Herbert Wrigley Wilson

Ironclads in Action - A Sketch of Naval Warfare From 1855 to 1895

Vol. II

Herbert Wrigley Wilson

Ironclads in Action - A Sketch of Naval Warfare From 1855 to 1895
Vol. II

ISBN/EAN: 9783337014155

Printed in Europe, USA, Canada, Australia, Japan

Cover: Foto ©berggeist007 / pixelio.de

More available books at **www.hansebooks.com**

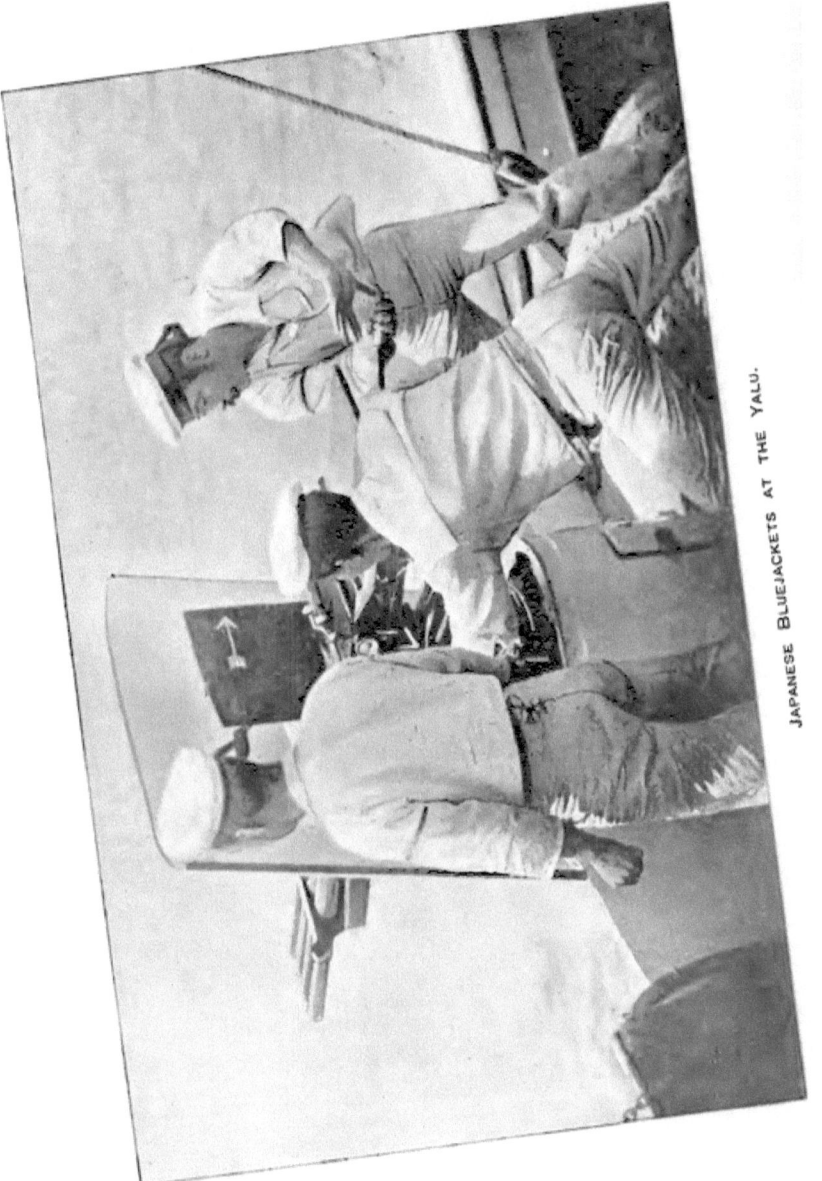

JAPANESE BLUEJACKETS AT THE YALU.

IRONCLADS IN ACTION

A Sketch of Naval Warfare

FROM 1855 TO 1895

WITH SOME ACCOUNT OF THE

DEVELOPMENT OF THE BATTLESHIP IN
ENGLAND

BY

H. W. WILSON

WITH AN INTRODUCTION BY

CAPTAIN A. T. MAHAN, U.S.N.

AUTHOR OF "THE INFLUENCE OF SEA POWER ON HISTORY," ETC.

WITH MAPS, PLANS, AND ILLUSTRATIONS

VOL. II.

LONDON
SAMPSON LOW, MARSTON AND COMPANY
Limited,
St. Dunstan's House,
FETTER LANE, LONDON, E.C.
1896.

LONDON:
PRINTED BY HORACE COX, WINDSOR HOUSE, BREAM'S BUILDINGS, E.C.

MAPS AND PLANS.

Map		Face page
XIX.	The Bombardment of Sfax and the Battle of Foochow	2
XX.	Descent of the Min and Torpedo Action of Sheipo	12
XXI.	Caldera Bay. Insets, Chilian Littoral, Valparaiso Bay, Torpedo Action	21
XXII.	Rio de Janeiro Bay	36
XXIII.	Sta. Catherina Bay	44
XXIV.	Attack on the *Aquidaban*. Elevation of *Aquidaban*, showing Injury.	46
XXV.	Theatre of the War in the East and Wei-hai-wei	52
XXVI.	The Yalu, I.	88
XXVII.	The Yalu, II., III.	90
XXVIII.	The Yalu, IV., V.	92
XXIX.	Naval Formations.	156
XXX.	Accidents to the *Grosser Kurfürst* and to the *Victoria*	194

ILLUSTRATIONS AND ELEVATIONS.

Plate		Face page
XIX.	Japanese Sailors at the Yalu	*Frontispiece*
XX.	The *Wei Yuen* and *King Yuen*	14
XXI.	Elevation of *Blanco Encalada*	28
XXII.	The *Itsukushima*	58
XXIII.	Elevation and Deck Plan of *Ting Yuen* and *Chen Yuen*	62
XXIV.	The *Yoshino*	68
XXV.	The *Tsi Yuen's* Conning Tower	70
XXVI.	Elevation of the *Naniwa*	74
XXVII.	Admirals Ting and Ito	82
XXVIII.	The *Matsushima's* Officers.	94

ILLUSTRATIONS AND ELEVATIONS.

Plate		Face page
XXIX.	The *Chen Yuen* in Battle	100
XXX.	The *Chen Yuen's* Side after Battle	110
XXXI.	The *Chih Yuen*	114
XXXII.	The *Ting Yuen*	122
XXXIII.	The End of a Battleship	172
XXXIV.	H.M.S. *Victoria*	197
XXXV.	Diagram of the *Victoria* just before she capsized	202
XXXVI.	The Last of the *Victoria*	204
XXXVII.	English Ironclads, I.	220
XXXVIII.	Systems of Protection	228
XXXIX.	English Ironclads, II.	232
XL.	Early and Modern Breech-loader	246
XLI.	Eight-inch Quick-firer	250
XLII.	French Ironclads	262
XLIII.	The Battleship *Neptune*	264
XLIV.	The Battleship *Formidable*	266
XLV.	The Cruiser *Alger*	268
XLVI.	The Sub-marine Boat *Gustave Zédé*	270

TABLES.

(PAGES 291—314.)

I. United States' Naval Ordnance, 1861-5.
II. Union Fleet at New Orleans. Confederate Forts and Squadron.
III. Union Fleet at Fort Sumter, April 7, 1863, and Confederate Forts.
IV. Union Fleet at Mobile.
V. Union Fleet at the Second Bombardment of Fort Fisher.
VI. The Southern Commerce Destroyers and their Prizes.
VII. Italian Fleet at Lissa.
VIII. Austrian Fleet at Lissa.
IX. Comparison of Fleets at Lissa.
X. Types of French Ironclads in 1870. German Ironclads.
XI. Fleets of Chili and Peru, 1878.
XII. Ships which took Part in the Bombardment of Alexandria.
XIII. Armament of the Alexandria Forts.
XIV. Shot and Shell Expended at Alexandria by the British Fleet.
XV. French and Chinese Ships at Foochow.
XVI. Congressional and Balmacedist Squadrons, 1891.
XVII. Fleets in Brazilian Civil War.
XVIII. Chinese Fleet at the Yalu.
XIX. Japanese Fleet at the Yalu.
XX. Comparison of Fleets at the Yalu and Notes on Guns.
XXI. Details of Japanese Losses at the Yalu.
XXII. Leading Types of English Battleships.
    1. Progress in Size, Dimensions, and Armour.
    2. Progress in Armament.
XXIII. Progress in English Cruisers.
XXIV. English Heavy Guns.
XXV. Summary of Torpedo Operations.

CONTENTS.

CHAPTER XVI.
FRENCH NAVAL OPERATIONS IN TUNIS AND THE EAST.

	PAGE
Difficulties between France and Tunis	1
Sfax bombarded. July 10-16th, 1881	2—3
Capture of Sfax. July 16th, 1881	4
Hostilities between France and China	4
Courbet at Foochow	5—6
The French attack the Chinese. Aug. 23rd, 1884	7
The Chinese squadron destroyed	8—9
Descent of the River Min. Aug. 25-28th, 1884.	11—12
Torpedo affair of Sheipoo. Feb. 15th, 1885	13
The Chinese ships sunk	15
Rice contraband.	15

CHAPTER XVII.
NAVAL EVENTS OF THE CHILIAN CIVIL WAR.

The revolt of the Chilian fleet. January, 1891	16
Balmacedist and Congressional fleets.	17—18
Physical features of Chili	18
The *Blanco Encalada* hit at Valparaiso	20
Balmacedist torpedo-vessels leave for Caldera	22
The attack on the *Blanco Encalada*. April 23rd, 1891	23
Congressionalist account. The *Blanco* sunk	24—6
Captain Goñi's report	27
Action between the torpedo vessels and the *Acongagua*. April 23rd, 1891.	29—30
Subsequent torpedo operations	31
Fall of Balmaceda	32
The case of the *Itata*.	33—4

CHAPTER XVIII.

THE CIVIL WAR IN BRAZIL.

	PAGE
Revolt of Admiral Mello. September 7th, 1893	35
Ships and resources of Mello and Peixoto	36—7
The Melloist ships pass the Rio forts	37—40
Peixoto acquires a fleet	40
Worthlessness of his fleet, Collapse of the revolt at Rio	41—2
Torpedo attack upon the *Aquidaban*. April 15-16th, 1894	43
The *Aquidaban* discovered	44
She is torpedoed by the *Gustavo Sampaio*	45
Melloist version	46—7
The value of the battleship	49
Lessons of the war	50

CHAPTER XIX.

THE STRUGGLE IN THE EAST.

Outbreak of war between China and Japan. July 29th, 1894	51
The Japanese revival	52—3
The effeteness of China	54—5
The state of the Chinese Navy	56
Renders tactical conclusions uncertain	57
The ships of the Japanese	57—61
Japanese merchant marine, and docks	61
Organization of the Chinese fleet	62
The *Chen Yuen* and *Ting Yuen*	63—4
Other Chinese ships	64—5
Chinese Merchant Marine, and Docks	65—6

CHAPTER XX.

THE ACTION OFF ASAN AND THE SINKING OF THE *Kowshing*.

Chinese and Japanese ships off Asan	67
The *Naniwa* engages the *Tsi Yuen*. July 25th, 1894	68—9
The state of the *Tsi Yuen* after the action	70
The *Kuwan-shi* and *Tsao Kiang* destroyed or captured	71
The *Naniwa* meets the *Kowshing*	72
The *Naniwa* sinks the *Kowshing*. July 25th, 1894	74
Violation of International Law	76
The position of the *Kowshing*	77
Admiral Ting puts to sea	79—80
Orders of Li Hung Chang to Admiral Ting	81

CHAPTER XXI.

THE YALU AND ITS LESSONS.

	PAGE
The Chinese despatch transports to the Yalu River	82
Ting with the fleet convoys them	83
Japanese fleet in the Gulf of Korea	84
The two fleets sight one another	85
Preparations of the Chinese	86—87
The battle opens. September 17th, 1894	89
The Japanese defile past the Chinese line abreast	90—91
Heavy losses of the Chinese	92
Both sides draw off	93
The *Matsushima's* share in the battle	94
Fortunes of other Japanese ships	95—6
The Chinese ships in detail	96—99
Alleged misbehaviour of the *Tsi Yuen*	100
The Yalu compared with Lissa	101—3
Losses of the Japanese	104—5
Losses of the Chinese	105—6
Does naval warfare grow bloodier	106—7
Guns of the Chinese	108
Broadsides of the two fleets	110
The Japanese Canet guns	111
Size and speed in the two fleets	112
The ram and the torpedo not used	114—5
Value of deductions from the Yalu	116
Line abreast	118
The value of armour	119
The gun still the predominant factor	123
Tactical value of speed in the battle	124
Training and discipline the secret of victory	125

CHAPTER XXII.

NAVAL OPERATIONS AT PORT ARTHUR AND WEI-HAI-WEI.

The Chinese fleet retires to Port Arthur	126
The Japanese capture Port Arthur. Nov. 21st, 1894	127
Wei-hai-wei bombarded	128
Torpedo attack upon the Chinese Fleet. Feb. 4th, 1895	130
The *Ting Yuen* torpedoed	131
Torpedo attack of Feb. 5th, 1895	132
Surrender and suicide of Admiral Ting, Feb. 12th, 1895	133

CHAPTER XXIII.
THE NAVAL BATTLE OF TO-MORROW.

	PAGE
Little material for induction	136
A special class of ships to fight in the line necessary	138
The cruiser and the battleship	139
Inferiority of the cruiser	141
Cruisers are of three classes	143
Weak ships in line a disadvantage	144
Examples of the division of a fleet	146
Sphere of torpedo craft in battle. The torpedo-gunboat	147
The torpedo-boat in battle	148
The ram in battle	150
The pneumatic gun	150
The position of the Commander-in-Chief	151
Battle dispositions	153
Line-abreast and bow-and-quarter line	155
Groups	156
Line ahead: its advantages	157
Manœuvring before and during battle	158
The value of the ram	160
The value of the torpedo	161
The warship's top-hamper	162
Effect of the long-range fire	163
Fires in action	165
Percentage of hits in battle	166
Maintenance of internal communication on board ship	167
The conning-tower	168
Perforation of armour in action	170
The encounter at close quarters	172
Loss of life in battle	173
Duration of battle	175
Losses of ships	176
The type of battleship best adapted to this forecast	178
The armament of the ideal battleship	179
Devolution of command during battle	181
Summary	182

CHAPTER XXIV.
IRONCLAD CATASTROPHES.

The *Captain*	183
She goes to sea with the Channel Squadron	185
She founders during the night. September 6—7th, 1871	168

CONTENTS.

	PAGE
Story of the survivors	187
Verdict of the court-martial	189
The *Vanguard* is rammed by the *Iron Duke*. September 1st, 1875	190
She sinks	191
Verdict of the court-martial	192
The *Grosser Kurfürst* rammed by the *König Wilhelm*. May 31st, 1878	193
Effect of the collision: the *Grosser Kurfürst* sinks	194
The *Victoria*	196
A dangerous manœuvre ordered	197
The manœuvre executed	199
The *Camperdown* rams the *Victoria*. June 22nd, 1893	200
Efforts to save the *Victoria*	202
Splendid behaviour of the *Victoria's* crew	203
The ship capsizes with grievous loss of life	204
Finding of the court-martial	206
Loss of the *Reina Regente*. March, 1895	207

CHAPTER XXV.
THE DEVELOPMENT OF THE ENGLISH BATTLESHIP.

Great changes in the implements of naval war	209
These changes due to steam	210
Growth of displacement	212
Number of engines on board ship	214
Specialization in the type of ship	215
The application of armour to ships	217
The *Gloire* and the *Warrior*	219
Captain Coles invents the turret	220
The central-battery ironclad	221
The *Sultan*	222
The *Alexandra* and *Téméraire*	223
Early turret-ships	224
The *Devastation* class	226
The armour deck	227
Unarmoured ends: the *Inflexible*	228
The "écheloned" turret-ship	229
The "Admirals"	230
The *Victoria* and *Sanspareil*	232
The *Nile* and *Trafalgar*	233

CONTENTS.

	PAGE
The *Royal Sovereign* class	234
The *Centurion* and *Barfleur*	236
The *Renown*	237
The *Majestic* class	238
Summary of progress in ironclad construction	239
Reappearance of moderate armour	241
English and foreign types of battleships compared	242
The development of artillery	245
The rifled gun and the early breech-loader	246
Resistance of early ironclads to artillery	246
The monster gun	247
The muzzle-loader abandoned in England	248
The new breech-loader	249
The quick-firer	250
Improvements in projectiles	252
Improvements in armour	253
Progress in engineering	254
The first cruisers	255
The *Esmeralda* and fast cruisers	255
The belted cruiser	257
The torpedo-boat	257
The torpedo	258
Appendix I. The Development of the French Navy, 1855—1895	260—274
Appendix II. Report of the French Committee in 1870 upon the practicability of attacking the Prussian littoral	275—76
Appendix III. British Ironclads	277—78
Appendix IV. Leading Authorities consulted	279—288
Appendix V. Illustrations	289
Tables I—XXV.	291—314
Index I. Actions	315
,, II. Names	317—357
,, III. Subject-Matter and Technical Terms	359—374

IRONCLADS IN ACTION.

CHAPTER XVI.

FRENCH NAVAL OPERATIONS IN TUNIS AND THE EAST, 1881—1884-5

IN 1881, difficulty with the Khroumirs, a Tunisian tribe on the Algerian frontier, and the fact that the Bey of Tunis was in secret abetting them, led France to take vigorous measures against Tunis. In the last week of April the army on land commenced operations whilst the fleet supported it. On April 25th, Tabarka, an island on the coast of Tunis, which was protected by an antiquated castle, was bombarded by the gunboat *Hyène*, supported by the *Surveillante*, *Tourville*, *Chacal* and *Léopard*. A very feeble resistance was offered by the Arabs, and next day a detachment of troops landed and occupied the place. A week later Bizerta was seized to serve as a base for the French operations. On May 7th, Béja was captured, and it was supposed in France that the war was over. A few days later the Bey practically accepted the protection of France, and about the middle of June the French army was recalled. But as a matter of fact the Arab population was by no means ready to submit to France. Fostered by Mussulman agitators, an insurrection against the Bey, who was accused of selling his country, broke out in the south. Sfax was seized and occupied by the insurgents, and the foreign residents in the country were in grave danger.

The French did not quietly acquiesce in these proceedings. On the contrary, as might be expected, they at once made preparations to re-conquer Tunis. For operations on the coast they had a formidable squadron ready in the Mediterranean under Vice-Admiral Garnault and Rear-Admiral Martin. The first had for his flagship the *Colbert*, the second, the *Trident*. The other ironclads of the squadron were the *Galissonière*, *Friedland*, *Marengo*, *Surveillante*, *Revanche*, *Alma*, and *Reine Blanche*. With these were the unarmoured vessels *Tourville*, *Hirondelle*, *Desaix*, *Voltigeur*, *Hyène*, *Chacal*, *Léopard*, and *Gladiateur*. To the force thus constituted the *Intrepide*, *Sarthe*, and *Pique* were added.

On July 5th, the *Reine Blanche* and *Chacal* appeared off Sfax, and the latter reconnoitred the place, and bombarded it at a range of 5000 yards, directing her fire especially upon the water battery, which was some little distance from the town wall, and making a breach in it. Only eighteen shots were fired in reply. Next day two more formidable vessels, the *Reine Blanche* and *Alma*, opened upon the town, early in the morning. They continued to shell it till midday when they were joined by the gunboats *Pique* and *Chacal*. As the reply of the Tunisians was exceedingly feeble and ill-directed, the gunboats ventured in to 2400 yards; in the evening the *Hyène* arrived. This day the batteries only fired thirteen shots. On the 7th, the *Reine Blanche* and *Alma* resumed their slow bombardment of the town, assisted by the *Hyène* and *Chacal*. A lighter, mounting one 14-centimètre smooth-bore, was also employed, and was of great value, since it drew very little water and could therefore be taken close in to the town. The depth of water in the Bay of Sfax was indeed the chief difficulty which the French had to face, as it prevented their heavier and more powerful ships from playing that part in the operations which would have been expected from vessels of their size and strength.

On July 8th, a boat attack was made upon the place. The boats approached within 1000 yards of the land and opened a vigorous fire upon the Arab trenches and lines. They were supported by the *Chacal* and *Hyène*. After this some days

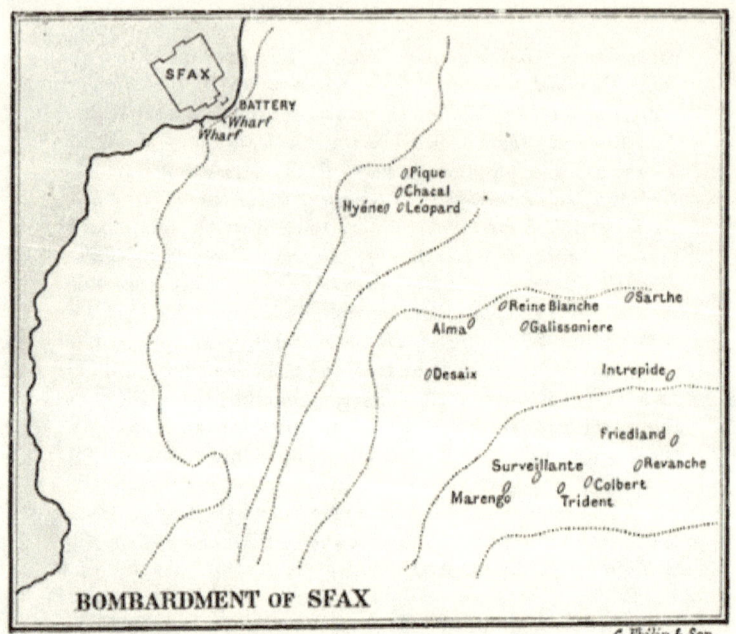

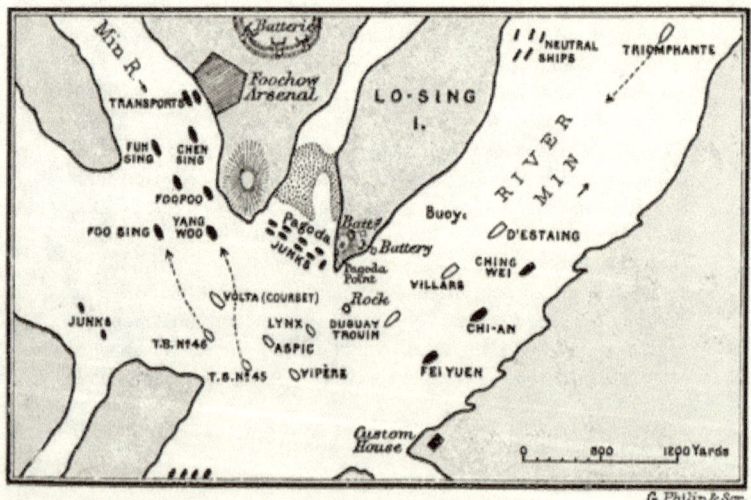

Map XIX.

of inaction followed, during which the two ironclads shelled the town from time to time.

On July 14th, the rest of the French squadron arrived, and on the following day the bombardment was resumed with increased vigour. The attack was delivered by the ironclads, gunboats, and boats of the squadron. The ironclads anchored according to their draught at a mean distance of 6500 yards from the town. They were in two groups:* in the first, furthest out, were the *Marengo, Surveillante, Colbert, Revanche* and *Friedland*; in the second, the *Alma, Reine Blanche* and *Galissonière*, with the unarmoured cruisers *Desaix, Sarthe* and *Intrepide*. Nearer in, at a distance of 2200 yards, were the gunboats. Finally, the boats of the squadron, armed with Gatlings and machine-guns, moved to and fro only 500 yards from the shore. The boats were supported by two lighters on one of which was a 16-centimètre rifle, and on the other a 14-centimètre smooth-bore. The fire maintained by the ships was slow and steady, and did much damage to the town.

On the next day, the 16th, it was decided to effect a landing. The beach was of soft yielding mud and the shallowness of the water made approach difficult, but this was overcome by the construction of a temporary floating jetty. Six of the ironclads had been ordered each to prepare a raft of spars and topmasts on the evening of the 15th. When ready, these rafts were towed in separately by launches and fastened together as close to the shore as possible. The boats laden with men were to be ranged as near in as they could go by four o'clock in the morning, and, at a given signal from the gunboats, were to dash in and land the men on the jetty. There were eighteen boats in all, armed with four 12-centimètre and one 4-centimètre gun, besides thirteen various machine-guns. By daylight the preparations were completed, and at 4.30 a shot from the *Colbert* opened an unusually fierce bombardment.

* *See* Plan.

The boats were massed 500 yards from the batteries and only waited the signal to dash in. The landing party numbered 3000 men, and was composed of 1600 sailors and 1400 soldiers. At three in the morning the raft was secured firmly inshore about 400 yards from the mole; at six, the signal for the troops to land came, and they made for the shore with some little confusion. The Arabs, however, had been unable to hold the trenches under the fire of the ships and boats, and to add to their discomfiture a great quantity of esparto grass near their lines was set on fire by shells, and burnt fiercely. The French sailors and soldiers quickly carried the water-battery, and forced the gates of the town. The loss of the fleet was not heavy. In all it amounted to eleven killed and fifteen wounded.

The attack upon Sfax was well conceived and well executed, but, of course, little resistance could be offered by the Arabs to the powerful artillery of the French squadron. Here, even more than at Alexandria, the attacking force was so much stronger than the defence that the result of the fighting was bound to be very one-sided. No damage was done to any of the ships.

In 1884, France, provoked by the attitude of China towards the corps of Black Flags or freelances on the Tonkin frontier, proceeded to acts of hostility against that power. Without any previous declaration of war, the port of Kelung in the island of Formosa was bombarded and occupied on August 6th. Nine days later the Chinese Government issued a declaration of war against France.

A French squadron under Rear-Admiral Courbet had, before the declaration, ascended the River Min, on which stands the city of Foochow, where was the most important of the Chinese naval arsenals. The ships which Courbet commanded included the *Duguay-Trouin,* a large composite cruiser, the *Villars* and *D'Estaing,* both wooden third-class cruisers, the *Volta,* a wooden sloop which carried the admiral's flag, and three gunboats, the *Lynx, Aspic,* and *Vipère,* all composite, and all of about 450 tons displacement. In addition, at the mouth of the Min, lay the armoured cruiser *Triomphante*, a central-battery and barbette vessel protected by iron armour, 6 inches to $4\frac{3}{4}$ inches thick. Her commander was busy lightening her for the ascent of the river, as he had received Courbet's permission to attempt it. Including this ship, the squadron mounted six 9·4-inch, eight 7·6-inch, fifty-one 5·5-inch, and nine 4-inch breech-loading rifled guns. The weight of one discharge from these would amount to a little over 6,000lbs. Two torpedo-boats were present with the squadron; their numbers were 45 and 46, their speed was sixteen knots, and they were armed with spar-torpedoes, containing a bursting charge of 28lbs of gun-cotton. The total of the French crews was 1830 men.

Ascending from the coast, the River Min at first takes a south-westerly direction, but about a mile below Foochow, at Pagoda Point, makes a sharp bend to the north-west. At this point no less than seven channels or waterways converge, the main channel of the river alone having sufficient depth of water to permit the movements of large ships, and then only at flood-tide, which lasts at this point four hours. Admiral Courbet anchored at his ships at the angle of the stream, the *Volta* being just abreast of Pagoda Point, the two torpedo-boats on her port quarter, and the three gunboats astern. The other ships were lower down, in the centre of the stream. On Pagoda Point were Chinese batteries, whilst other works protected the arsenal, which is 2000 yards above the Point.

In the River Min was moored a considerable Chinese squadron.[*] This included one moderate composite cruiser, the *Yang Woo*, which acted as the Chinese flagship; six very indifferent wooden sloops, *Foo Poo, Chi-an, Fei Yuen, Ching Wei, Foo Sing* and *Yu Sing*; two transports; two Rendel gunboats, each mounting one heavy gun in the bow, the *Chen Sing* and *Fuh Sing*; seven launches fitted with spar-torpedoes; and eleven war junks, sailing vessels, armed only with smooth-

* *Vide* Table XV. for details.

bore guns. This flotilla was as badly manned and commanded
as it was armed. The eleven steamers of any size included in
it, carried only 1190 men, whilst the artillery on board, excluding smooth-bores, comprised one 18-ton, two 16-ton, one
6½-ton, eleven 3½-ton, twenty-four 45-pounder, and two
40-pounder guns, mostly muzzle-loaders, and of inferior power
and penetration to the French breech-loaders. The weight of
metal discharged by them was under 4500lbs. The French
had thus a superiority of one-third, which would probably have
given them the victory over well-trained opponents, but the
men whom they were to fight, a disciplined force themselves,
had neither skill, discipline, nor courage. Add to this that
the Chinese captains, with one or two exceptions, were as
ignorant and as cowardly as their men, and it will be seen
that the work before Admiral Courbet was not of a very
difficult nature.

The Chinese flotilla was thus disposed:* Abreast of the
D'Estaing, and lowest down the Min, lay the *Ching Wei*;
abreast of the *Villars*, the *Chi-an*; and abreast of the *Duguay
Trouin*, the *Fei Yuen*. In a backwater of the river between
Pagoda Point and the mainland, lay nine of the junks; two
others with the torpedo launches were on the opposite side of
the river, thus flanking the *Volta* and the French gunboats.
The rest of the squadron lay above the *Volta*, between her
and the Arsenal, and the leading ships *Yang Woo* and *Foo
Sing* were 400 to 600 yards ahead of her. In the river were
several French mercantile steamers, the English warships
Champion, Surprise, and *Merlin,* and an American squadron
composed of the *Enterprise, Juniata, Trenton,* and *Monocacy.*
These vessels were either above or below the hostile
squadrons, and kept well out of the way, in view of a French
attack upon the Chinese.

For day after day the two enemies confronted each other,
whilst the neutrals expected every hour to see the attack
begin. The French were always cleared for action; their

* *Vide* Plan, p. 2.

upper masts were struck, their cables ready for slipping, and their crews, in watches, relieved one another at the guns Each day brought rumours that this was the appointed time, and as each day passed without incident, the vigilance of the Chinese, never very remarkable, was relaxed more and more. The French were waiting for the arrival of the *Triomphante;* the Chinese close under their enemy's guns, perhaps, imagining that the French did not really mean business. They were to be terribly undeceived. On Friday, August 22nd, in the evening, Courbet summoned his officers to his flagship, and there made known to them his plans. Next day, just before two o'clock, when the tide ebbed, the ships were to weigh and get to work. A signal from the Admiral would tell the torpedo-boats to steam forward and attack, No. 45, the *Yang Woo,* and No. 46, the *Foo Sing.* A second signal, and the ships were to open fire, the *D'Estaing, Villars,* and *Duguay-Trouin* upon the Chinese steamers which lay abreast of them, with their port batteries, whilst their starboard guns played upon the junks off Pagoda Point. The *Duguay-Trouin,* when this work of destruction was completed, was to settle the launches and support the *Volta,* which in the meantime would cover the torpedo-boats, and with either broadside assail the junks. Finally, the three gunboats were to steam forward and sink the Chinese vessels off the Arsenal.

The morning of the 23rd dawned cloudless and intensely hot, the strong sun causing the men at the guns of the French ships no little discomfort. Courbet early gave notice to the consuls and others, that he intended to attack the Chinese soon after noon, so that he gave his enemy some warning. To the watchers on board the *Enterprise,* the hours seemed to go slowly beyond endurance, and it appeared as though the French were never going to begin. At 9.30, the flood-tide was in, and soon after, steam was up, and the French crews took a meal. At 1.30, silently the men went to quarters, and the Chinese followed suit. A quarter-of-an-hour later, anchors were weighed, and the preparations were completed. At 1.50, the *Triomphante* came in sight, and there passed some signals between her and Courbet. Six minutes later, the stillness was broken by the rapid discharge of the Hotchkiss in the *Lynx's* top, and almost simultaneously the *Ching Wei* replied with a broadside. At the signal, the torpedo-boats had dashed forward, and only twenty-seven seconds after the first shot, the sound of a terrific explosion dominated the uproar of the engagement. No. 46 had exploded her torpedo under the *Yang Woo's* side, amidships. Of 270 men on board the Chinese vessel, only fifteen escaped, and so great was the force of the explosion that, it has been stated by an eye-witness, mutilated bodies were found after the engagement on the roofs of houses a mile away on shore.* The *Yang Woo* drifted away, an utter wreck, and as the French continued to pour in a hail of shells, caught fire and sank. The boat which had dealt the blow, reversed her engines, but was struck by a shell and disabled. Having lost one sailor, killed by a bullet, she drifted slowly down stream and anchored. No. 45 was not so successful. As she ran forward to attack the *Foo Sing*, a Chinese torpedo-launch encountered her, and forced her to swerve to one side. Thus she failed to strike the *Foo Sing* amidships, and, embarrassed by the Chinese boat, caught her bow in the *Foo Sing's* side. A hot fire was poured in upon her by the Chinese. Her commander was dangerously wounded, and one of her crew had his arm broken. At last she got clear, and retired down stream towards the *Enterprise*, whose surgeon went to the help of her wounded. Her crew were bespattered with blood from head to foot, but their boat had received no serious injury. They had not, however, disabled the *Foo Sing*, which was slowly steaming ahead, when a second torpedo attack was made by a launch from the *Volta*. The torpedo this time exploded close to her screw, and completely disabled her. She drifted slowly down stream

---

* Messrs. Roche and Cowen. "The French at Foochow," p. 25. The statement seems to me most improbable, but I give it for what it is worth.

upon the French vessels, which received her with a terrible fire. A minute or two later she was boarded and carried, but she was too much injured to float, and quickly went to the bottom.

Meantime, the French fire upon the Chinese was terrific. In the stillness of the air the smoke hung heavily about the ships, but it could not save the unfortunate Chinese. The *Volta* was hotly engaged with the junks, and had a shot through her chart-house, which killed her pilot, and all but killed Admiral Courbet. She retaliated by sending the junks to the bottom, and even when they were helpless and sinking, continued to fire her machine-guns into them. The *Chi-an* and *Fei Yuen* could offer no resistance to the *Villars* and *Duguay*, and they were quickly on fire and sinking, whilst the Chinese crews fled ashore. The *Ching Wei* alone showed heroism, and for a time, faced the *D'Estaing* bravely. But now the *Triomphante* was coming on, having passed the Chinese batteries with the interchange of only a few shots, and as she neared the *Ching Wei*, fired her 9·4-inch guns at her. A shell from one of these struck the Chinese vessel in the stern, and, passing the whole length of the ship, burst with a tremendous cloud of smoke in her bow, lifting her in the air and setting her on fire. The gunners thereupon jumped overboard, but the officers stuck to the ship, working her guns with their own hands. They made a desperate effort to run her alongside of the *D'Estaing* and sink their enemy with them, by exploding the *Ching Wei's* magazine. Attempting this, they received a tremendous broadside from the *Villars*; fresh fires broke out; the small-arms' ammunition exploded; and the *Ching-Wei* went to the bottom, but not before she had fired one last parting shot as she vanished in the water. In seven minutes from the first shot the action was virtually over, and every Chinese ship was sunk or sinking. The *Foo Poo* had run away up stream at the commencement of the fight. Her captain drove her ashore, breaking her back, gave her crew leave of absence, and then fled himself. A second Chinese captain, after firing one broadside, left his ship with his crew, having first set her on fire. The *Lynx, Aspic,* and *Vipère,* proceeding up stream after the Chinese sloops, were now busied in shelling the Arsenal and the shore forts, and traces of their handiwork soon began to drift down the river towards the heavier ships. At 2.8 a Rendel gunboat came round Pagoda Point, and fired her 16-ton gun at the *Duguay,* but missed her. Immediately the guns of the fleet were concentrated upon this luckless craft, and the torrent of descending and exploding shells was so great that it literally stopped her way. Two minutes she remained almost stationary, a helpless target ; then, with a crash, her magazine exploded, and she dived headlong to the bottom. At 2.20 a mine the Chinese had laid under the dock at Foochow exploded, fired either by inadvertence or the French shells. At 2.45 the French fleet, which had slackened its fire, again opened on the Chinese forts. All the while burning Chinese vessels were drifting down past the neutral ships. One sloop was seen in a great blaze with crowds of Chinamen close under her stern clinging to her hissing rudder-shaft ; one of these had had his thigh shot close off, and many were terribly wounded. The water was full of wreckage and Chinamen, six or seven together, clinging desperately to masts or timbers. Many were rescued by the English and American launches.

The French machine-guns were extremely effective. "The continued hail of shell from Hotchkiss cannon in the tops of the French men-of-war upon their antagonists, swept them down like wheat before the mower. Relays of men could not be brought up from below fast enough to fill up the gaps in the gun-crews. The diminutive shell came crashing through the sides and bulwarks of the ship. Splinters, flying in every direction, killed many more."*

About four o'clock the French ceased their fire, when at once the land batteries redoubled their exertions. Flaming

* Roche and Cowen, 43.

rafts, set loose by the Chinese, came drifting down upon Admiral Courbet's ships. Courbet re-opened the duel with the forts, whilst the rafts were caught and towed aside. At 4.55 the fleet anchored for the night out of range. The ships were not much injured. The *Volta* had one hole a little above the water-line. Her pilot and two men at the wheel were killed, whilst six powder passers, on the berth deck, had been cut down by the shot which hulled the ship on the water-line. The *Duguay* and the other ships had the most trivial hurts. The French loss is reported, by eye-witnesses, to have been twelve killed, though in the official despatch of Courbet it is only given at six killed, with twenty-seven wounded. The Chinese loss is returned at 521 killed and 150 wounded; and, in addition, there were a large number of men missing.

This fight, if fight we can call it, has been described by French writers as a most splendid achievement. "The great glory of Foochow," is the title given. As a matter of fact, it was little more than slaughter—necessary no doubt, but yet deserving no extravagant laudations. It may be placed in the same class with the bombardment of Alexandria. Both operations were undertaken against men who lacked training, and in each case very heavy loss was inflicted by the Western force upon its Oriental opponent. Courbet's great reputation rests rather upon his professional ardour, and the skill with which he formed his plans, than upon great performances on the scene of action. His enemy was contemptible.

The day after the battle the French once more bombarded the Arsenal. That night the Chinese attempted a torpedo attack, but on being discovered by the search-lights of the fleet ran ignominiously. On August 25th Courbet moved his flag from the *Volta* to the *Duguay*, and prepared to descend the river, his ships being placed in the following order: *Triomphante, Duguay, Villars, D'Estaing, Volta*, and the three gunboats. A succession of Chinese batteries fringed the banks, but most of the guns pointed down stream, and little preparation had been made against the possibility of an attack from above. The French ships, except the *Triomphante*, had ascended the river before the declaration of war, and thus they were in a peculiarly favourable position to effect the destruction of the forts. For this they used their 9·4 and 7·6-inch guns, the 5·5 and 4-inch weapons, being too small to make much impression. Whilst they were at work the cannon of the *Galissonière* and *Bayard*, two ironclads similar in type to the *Triomphante*, could be heard bombarding the forts on the lower reaches. On the 25th a battery on Couting Island was silenced by fire from the rear, and its one 8-inch gun burst by a landing-party. The fleet then entered the narrows of Mingan, on which a number of works bore. On the 26th these were attacked in succession by the *Duguay* and *Triomphante*, and silenced one by one. On the 27th landing parties destroyed the guns with gun-cotton, after which the ships weighed and proceeded to the Kimpai Narrows. On the afternoon of that day a battery there was silenced, and during the night a number of junks loaded with stone, which had been placed in readiness to block the channel, were attacked and destroyed by the French boats, covered by the gun-vessels.

Next day the squadron, reinforced by the *Sâone* and *Château Renault*, proceeded to force the narrows. The passage here is little over 400 yards wide, and on either side of it wooded hills rise steeply. Two strong works, armed with 7-inch and 8¼-inch guns, had to be silenced. This operation was very skilfully conducted. The *Duguay* and *Triomphante* anchored with springs to their cables. These were paid out, till, dropping down stream, the ships' batteries would bear upon the first embrasure. On this the whole fire was concentrated, and, needless to say, at such short range, the work quickly crumbled away, and the gun was silenced. These tactics were repeated with embrasure after embrasure, till at noon on the 28th the Chinese had abandoned their forts. Courbet then led his fleet through the narrows, and after forty days' absence rejoined the ships which had remained below.

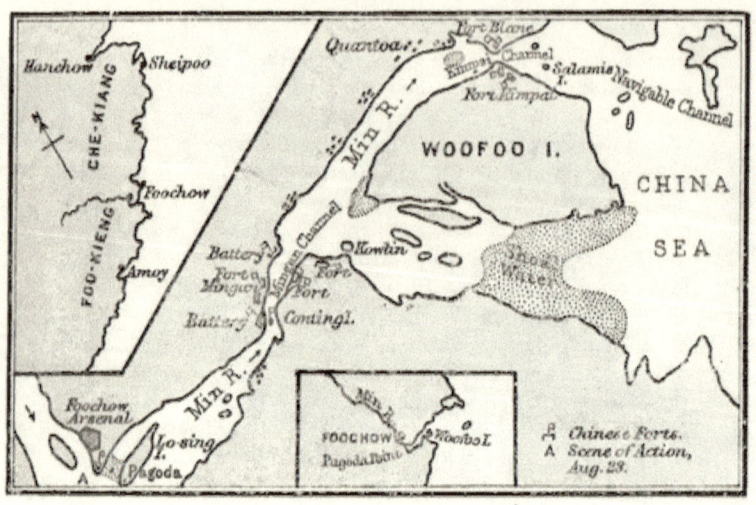

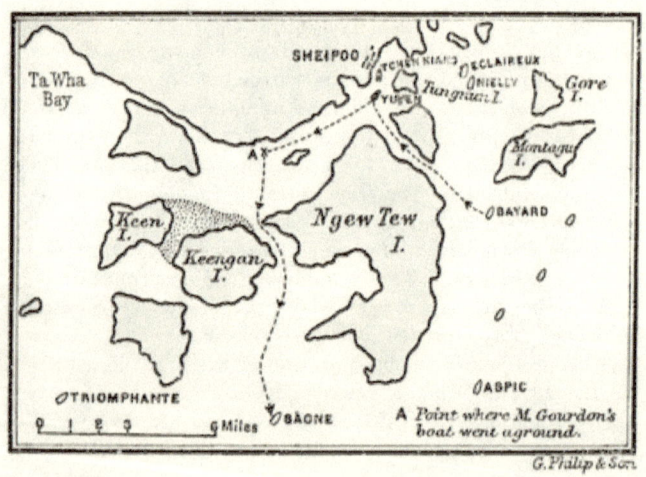

Map XX.

The next important action in the East was the affair of Sheipoo, which took place on the night of February 14th—15th, 1885. Two Chinese vessels, the cruiser *Yu-yen*, mounting one 8¼-inch, eight 6-inch, and twelve 4·7-inch Krupps, and the despatch-boat *Tchen K'iang*, carrying one 6·3-inch, and six 4·7-inch guns, had been cut off by the French, and were lying between Sheipoo and Tungnun Island. They were watched by the French warships *Eclaireur, Nielly, Bayard, Aspic,* and *Sâone*, which effectually prevented their escape. As the navigation of these waters is intricate and difficult for heavy ships, Admiral Courbet decided to attack them with torpedo-boats, on the night of the 14th. For the command of the two launches selected, he designated Lieutenants Gourdon and Duboc. The boats were 30 feet long, carrying spar-torpedoes loaded with 28lbs. of guncotton; the engines were constructed to work silently, and picked coal was burnt. The colour of the hulls was black.

At eight o'clock the boats were made ready, and at 11.30 M. Gourdon started from the *Bayard*. At 12, M. Duboc was to follow. M. Gourdon* gives us the following account of the expedition. "The moon was new, and the night absolutely dark, so that it was a matter of great difficulty for the vidette boat and launch, which were accompanying my torpedo vessel, to keep together. We repeatedly lost and found one another. Our difficulties were increased by the fact that, from fear of injuring the spar-torpedo, I could not go close to the vidette boat, and the strong south-easterly current was another disturbing factor. A turn brought us clear of the Ngew Tew Straits; then leaving these narrows we tested our conductors and their insulation, and ran the spar in and out. All worked well. And now we go straight for the Chinese, but they are not at the anchorage which they occupied during the day. They have disappeared. It is 3.15 a.m., and we steam in search of them. At 3.30, I see a great black mass in front

---

* I have compressed the account given, but without, I trust, losing the spirit of the narrative or impairing its accuracy.

of Sheipoo, with four or five lights on shore. The vidette
boat is detached to inform Duboc of her whereabouts. I get
her three masts in line and move slowly on. On shore there
are numerous lights; are they signals? The loud rattle of
my boat's engines prevents me from hearing anything. When
200 yards from the Chinese, I run out my spar, and connect
the firing wires with the battery, then 'Full speed ahead.'
The frigate lights up; on port and on starboard there are
spurts of flame. Are these the Nordenfelts? On we go
swiftly. Then the order 'Astern,' and a violent shock. The
torpedo has exploded; the boat is violently lifted and strikes
the enemy's side, catching in it. 'Astern quickly,' is my
order. The quarter-master is trying to push us clear, when a
Chinaman looks out of a port-hole and gets his fist in his eye.
All the while the boat is stationary, and steam escaping from
her valves. The oil feed has been smashed by the shock; I
plug it with a bayonet; still the boat does not move. Our
spar has caught and must be abandoned. It is freed and falls;
the boat goes astern at last, and leaves the lights of the frigate.
Duboc's boat has come up now, and I wait to help it, in case
it needs assistance. Meanwhile, the Chinese ships are firing
on each other, and the men ashore on both. A marine falls
killed by a bullet. Through the hail of projectiles, Duboc
advances, passes to starboard, and explodes his torpedo, then
retires. We meet: 'What news? I have a man killed.'
'We've not a scratch.' We cannot find the vidette, which
was to show a red light. I take Duboc in tow, and we are
off."* Whilst retreating, the boats experienced some little
delay through the towing rope getting caught in the screw of
the leader, and then through Gourdon's craft running aground.
He got her off, and rejoined the *Sâone* soon after ten o'clock
that morning. The vidette-boat saw the explosions, and
waited till six o'clock, but then as the boats did not re-
appear, gave them up for lost. It was a welcome surprise
when they returned safe and sound.

* Loir, 215-8.

THE CHINESE SLOOP WEI YUEN.
See p. 132.

THE CHINESE CRUISER KING YUEN.
See p. 64.

PLATE XX.

On reconnoitring the two Chinese ships, it was discovered that both had sunk. As only one had been torpedoed, this must have been due to the carelessness of the Chinese, who doubtless fired into friends in the hurry and excitement of the attack. This is a serious danger in all night engagements with torpedo-boats, as the English manœuvres have shown.

One or two questions of importance in international law were raised during the war. On October 20th, 1884, Courbet proclaimed the blockade of all the ports and roads of southern Formosa. England, however, protested against this blockade as inefficient, and the proclamation was withdrawn. On February 20th, 1885, the French Government declared that it would treat rice, when bound for open Chinese ports, as contraband. Four days later this declaration of contraband was restricted to cases where the rice was being conveyed to the northern ports of China. In a note the French Government explained that the stoppage of supplies would bring China to reason with less injury to neutral trade than would be inflicted by a close blockade of the Chinese ports. Protests were made both by the English minister at Pekin and the English Government at home, but the war came to an end on April 7th, 1885, before the point at issue had been settled. In the course of the war French cruisers seized lead on English ships as contraband, though it was a regular article of trade with China. They also used Hong Kong as their base, coaling and refitting there.

CHAPTER XVII.

NAVAL EVENTS OF THE CHILIAN CIVIL WAR.

January—August, 1891.

IN January, 1891, the Chilian fleet, which was lying off Valparaiso, declared against the government of President Balmaceda, who was accused by the Congressional party in Chili of aiming at a dictatorship. On the 6th, the ironclad *Blanco Encalada*, which we have met before, the fast Elswick-built cruiser, *Esmeralda*, which had replaced Arturo Prat's gallant little craft, the *O'Higgins* and *Magallanes*, put to sea. That night they were joined by the *Cochrane*, and returning on the next day, took possession of the *Huascar*, which was lying, out of commission, in the harbour. Having prepared her for sea, they added her to their squadron, of which Captain Montt took command. They also laid their hands upon every steamer which carried the Chilian flag, including some mail steamers of the South American Steamship Company. The *Imperial*, however, the fastest vessel of the line, happened, with another vessel, to be laid up, and consequently escaped seizure.

President Balmaceda was thus left without a sea-going warship on the coast. He had the *Imperial*, which he armed, and which could perhaps accomplish fifteen knots an hour, and he had a dozen torpedo-boats of various patterns, mostly equipped with spar-torpedoes. Two torpedo-gunboats of the most recent design, were on their way out from Europe. Their names were the *Almirante Condell*, and the *Almirante Lynch*. They were steel vessels of 750 tons displacement, built by Messrs. Laird, of Birkenhead, and launched in the previous year. With water-tube boilers they had steamed twenty-one knots under forced draught on the measured mile, and with a bunker capacity of 175 tons, could carry coal enough for 2500 knots at economical speed. Their armament consisted of two 14-pounder Hotchkiss quick-firers placed *en échelon* forward on their deck, and one aft, besides four 3-pounder quick-firers and two Maxims. They carried four torpedo-tubes for the discharge of the 14-inch Whitehead torpedo. The *Capitan Prat* was building for the Chilian Government in Europe, but was not ready at the beginning of the war; and two other cruisers, the *Presidente Errazuriz*, and the *Presidente Pinto*, were detained at La Seyne, in France, till the French Supreme Court should decide whether they were to be permitted to depart. Week by week the Balmacedist papers reported the movements of these ships, as if they were actually on their way to Valparaiso, and thus, perhaps, deceived the people of Valparaiso and Santiago, if not the Congressionalists. Balmaceda held the forts of Valparaiso, where numerous heavy guns were mounted, and had at his back an army of 40,000 men.

The Congressionalists had the ironclads *Blanco*, *Cochrane*, and *Huascar*, of which full details have already been given. All three had been re-armed with the 8-inch Armstrong breech-loader, and the two central-battery ships carried each four 6-pounder quick-firers four Nordenfelts, and two Gatlings, in addition to their heavy guns. The speed of the three cannot have exceeded eleven knots. Perhaps the most formidable vessel of the squadron was the *Esmeralda*, the first fast protected cruiser, launched at Newcastle in 1884, of 3000 tons displacement, and 18·3 knots speed on the measured mile. She carried an end-to-end 1-inch steel protective deck, a little below the water-line, under which were boilers, engines, and magazines. Above the water-line she had considerable protection from her coal-bunkers. Fore and aft were mounted 10-inch 25-ton breechloaders, and amidships, six 6-inch breech-loaders, three on each side in sponsons, with steel bullet-proof shields. In her bunkers she carried 600 tons of coal, which would enable her to steam 2000 miles at ten knots. She was a good sea-boat though so heavily gunned, and her speed, which was in 1891 sixteen knots or a little less, made her a dangerous enemy to the craft which Balmaceda possessed. The other Congressional ships took little part in the struggle, or have been already described. The *Aconcagua*, however, deserves a word. She was a mail steamer of 4100 tons gross, and had been armed with two 5-inch guns, one 40-pounder, and several machine guns. Her speed was twelve knots.

The physical configuration of Chili, to which we have already alluded,* rendered sea power of peculiar importance in this civil war, as in the struggle between Peru and Chili. Since 1880, the Chilian frontier had been moved north 450 miles by the annexation of the arid and waterless Peruvian provinces of Tacna and Tarapaca, and the Bolivian department of Antofagasta, territories which are rich in nitrate and guano deposits, and which for that reason are a source of great wealth to the country. Dues levied on the export of these commodities formed no inconsiderable part of the Chilian revenue, so that whoever held the north held the purse strings. Land communication between the nitrate ports and Valparaiso or Santiago there is none; in fact, the towns along the coast might be regarded strategically as a series of small islands, each separated from the other by seas which are represented by deserts. The Congressionalists with their mobile naval force were thus at an enormous advantage, of which they proceeded to make the fullest use. The small Balmacedist garrisons in the various nitrate ports were attacked one by one and compelled to surrender.

The first interchange of shots was at Valparaiso on the 9th, when pickets on shore fired on a boat in the harbour, to which the warships replied from their tops. The Congres-

---

* Vol. i. 314.

sionalists procured their supplies from the ships which they had seized, and from the coast to the north and south, whither their vessels were despatched to raise recruits and obtain food. At a later date they actually drew coal, stores, and provisions from Valparaiso itself, by a most ingenious use of neutral ships. These would come into the harbour and load the stores most required by the insurgents, and then proceeding up the coast would be brought to by Congressional warships, and would, under the pretext of compulsion to satisfy their consuls, sell coal and food. President Balmaceda knew perfectly well what was happening, but owing to the presence of strong foreign squadrons on the coast, found it expedient to shut his eyes.

On January 16th the only Congressionalist ship in Valparaiso Bay was the ironclad *Blanco*, which was quietly lying at her moorings. Her business off the port was to procure supplies and forward them north. There was no blockade as yet, and indeed, when the Congressionalists endeavoured to close the harbour and to stop all trade at a later date, the foreign consuls refused to permit such action. She was distant from Fort Bueras, which was on her starboard beam 150 feet above the water level, 600 yards, and from Fort Valdivia 1200. Suddenly, at 5 a.m., these two forts and Fort Andes fired each a round at her. In Fort Bueras a 20-ton 10-inch muzzle-loader was used, firing a common shell of 450 lbs. weight, filled, and fitted with a time and percussion fuse. The charge on this occasion was 130lbs. of pebble powder, a very heavy one. The shell struck the 8-inch armour on her starboard battery just upon a bolt, and bursting outside drove the bolt through. It disabled an 8-inch gun and made a large hole in the deck above, but did no other damage. The shell from Valdivia was fired from a Krupp 8·2-inch 10-ton gun, with a charge of 100lbs. powder, and weighed 250lbs. It struck the *Blanco's* stern outside the armour, and, entering a compartment, where a number of the crew were sleeping on deck or in hammocks, was shattered by the 5-inch iron bulkhead which

protects the central battery from raking fire. Fortunately for
the forty men asleep in the compartment it was charged, not
with powder, but with sand, yet even so it killed six men and
wounded six others, three mortally. One of the killed, who
was in the line of fire, was horribly mutilated. On its way
into this compartment the projectile had passed through the
captain's cabin, and carried the pillow of an officer sleeping
there from under his head, without hurting him. The third
shell missed. After this the *Blanco* slipped her moorings and
stood in to the town, making fast to a buoy off the Custom
House. Here she could not be fired at without the risk of
projectiles dropping into the town. At night she left the
harbour, and henceforward watched it from a distance. The
O'Higgins joined her, and pitched some shells from time to
time at the torpedo depôt outside the bay. Meanwhile
Balmaceda fitted out his torpedo-boats, which were not much
the worse for the *O'Higgins'* fire, and armed the *Imperial*
and *Maipo*, both merchant steamers.

About the same date the rest of the Congressionalist squadron was busy in the north laying its hands upon the nitrate
ports. Iquique was the first place to be attacked, but in it was
Colonel Soto, with a Balmacedist garrison, who gave no signs
of willingness to surrender. A blockade was maintained by
the *Cochrane* and other ships for some days, and the Congressionalists, concentrating their troops, marched upon it by
land from Pisagua. After a check the town was captured on
the 16th of February, but three days later the Balmacedists
re-entered the town, driving the Congressionalists to the
Custom House. The rebel ships supported their men ashore,
and vigorously bombarded the town, the possession of which
was of vital importance to them. In the afternoon a dynamite
magazine blew up and the town took fire. On this the English
commander on the Pacific Station, Rear-Admiral Hotham,
invited the Balmacedist and Congressional commanders to a
conference on board his flagship, and succeeded in arranging
an armistice. Next day Soto evacuated the town, though he

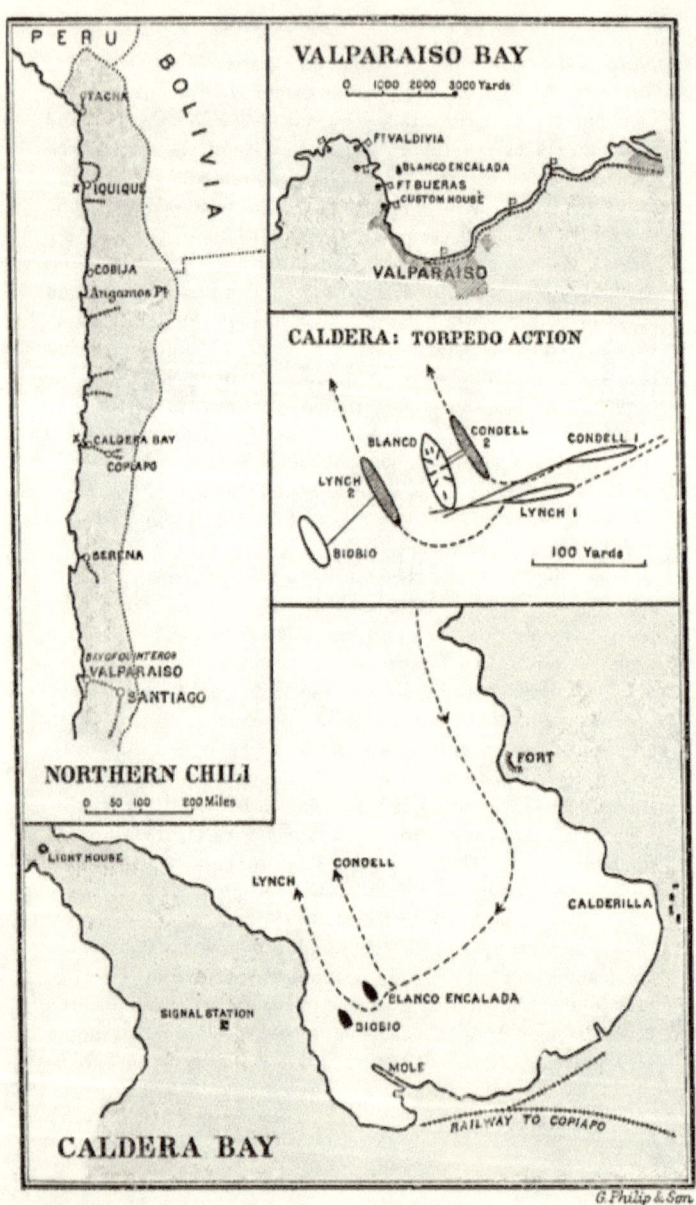

Map XXI.

should have held to it to the last; and the insurgents obtained control of the richest of the nitrate ports.

On January 27th an attempt was made by a steam-launch of the *Blanco* to destroy the armed steamer *Imperial* with a Whitehead. The torpedo missed the Balmacedist vessel, passing unpleasantly near the British mail-steamer *Britannia*. After this the torpedo boats in the harbour, which had so far remained inactive, in spite of the entreaties of Balmaceda's officers to be permitted to take them out against the enemy, were more on the alert and patrolled the harbour. On March 8th the *Maipo*, an armed steamer, was carried off by her crew to the Congressionalists. On March 21st the torpedo-gunboats *Lynch* and *Condell* arrived. They were in very bad order, as their English engineers had left them at Buenos Aires, not caring to face the risks of war, and the men who had brought them on had damaged their boilers, burning through many of the tubes. These were replaced at Valparaiso, and a French torpedo artificer was engaged to handle and adjust the Whiteheads.

At last they were ready, and on April 16th left Valparaiso in the company of the *Imperial*. On board this ship was Mr. Hervey, the "Times" correspondent, whilst Captain Moraga, of the *Condell*, was in command of the small squadron. Under him was Captain Fuentes, in charge of the *Lynch*. Both these officers had passed through the torpedo school at Valparaiso, and both were admirable disciplinarians, men of dash and determined courage, beloved by their crews. Just before he left Valparaiso, Mr. Hervey had somewhat indiscreetly telegraphed to the "Times" that he was leaving with a torpedo expedition, and the news had been cabled back from Europe on its appearance to Chili, and on April 21st it was published in the Congressionalist journal "Patria." Hence the insurgents received some warning of what was to be attempted.

After some days of constant small-arms and torpedo practice at sea, the three vessels left Quinteros Bay on April 21st, the *Condell* and *Lynch* hugging the shore, and the *Imperial* further out. No lights were carried at night, the better to escape the Congressionalist ships, though of these there was only one, the *Esmeralda*, faster than the *Imperial*, and even she would, in fine weather, have been distanced by the torpedo-gunboats. The destination of the squadron was Caldera Bay, 450 miles north of Valparaiso. According to the information which Captain Moraga had received, the revolted fleet was at anchor there, without its torpedo-nets, which had been left at the outbreak of the insurrection in store at Valparaiso. On the 22nd the *Imperial* lost sight of the gunboats at dusk, and heaving-to off Caldera waited for signals from them. The night passed without any being made, but an officer on board believed that he heard firing towards Caldera.

The gun boats reached Huasco on the afternoon of the 22nd, where news was received that at Caldera were the *Blanco*, *Cochrane*, and *Huascar*, with a corvette and three transports. Captain Moraga, therefore, summoned Captain Fuentes on board his vessel and arranged with him the plan of attack. He did not stop to take Mr. Hervey on board, as had been agreed, but went straight on to Caldera. A little later the *Lynch*, which had been running her torpedoes, spoke a boat and received the intelligence that three Congressionalist vessels had already left Caldera, and that if the torpedo-gunboats wanted to catch the others they would have to make haste. The two boats increased their speed, steaming abreast, and at half-past three in the morning found themselves well to the north of Caldera. The plan decided upon was as follows:—Entering the bay from the north, the *Condell* was to lead by 200 yards. Both vessels were to make for the insurgent ships, the *Condell* upon the starboard side and the *Lynch* upon the port side. They were to creep up as close as possible, and, first of all, to use their bow tubes; then afterwards, the training tubes abeam. The night was dull and cloudy, and though there was a moon, it was, from time to time, hidden by the clouds.

At 3.30, just as the day was dawning, Moraga led the way into the harbour. In the uncertain light the hostile ship could be discerned from the bridge of the *Condell*, lying at anchor in the western curve of the bay. The *Lynch* was very close astern, not more than twenty yards off her leader. When the position of the enemy had been ascertained by Moraga, he headed at half-speed for the larger of the two vessels, which he imagined to be either the *Blanco* or *Cochrane*. Astern of her was the *Biobio*, a small transport, which he mistook for the *Huascar*. When only a hundred yards off he fired his bow tube at the ironclad, but the torpedo just missed her, going astern, and passed very close to the *Biobio*. An instant later he turned his helm hard to starboard, and ordered Lieutenant Vargus, who was in charge of the port tube, to fire a second torpedo, which, he thinks, hit the target.* Just at this moment the ironclad opened a sharp fire, using small-arms, quick-firers, and heavy guns. The *Condell*, going full-speed, now discharged her third torpedo, which missed. But the *Blanco's* gunners had not noticed the *Lynch* creeping up behind the *Condell*, and had concentrated all their fire upon the latter. The *Lynch* was able to come on unmolested till within fifty yards of the big ship, when she fired her bow torpedo, which missed. Turning, she fired her starboard tube and hit the *Blanco* amidships. The torpedo exploded with great violence, and two minutes later the *Blanco* went to the bottom. The *Biobio* lowered her boats, and others put off from the shore, saving between them ninety-six officers and men. When the ironclad opened fire, and it was clear that the torpedo vessels were discovered, they replied to her discharge with deadly effect from their Hotchkiss 12-pounders and 3-pounders. Seven minutes only elapsed between the discharge of the first and last torpedoes. After the *Blanco* had sunk, the two torpedo-gunboats left the bay at full speed.

---

* In this he was probably mistaken. The Congressional account mentions only the bow torpedoes of the *Lynch*, and does not allude to these two, which perhaps dived. But inconsistencies in such a matter are only to be expected.

The above is Captain Moraga's official account of the action. Mr. Hervey adds some interesting details. The first torpedo fired by the *Lynch* went straight to the bottom; the second steered wide; and the third, which had always run badly in practice, alone struck the *Blanco*,* A torpedo was afterwards picked up in the bay, and was found to be a Mark IV. Fiume Whitehead, set for 600 yards and to sink. The pistol had been altered unskilfully.

What was the *Blanco* doing to be thus taken off her guard? we may ask; and to her proceedings and the Congressionalist accounts we may now turn. To begin with, her Captain, Señor Goñi, was under the impression that he was outside the range of Balmaceda's torpedo-boats, and never, apparently, gave the least thought to the fast and formidable gunboats. He may also have imagined, as is said to be the case, that the President's boats would never, from patriotic motives, go to the length of sinking a Chilian ship. If so, he was singularly deceived. On this particular night he was on shore with a force which was attempting to capture Copiapo, according to one account; according to other stories, at a banquet. In his official report he speaks as if he had been on board, and his own evidence must be accepted. The following is the drift of the Congressional account: The ship was left moored to a buoy, with enough steam up to move her engines. Her armament, in addition to her six 8-inch breechloaders, consisted of three 6-pounder Hotchkiss quick-firers, mounted one on each side of the forecastle, and one on the poop. She had four 1-inch Nordenfelts, which were placed, two on the fore-bridge, one on the after-bridge, and one on the poop; one Hotchkiss quick-firer in her top; and two Gatlings. Of her crew, which numbered nominally 300 officers and men, the greater portion were ashore, leaving only eighty of her original complement. The vacancies on board had been made up to 288 by adding raw recruits, who neither knew the ship

---

* Hervey, 322. According to Moraga, the *Lynch* only fired two torpedoes.

nor her weapons, and could not therefore be expected to perform well in the trying emergency of a torpedo attack. Of her two steam-launches one was undergoing repair on deck, and the other was in bad order. Neither was employed on patrol duty on the night of the 22nd, though this is stated to have been exceptional. Seven men formed the total of her watch, and the officers on deck were Commander Gonzalez and Aspirant Aguilar. At about 4 a.m., or probably before, the torpedo vessels were seen at a distance of no less than 2000 yards. The alarm was given at once, but the crew were slow in going to quarters, as they thought that this was merely a false alarm to practise them. It is also said that the bugler made a mistake, and sounded the ordinary morning call, instead of the summons to go to quarters. In any case the leading torpedo vessel was 500 yards off before the *Blanco* opened fire. This does not altogether agree with Captain Moraga's version, according to which the *Blanco's* guns were not fired till the *Condell* was well within a hundred yards. Either then the Congressional officers over-estimated the distance, or having mistaken the *Condell* for a friend, and allowed her to approach, they waited till her torpedo showed her to be an enemy. Their inaction is certainly hard to explain Accepting the Congressional statement, an interval of five minutes at least, the time required by the *Condell* to cover 1500 yards at half speed, must have elapsed between the giving of the alarm and the commencement of fire. The *Blanco* did not use her search-light, either because it was out of order, or because the growing daylight was strong enough to enable her men to see their enemy. Nordenfelts and 6-pounders were the first to open, and the assailants at once replied. On giving the order " Clear ship for action," the port engine had been ordered to go ahead and the starboard one to go astern, to turn the ship a little. The engines were just in motion when the torpedoes struck her. The *Lynch* and *Condell* having now drawn very close, the *Condell* fired her bow tube, but the torpedo ran ashore and exploded, after which the two seemed to stop their engines. Probably, this was at the moment when they were turning. An instant later, the *Lynch* went on full speed, passing the *Blanco's* starboard quarter, and as she passed, fired simultaneously two torpedoes at a distance of 100 yards.* The *Blanco* was struck on the starboard side, near the dynamo room. The shock was tremendous. Every light in the ship was extinguished, one of the 8-inch guns was thrown off its trunnions, and a large number of men were killed. Portions of iron and machinery flew about in the engine-room, and killed or wounded six engineers. The only one who escaped, was carried by the violent rush of water up a ventilator. Though orders had been given to close all water-tight doors, it is almost certain that this had not been done. The *Blanco* began to heel heavily to starboard, exposing her decks to the pitiless hail from the quick-firers of the torpedo-gunboats, which mowed the men down as they poured up from below. A shell from a 14-pounder Hotchkiss burst in her fore compartment, killing Lieutenant Pacheco. The ironclad fired one of her heavy guns in reply, but the shot passed over the torpedo gunboats. The ship began to go down very fast, and a minute later the order to abandon her was given. At that moment, the sills of her battery ports were level with the water. Of her total crew, which mustered 288 officers and men, eleven officers and 171 men were killed or drowned. Forty of these were said to have fallen victims to the machine-guns of the torpedo vessels.

As the *Blanco* sank, her men saw the *Lynch* steam round under her stern and fire a torpedo at the *Biobio*. This missed, as it was said to have passed under her. The gunboats then left the bay apparently uninjured. The only damage done to

---

* In the plan of the action the final torpedo, which settled the *Blanco*, and which was fired by the *Lynch*, is that which was discharged just as she turned to go under the ironclad's stern. Apparently from Captain Moraga's account she only fired one torpedo at the *Blanco* in this position, and not two as the Congressional version runs, or three as Mr. Hervey's.

either of them by the *Blanco*, was the cutting of the electric firing apparatus of one of the tubes just after the torpedo had been discharged. Both vessels had gone into action with five loaded torpedoes on their decks, so that a hit before they had got rid of any of them would almost certainly have been fatal. The worst damage was self-inflicted, caused by the firing of their 14-pounder Hotchkisses, which jarred them so much that some of their boiler tubes burst.

The official report of Captain Goñi was as follows:

"To my regret, I have to inform your Excellency that the ship under my command was sunk this morning, at half-past four o'clock, by the combined attack of the torpedo vessels *Lynch* and *Condell*, which succeeded in hitting the *Blanco* with six out of seven torpedoes fired. We have lost one-half of our crew, including amongst the officers, Paymaster Guzman, Lieutenant Pacheco, Aspirants Soto, and Aguilar, and numerous engineers. I have also to lament the loss of Don Enrique Valdes Vergara. The torpedo-gunboats received a heavy fire from the *Blanco* before she went down, and afterwards from the *Aconcagua*, when, on leaving Carrizal, she entered the harbour. We are, however, unable to ascertain what damage they have suffered. The ship is lying on her starboard side, with the tops of her bridges out of water. I believe we shall be able to recover the guns on the upper deck, and the others later. I have a diver here."

There is some discrepancy as to the number of torpedoes which hit the *Blanco* and exploded. According to Moraga six were fired and three exploded. Goñi's statement that six struck her is probably a loose exaggeration. It is even doubtful whether more than one torpedo struck the *Blanco*. Her hull was carefully examined by the diver of the *Champion*, a British cruiser which was present on the coast. It was found in about sixty feet of water, lying on the starboard side. The port side was intact, but on the starboard side was a large hole blown through the bottom of the ship, fifteen feet long and seven feet broad. The ship, having been wood-sheathed and cased with zinc, the iron skin had been detached near the hole from the wood-sheathing, and much of the zinc was torn off. The scuttles, on the port side, were found open, only the officers' ports being closed. From this account it looks as if the injury had been inflicted by one torpedo alone, unless two or more happened to strike exactly in the same place—a most unlikely supposition. The uproar and confusion which prevailed can readily explain mistakes in the accounts of eye-witnesses.

This is the first occasion on which the Whitehead was successfully employed against an ironclad, where we possess full details. Once before, during the Russo-Turkish war, a ship had been sunk by it according to the assertions of the Russians; but whereas then the Turks denied their loss, now the fact was beyond dispute, and the hull of the sunken ship could be seen and examined. Once more, as so often in these minor struggles, we have exceptional circumstances which would not be likely to recur in a European conflict. For an ironclad to be at anchor without nets out, without launches to protect her, without a search light, and with a very insufficiently trained crew on the top of all this, points to singular carelessness on the part of her commander, who had had warning, be it remembered. If he had chosen to trouble himself he had ample time, between the publication of the intended attack at Iquique on the 21st, and the morning of the 23rd, to have constructed a boom, or again he might have taken his ship to sea for the night and cruised in the offing with lights out, when he would have been comparatively safe. The torpedo-gunboats were handled with courage and coolness, but the task before them was not difficult. Smaller and weaker vessels could have destroyed, with less risk, this ironclad at anchor. If captains choose to imperil their ships as Captain Goñi imperilled the *Blanco* they will lose them. But French or English ironclads are not likely to lie in open harbours without taking the most elementary precautions. We may notice that the inexperience or want of smartness of

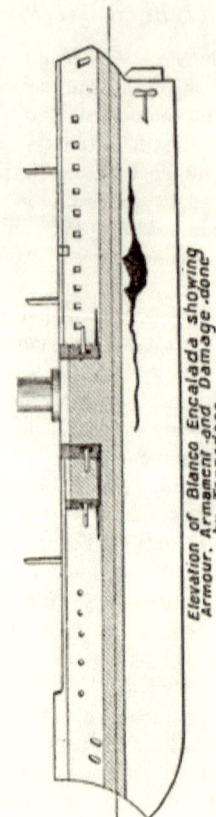

Elevation of Blanco Encalada showing Armour, Armament and Damage done by Torpedoes.

PLATE XXI.

the *Blanco's* men rendered her quick-firers and machine-guns almost useless. If they had stood smartly to their Nordenfelts and Hotchkisses they might have sunk or disabled the gunboats before they got close, and we may be certain that a hail of shells would not have improved the aim of the men at the torpedo-tubes.

For the *Lynch* and *Condell* the fighting was not yet over. As they stood out of the Bay they saw what they thought was the *Imperial*, and went towards her. When they got closer they discovered that it was not the Balmacedist vessel, but the Congressional armed transport *Aconcagua*. Her decks could be seen crowded with troops. She carried no flag, and was at once attacked by the two gunboats with their Hotchkisses. They poured into her such a furious fire, that the roll of it resembled the rapid discharge of small-arms rather than heavy guns. At first they separated and prepared to assail her one on each beam, but realising that this would enable her to bring both her broadsides into play they drew together again and fought her, keeping on her starboard beam. The *Aconcagua* replied vigorously with her 5-inch guns and small quick-firers. After some minutes, finding that she was getting the worst of it and that from the great speed of the gunboats she could not well retreat, she headed towards Caldera, probably hoping that the *Blanco* would come out to her support, but the *Blanco* was, of course, at the bottom. At last, after an hour-and-a-half, she stopped her engines, according to Captain Moraga, and had apparently surrendered. The gunboats were on the point of steaming in when a large cruiser was seen on the horizon. This they supposed to be the much dreaded *Esmeralda*, and therefore at once retreated. The tubes of one of the *Condell's* port boilers were leaking heavily and neither vessel was in condition for a fresh action with a powerful cruiser. Soon afterwards, when their prey had got safe into Caldera, they discovered that the approaching vessel was the British flagship *Warspite* and not the *Esmeralda*, and the chagrin of Captains

Moraga and Fuentes was unbounded. Neither of the torpedo-gunboats was much the worse for the fighting, though considering the loaded torpedoes which they had about, it was a most foolhardy adventure to attack the *Aconcagua* in broad daylight. They fired more than 400 projectiles, of which only eight, or two per cent., hit her. All these struck her above the water-line, and the damage they did was most trifling. Of the crowd of men on board only four were slightly wounded, two by a shell which struck the upper part of the funnel casing. The *Aconcagua* fired 137 rounds, seven from her 5-inch guns, and with these probably scored an even lower percentage of hits than the *Lynch's* and *Condell's* guns. The captain of the *Aconcagua* in his official report drew attention to the worthlessness of the torpedo-gunboat for open and straightforward fighting. She is built to act by surprise, and except when so acting is of little value. Had these boats, however, been armed with a heavier quick-firer than the 14-pounder, the 4·7-inch gun, for instance, as in the British service, the hits might have been as numerous, since the men would have fired more carefully, whilst the damage done would have been infinitely greater. The medium quick-firer, for such we may call the 12-pounder and 14-pounder, is not of much value except for the attack upon torpedo-boats, but the larger weapons of that type can do a great deal of damage. The Yalu confirmed this conclusion, as the small projectiles fired on that occasion had little effect. This action would also point to the comforting fact that merchant steamers, well equipped with medium quick-firers, have little to fear from torpedo-craft. This has an important bearing upon the defence of our commerce.

The *Imperial*, after waiting hours for her two consorts, returned to Valparaiso, where she found them all safe. The next occurrence was an attempt of the torpedo-boat, *Guale*, which was employed by the Balmacedists to patrol Valparaiso Harbour, to run off and join the Congressionalists. She did not, however, get very far, as the *Lynch* was sent in pursuit, and quickly overtook her. She was taken back to Valparaiso, where her crew were shot for their disloyalty.

On May 14th, the *Condell*, with the *Imperial* and *Lynch*, was off Iquique, endeavouring to repeat her performance at Caldera. The Congressionalists, however, had learnt by experience, and were quite ready for them. Though there were no warships in the roads, booms and numerous sunken torpedoes had been placed about the anchorage, so that there was nothing to be done. The Congressionalists, since the loss of the *Blanco*, steamed out to sea every night. Thence the *Condell* proceeded to Caldera, off which port she arrived on the night of the 16th. It could be seen that there were a large number of vessels inshore, and with everything ready for the attack, she ran in. Mr. Hervey, the "Times" correspondent, was on board, and has given a vivid picture of the suspense, the breathless anxiety, of those whose fate it is to be boxed up in these fragile cases of machinery with hundreds of pounds of gun-cotton, in the shape of torpedoes, covering the deck. When the *Condell* got close in, she found that once more the insurgents had been too clever for her. Two lines of ships were moored in the bay; the inner line towards the shore, composed of armed transports; the outer line, of neutral sailing-ships; and the worst of it was that the outer line was so disposed, as to cover the transports. It was therefore impossible to use the torpedo, and though Moraga was anxious to give his tubes an airing, and neutrals a lesson, he was dissuaded from a course which would have certainly embroiled his government with powerful enemies. A few days later he had an opportunity of sinking the *O'Higgins*; he had caught her in harbour at night,* and apparently off her guard, and was advancing to the attack upon her, when one of his crew told him that he had a brother on board. With touching tender-heartedness, Moraga abandoned the attack and with-

---

* At Pacocho. Hervey, 213-4.

drew; the act deserves to be remembered, though in war there is small room for consideration such as this.

The torpedo gunboats made one more attempt to ensnare their opponents. On June 4th the *Imperial* started north. At night she steamed ablaze with lights, whilst astern of her in absolute darkness followed the gunboats. It was hoped that the Congressional cruisers would see her, and stand towards her when she was to surrender. Whilst the Congressionalists were lowering boats the torpedo craft were to attack them off their guard, and sink their ship or ships. But the plan came to nothing as no Congressionalists showed themselves.

The final downfall of the Balmacedist Government came in August, and was the result of a combined military and naval expedition against Valparaiso. Having week by week collected more and more troops in the north, armed them with repeating rifles, and obtained German generals, the Congressionalists were ready. The whole insurgent fleet collected 100 miles north of Valparaiso, and as there was little to fear from Balmaceda's torpedo-vessels, steamed to Quinteros Bay. It was off the bay on the night of August 19th, and next morning, after a search for mines conducted by the smaller vessels, the troops carried on board were landed. Meanwhile, on the 19th, the *Esmeralda* had engaged the Valparaiso forts to divert the attention of the Balmacedist troops. On the 20th, in the morning, two of Balmaceda's torpedo-boats came out to see what was happening, but were quickly driven back. The troops having been put ashore by means of large punts, which had been conveyed from the north, fastened bottom outwards to some of the ships, at Concon on the 21st, supported by the fire of the ships, drove back the Balmacedists, and a week later captured Valparaiso. Thus ended the naval struggle on the Pacific.

The Chilian Civil War is important for two reasons. Firstly, because then the Whitehead for the first time sank an ironclad; secondly, because of the admirable strategy of the insurgents. They used their fleet sparingly against fortifications, making no attempt to capture Valparaiso by bombardment from the sea. They recognised a truth which is sometimes forgotten, that fleets cannot act on land, though they do exercise a very marked influence on land actions. They were confronted by a naval force which lacked all capacity of action except by surprise, and they showed that such a force is powerless to change the issue of a war, though it may destroy individual ships. We may, perhaps, wish that Balmaceda had been able to place in line the two cruisers detained at La Seyne. In that case we should have seen what we are particularly anxious to see—a contest between fast, lightly-armed vessels on the one hand, and slow, heavily-armed vessels on the other. Such a contest would have taught us much concerning the value of speed, and the practicability of raiding the enemy's coast, in the face of a superior but slower force.

Efforts were afterwards made to raise the *Blanco*, but they were unsuccessful.

One other incident of the war deserves mention, though it concerns the lawyer rather than the sailor. At the outbreak of the struggle the Congressional party had despatched a commission to purchase arms and munitions of war in the United States. The *Itata*, a Congressional steamer, was sent to embark these at San Francisco, but through a mistake arrived too soon, and lay waiting in harbour there. The suspicion of Balmaceda's representative—who was, of course, the official representative of Chili, as the Congressionalists had not received recognition—was aroused, and a watch was kept on the *Itata*, to prevent her loading with contraband. Seeing that it would be impossible to take a cargo on board at San Francisco, the Congressionalists embarked the arms on board two American schooners, which were to meet the *Itata*, and tranship their cargoes to her, outside the three-mile limit. In the meanwhile the *Itata*, proceeding south, ran short of coal, and put into the United States' port of San Diego. As it was believed that she intended to take the military stores on board there, a United States marshal was sent on to her to detain her. This was on May 5th. On May 6th, early in the morning, she slipped her cable, went out to sea, met the schooners, and transhipped the arms. Two days later she put the United States' marshal ashore. Indignant at this outrage to their flag, the United States' Government issued orders to their officers on the Pacific Station to seize her when she reached the Chilian coast. She was surrendered at Iquique on June 3rd to the American Admiral Brown. The incident very nearly led the Congressionalists to come to blows with the United States. In the end the case of the *Itata* came before the United States' Supreme Court, which returned the ship and arms to their owners, on their giving bonds for 120,000 dollars. The Balmacedist Government had fallen by then, and the Congressionalists had become the legitimate rulers of Chili.

CHAPTER XVIII.

THE CIVIL WAR IN BRAZIL.
September, 1893—April, 1894.

HARDLY had the civil war on the Pacific coast come to a conclusion when a fresh struggle broke out upon the Atlantic. Since the fall of the Empire Brazil had been in a more or less disturbed state, and considerable ill-feeling had arisen between the army which supported Marshal Peixoto, and the navy which was in favour of Señor Custodio di Mello. The trouble came to a head on July 6th, when Admiral Wandenkolk of the Brazilian navy seized the merchant steamer *Jupiter* at Montevideo. With a large number of sympathisers he steamed to Rio Grande do Sul, captured two diminutive Brazilian warships, and issued a proclamation in which he invited his brother officers of the fleet to join him. His career was short and unsuccessful. On July 20th, the *Republica* and *Santos* came up with him and compelled him to surrender; but a more serious attempt had to be faced on September 7th, 1893, when Admiral Mello, assisted by thirty-six naval officers and six or seven members of the Brazilian congress, seized the warships in Rio Harbour. Amongst these were the following vessels: the *Aquidaban*, a sea-going turret-ship of a design similar to the *Ajax*, with two turrets amidships placed *en échelon;* she had been built in England and carried as her heavy armament four 9·2-inch guns; the *Javary*, a low freeboard, coast-defence, ironclad, carrying four 10-inch muzzle-loaders in two turrets placed fore and aft; the small river monitor *Alagoas*; the cruiser *Almirante Tamandare*, which had been disabled by injuries to her machinery; and was armed with quick-firers which were protected by thin armoured casemates; the Elswick cruiser *Republica*; the old cruisers *Guanabara* and *Trajano*; two ancient gunboats; two transports; three large sea-going torpedo-boats, of twenty-five knots maximum trial speed; four smaller ones; and a host of armed merchantmen.*

Though this very miscellaneous squadron only included two ships of any real power, the *Aquidaban* and *Tamandare*. Marshal Peixoto found himself quite unable to oppose it at sea at the outset of the war. The *Riachuelo*, which closely resembled the *Aquidaban*, was in Europe, as also was the cruiser *Benjamin Constant*. On the South American coast he had the Elswick gunboat *Tiradentes*, which was in dock at Montevideo, with a not too trustworthy crew; the *Bahia*, a very indifferent coast defence monitor, and five small gunboats or cruisers. The torpedo gun-vessel *Aurora* was, however, expected from Europe, where she had just been completed by the Armstrong Company. She was of 480 tons displacement, equipped with three torpedo-tubes, two 20-pounder quick-firers, and four 3-pounders. Her trial speed was eighteen knots.

The contest that followed was, in its initial stages, one between ships and forts, and almost entirely destitute of incidents of either tactical or strategical importance. The Melloist ships lay in the Rio Harbour, and day by day exchanged fire with the Peixotoist forts. The ships were short handed, as the total force at the disposal of the insurgents was only 1500 men,† and though Admiral Mello had numerous adherents in the southern provinces of Brazil, he was quite unable to equip a land army. He started with a considerable advantage in being able to control the sea, as it is needless to state that the vast extent of the Brazilian coast

* *See* Table XVII.

† The naval force of Brazil consisted of 3020 marines, 990 sailors, 3300 firemen and naval apprentices.

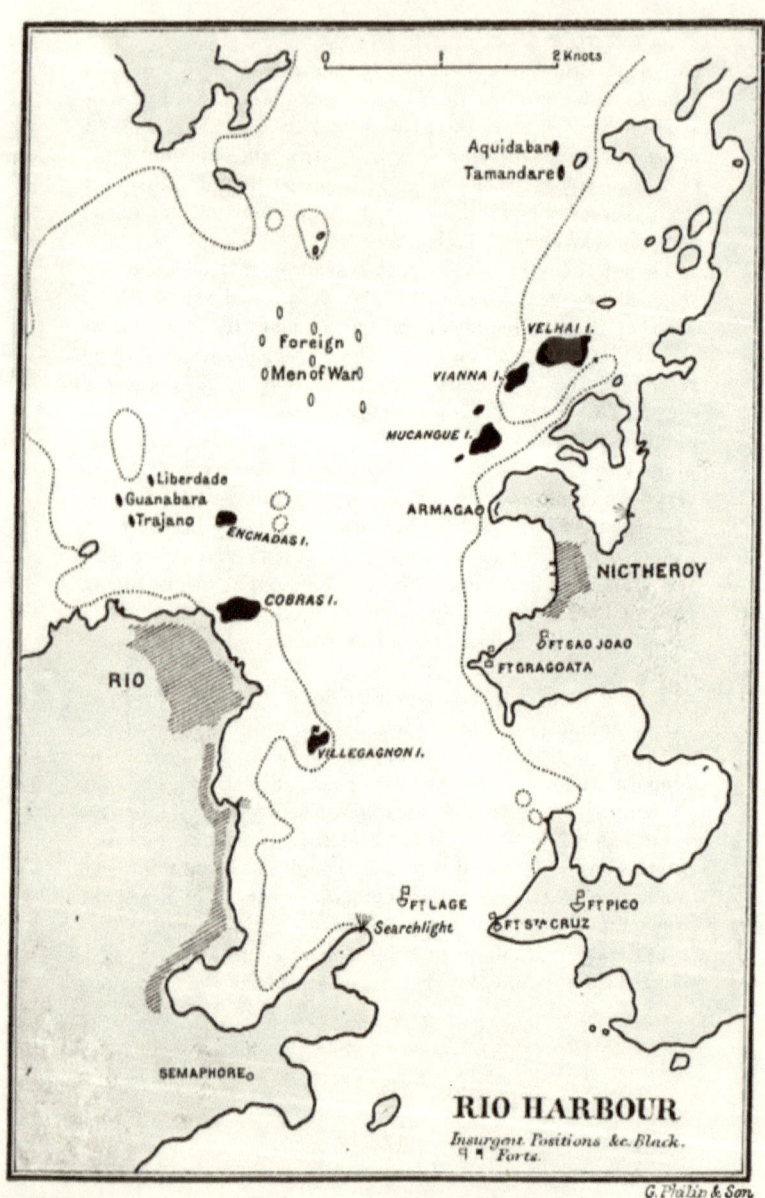

renders communication difficult between the various points by land. He endeavoured, following the strategy of the Chilian Congressionalists, to establish himself in the south, but did not meet with their success. Peixoto, on the other hand, had at his disposal considerable financial resources, which the Melloists lacked, a formidable military force of 24,800 officers and men armed with breech-loaders, and all the forts commanding the harbour of Rio, with the sole exception of that on the island at Villegagnon, by far the most heavily armed. This was at first neutral, but later declared for Admiral Mello. Numerous foreign ships of war were present in the harbour and their commanders prevented Admiral Mello from enforcing a strict blockade, which might perhaps have reduced Rio. They also did all in their power to avert a bombardment of the city.

On the morning of September 17th, the *Republica* and *Diaz* ran out of the harbour, past the forts, unharmed, and were followed next day by the *Pallas* and a second ship, with equal success. From September 14th onwards, there was constant firing between the forts and the ships, but the powder used by both sides was extremely bad, and though there was a liberal expenditure of ammunition, extraordinarily little damage was done. In this hot climate explosives deteriorate greatly by years of storage, and it was never certain how any gun would shoot with the stuff from the magazines. On September 23rd, Fort Santa Cruz fired eighty-five projectiles without harming anything but an old wooden hulk. On September 30th, the *Aquidaban* was under fire, and was hit several times, receiving very slight damage. A 6-inch shell came through her deck and exploded in the admiral's cabin, but did not set her on fire. A 6-inch shot entered her starboard side, and struck a shell which was standing in a shot-rack, ready loaded and fused. This exploded, wounding an officer. A third shell entered the admiral's bath-room, and a fourth burst in a coal-bunker, penetrating the unarmoured side, but only made a round hole of the diameter of the projectile. Fragments from this shell were driven through the inner wall of the bunker and wounded a man. A fifth projectile struck the shield of a 6-inch gun, but did it no harm. Other shots struck the armour, leaving only dents. On October 6th, Fort Villegagnon joined the insurgents, and co-operated with them in bombarding the other forts.

On the night of October 12th, an incident of more interest occurred. The armed merchant steamer, *Urano*, with 200 Melloist troops on board, attempted the run out to sea, past the Peixotoist forts. The mouth of the harbour is not more than a mile wide, and is commanded by three forts—Santa Cruz, which mounted two Armstrong 10-inch muzzle-loaders; São João, with one 10-inch muzzle-loader; and Lage, with three 6-inch Whitworths. As she went by, the forts fired into her, Santa Cruz hitting her several times, and she lost forty killed or wounded. She was riddled by machine-gun bullets, and the men on board her were demoralised, and fled on shore. A few days later the Melloists scored a success to counterbalance this loss, as the cruiser, *Republica*, rammed the *Rio de Janeiro*, which was conveying 1100 troops south, for Marshal Peixoto. Five hundred soldiers are said to have been lost, but details are wanting. On the 25th, a Melloist magazine was blown up by the fire of the forts, and on the 3rd of November, a depôt on Gobernador Island, containing 100 tons of powder, was destroyed by an emissary of Marshal Peixoto. Five officers and seamen from the British squadron happened to be near the magazine and lost their lives. On November 20th, a field-gun on shore struck a torpedo-boat which was lying in the harbour close to the *Aquidaban*, and sank it. On the 22nd, the *Javary* was hotly engaged during the morning with the forts. In the afternoon, whilst off Villegagnon, she was discovered to be sinking. Efforts made to tow her into shallow water were unsuccessful, and her crew were compelled to abandon her, though not before an audacious seaman had fired two of her heavy guns, which had been left loaded, at the forts, only a minute or two before she capsized and sank. She was an old vessel, and had probably been shaken badly by her own fire, though it is just possible that one of the enemy's projectiles had damaged her, as a shell from Fort São João fell either on her deck or close to it, just before she went down.

On November 29th, the *Aurora* arrived from Europe, and was taken over by Marshal Peixoto at Pernambuco. Her name was changed to *Gustavo Sampaio*, but as yet she was not brought round to Rio, probably through want of men to work her. On the night of November 30th—December 1st, at 12.30, the ironclad, *Aquidaban*, accompanied by the armed steamer, *Esperança*, ran past the forts. The *Aquidaban* manœuvred to draw the fire in the glare of their search-lights, and received the attention of all their guns, herself returning their fire. The gunners in the forts knew that she was coming and were ready for her, but in spite of this could inflict no damage upon her. Not a man on board was killed or wounded, and only a single shell struck her hull. This burst in a coal bunker. Some small projectiles also dropped on her deck. The *Esperança* was hulled by a shot which passed through her engine-room, and lost her chief engineer, killed, besides one or two men wounded; but was quite able to continue her voyage. Admiral da Gama was left in charge of the vessels in harbour, and on December 3rd engaged the forts with the *Tamandare*. On the 4th another armed steamer ran past the forts. On the 22nd the *Tamandare* was again in action, using ballistite for her quick firers. The new explosive performed well, and it was most demoralising to the men in the forts to find shells dropping amongst them with no premonitory warning, such as is given by the smoke from an ordinary charge of pebble or brown powder. The *Tamandare* was often hit but not damaged. About this time the *Republica* had a brush with the Peixotoist transport, *Itaipu*.

On the 12th of January, 1894, the *Aquidaban* ran past the forts at dawn without being touched. Steaming up the harbour she was hit twice with very trivial damage, and the loss of two men, wounded. Admiral Mello was not on board, having remained in the south to organise a military force. In this he was unsuccessful, as there were numerous insurrections on hand, each with a separate leader, competing for support. On January 16th Mucangué Island, which had been occupied by the Government, was recaptured by the Melloists. January, and the first days of February, passed without more than an intermittent cannonade, but on the 9th of February the insurgents made a desperate attempt to carry the Armaçao battery, on the east side of the harbour. They were repulsed with heavy loss, and the gunboat, *Liberdade*, was repeatedly hit whilst covering the landing party.

In the meantime Peixoto had been busy acquiring an improvised fleet. In the United States he had purchased a collection of naval curiosities, which he now prepared to use against the Melloists. One of his purchases was Ericsson's submarine gunboat *Destroyer*, renamed the *Piratiny*. This was a small steamer, with a gun built into the bow, from which a torpedo, thirty feet long, could be fired. Above the compartment which contained the gun, was a thin armour-deck, and above this again a compartment filled with inflated india-rubber air-bags. The engines and boilers were protected by two stout armour-plates, inclined at an angle of twenty-three degrees, forming an athwartship bulkhead above the water-line. This vessel had been built for the United States' Government, but, the Ordnance Department having changed heads, in 1881, the new chief refused to purchase her. A similar gun to the one carried on the *Destroyer* was sent to the Peruvians in 1880, but never used; the weapon was also experimented with by the British Government. Besides the submarine gun she carried a Howell training torpedo tube and two Hotchkiss 1-pounders. On the steamer *Nietheroy* (formerly *El Cid*, of the Morgan line), a Zalinski pneumatic gun had been fitted forward, for the discharge of aerial torpedoes containing 50lbs. of dynamite. This weapon is fifty feet long, and has a calibre of 15-inches. It is capable of training, and can throw its shell by compressed air to a distance of 3000 yards, but the time occupied by the shell in its flight, projected as it is with very low initial velocity, is so considerable that it must be difficult with it to hit a moving target. It was, however, an interesting experiment to mount it, and we may regret, in the interests of naval science, that it was not tested in action. In addition there were one 4·7-inch quick-firer, two 10-centimètre quick-firers, eight 6-pounder quick-firers, and several small guns. Besides these vessels there were the armed steamers *Advance, Alliança, Finance,* and *Segurança*, the *America*,* with a Sims-Edison controllable torpedo, the small steam-launch, *Feiseen*, re-named *Inhanduay*, which had the phenomenal trial speed of twenty-five knots, the twenty-two-knot launch *Nada*, five Schichau torpedo-boats, and a small Yarrow boat. The *Nada* and *Feiseen* were carried on the *Nictheroy's* deck.

Trouble was early experienced with this odd assemblage of ships. Their crews were untrustworthy, and on board almost all of them were Melloist emissaries, who lost no opportunity of damaging their machinery. Breakdowns were constant; there were neither swords nor cutlasses for the sailors who were to fight them, and trained men to handle the new engines of war were lacking. There was no one who understood the Sims-Edison torpedo. The dynamite gun had an altogether insufficient supply of ammunition. Five loaded shells, only one of which contained a full charge, were not enough to sink the Melloist squadron. In addition, there were, it is true, five unloaded shells, but the few fuses on board were untested and untrustworthy. The torpedo-boats had been very carelessly placed on deck, and could not be hoisted out, as the necessary tackle was wanting. Parts of the boats' gear were missing, and their torpedo tubes were in bad repair. The *Destroyer's* engines had broken down, and, in short, the improvised fleet was useless. The most serviceable part of it was the squadron of Schichau boats, which had safely crossed the Atlantic; but

* Re-named *Andrada*.

these were not used as yet. After numerous executions of untrustworthy Brazilians, some ships were at last ready for action. On February 18th, the *Nietheroy* appeared off Rio, and landed 300 troops. She had intended to try her dynamite gun on Villegagnon, but the gunner had disappeared, and at the critical moment the weapon would not work.

On February 23rd, the armed Melloist steamer *Venus* was struck by a shell, which exploded her boilers or magazine, blowing her in two, and killed thirty officers and men. A fortnight later came the end of the tedious warfare at Rio. On March 10th, Peixoto's squadron appeared off the harbour, and next day, notice was given by him of his determination to attack the insurgents. On the 12th, Gama proposed terms of submission, which were rejected. Despairing of his cause, he went on board a Portuguese cruiser which lay in the harbour, with all his superior officers, and landed his men on one of the islands. The warships were left to serve as a target for Peixoto's fleet. At noon on the 13th, the *Nietheroy, Andrada, Tiradentes, Paranahyba, San Salvador, Gustavo Sampaio*, two steamers and five torpedo-boats began the attack. For four hours, supported by the forts, they played upon the squadron in harbour, and then discovering at last that there was no reply to their fire, came in towards Rio. Next day, they seized the insurgent vessels. It does not appear that the *Nietheroy* used her dynamite gun.

Having recaptured the rebel ships, Peixoto demanded the surrender of the officers, which was refused him, the Portuguese vessel steaming away and landing them in neutral territory. Peixoto on this, broke off diplomatic relations with Portugal, but nothing more followed.

The insurrection, or one of the insurrections, still smouldered on in the south, where Admiral Mello, with the *Aquidaban*, was at large. In April, Peixoto decided to make a torpedo attack upon her, and in consequence, a squadron of six ships*

---

* The ships were the *Andrada, Nietheroy, Tiradentes, San Salvador, Itaipu,* and *Santos*.

and four torpedo-craft were sent against her. She was lying, short of men, stores, and ammunition, in the Bay of Santa Catherina, close to Desterro. Off this place, the Peixotoist flotilla arrived in April, and from Tijucas Bay, at a distance of four or five miles, kept a watch upon her. She lay close to the small island of Santa Cruz, upon which stands a fort, whence she drew her supplies. Between this small island and the larger one of Santa Catherina she had placed one line of mines in the fairway, and was preparing to place a second to the rear of the first. The islands and the mainland were in the possession of Melloist sympathisers. Four torpedo vessels were selected for the attack. They were the Elswick torpedo gunboat, *Gustavo Sampaio*, with three tubes, one fixed in the stem and the other two training, one on each beam, and the *Affonso Pedro*, *Pedro Ivo*, and *Silvado*, Schichau boats, of 130 tons and twenty-six knots' speed, with complements of twenty-four officers and men each, and three 16-inch torpedo-tubes. On April 14th the *Aquidaban* was reconnoitred by the boats, and was seen to be lying at anchor about two miles to the south of the Santa Catherina light, and under the lee of Santa Cruz. On the night of the 14th—15th the first attack was attempted. The boats ran in under the northern shore of the bay or strait, but, before they had got far, saw that they were discovered. Bengal lights were burnt by the insurgents on shore, and signals made to the *Aquidaban*. In consequence the boats had to return without effecting anything.

On the next evening at dusk the steamer *San Salvador* steamed in towards the bay, and on the Bengal lights being shown, as before, opened upon them with machine-guns. In this way she must have destroyed the look-out stations, or driven off the watchers, as there were no more signals made during the night. When the moon set the torpedo-boats started once more. The sky was cloudy and the night extremely dark. They entered the bay, keeping in the centre of the fairway, and crossed the line of mines without misad-

venture. The plan arranged was as follows: They were to
advance in line abreast till well within the bay, and then to
turn in succession to starboard, thus spreading fan-wise, and
steaming in a north-westerly direction.* By this disposition
they confidently expected to find the *Aquidaban*, although the
night was so dark that they could not see her at any distance.
They would, too, in the formation adopted, be less likely to
get in each other's way, and thus impede the execution of
their attack. The *Aquidaban*, as it turned out, had moved
from her anchorage of the 14th, further to the north-west. At
2.30 the *Sampaio* was in the bay, and almost at once lost sight
of the other boats. On drawing near the place where she
expected to find the ironclad, her speed was reduced to make
as little noise as possible, and silently, without showing any
lights, she glided through the still water. The obscurity was
impenetrable: nothing could be seen or heard. After passing
the *Aquidaban's* anchorage of the 14th, she turned in a north-
westerly direction, as had been arranged, but could still see
nothing, and began to suppose that in the darkness the *Aquidaban* must have slipped out of the bay. Her captain resolved,
however, to make the full circuit of the waters before retiring,
and had gone one mile to the north-west, when he made out
what he took to be a small tug upon his starboard bow. He
was passing it, not thinking it worth his attention, when it was
suddenly lighted up, a sharp fire of artillery was opened upon
him, and it became evident that here at last was the *Aquidaban*. At once he went full-speed towards his enemy,
circling so as to bring his bow tube to bear. When he
thought the range sufficiently short, he shouted down the
voice-pipe of the bow tube to fire, but the officer in charge
had already fired the tube, fancying that he heard the order
given. The first torpedo consequently missed the *Aquidaban*.
The captain now took his boat round the ironclad's stern,
passing very close indeed to her, and turning, with his star-

* *See* Plan.

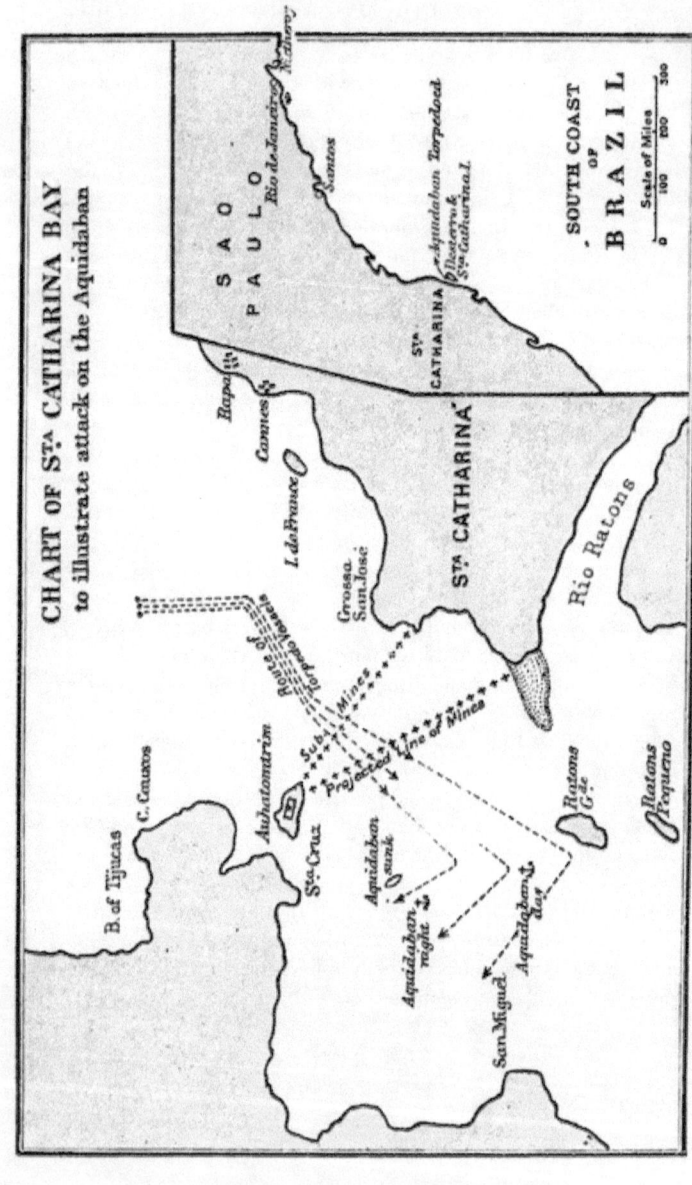

board to her port, stopped. The way on the boat took her slowly past the *Aquidaban*, at a distance of only 400 feet. When abreast of her funnel the order was given by the voice-pipe to fire the starboard tube, but nothing happened. The second officer, who was standing by the captain, seeing that there was danger of missing the *Aquidaban* in spite of all the risk that had been run, since the *Sampaio* was passing her fast, hurried aft, and seizing the lanyard of the tube, pulled it himself. The *Sampaio* had meanwhile travelled sixty feet, as three seconds had elapsed from the first giving of the order to fire the starboard tube. The torpedo ran straight, and in consequence of the delay, struck the *Aquidaban* very far forward instead of amidships.* The explosion was exceedingly violent, but to the surprise of the *Sampaio's* men, who saw a great uprush of water, and heard a terrific crash, did not change the ship's trim. The *Sampaio* moved ahead, and, just as the torpedo struck, the ironclad stopped her fire. The *Sampaio*, travelling at her fullest speed, covered 1000 yards before the Melloist gunners re-opened. Up to this point they had not used their search-lights, but these were now turned on.

Whilst the *Sampaio* was thus making her attack, the other torpedo-boats had been attracted to the sound of firing. The *Pedro Ivo*, indeed, had had to abandon the attempt, as the pressure in her boilers suddenly fell to 10lbs., and she could scarcely keep in motion; but the other two had gone forward. The *Silvado*, at first led, but when she got near the *Aquidaban* found the *Sampaio* between herself and the ironclad, and was forced to stop. She turned to starboard and passed the *Aquidaban's* bows at a distance of 1000 yards, without discharging her torpedoes, or taking any part in the attack. At the same time, a launch coming from Desterro, headed towards her, endeavouring to ram her, and to avoid this launch, she turned once more and retired. The *Affonso*

---

* It was of Schwartzkopf pattern, carrying 125lbs. of guncotton.

Pedro, as the *Aquidaban* opened, put on full speed and steamed past the ironclad's starboard side. When abreast of her, she fired two torpedoes from her training tubes, but it is uncertain whether either hit. According to the lieutenant in command of the *Affonso,* the first struck the *Aquidaban.* Having passed her enemy, the *Affonso* circled and retired, rejoining the other boats at Tijucas Bay. They were not aware of the success of their exploit; indeed, they seem to have supposed that the explosions had not harmed the ironclad, and were preparing for another attack on the night of the 16th, when the officers of the German cruiser, *Arcona,* informed them that the *Aquidaban* was deserted. The *Affonso Pedro,* though she went within 200 yards of the *Aquidaban,* was not touched. The *Silvado* was struck by one Nordenfelt bullet. The sailors of both boats heard the enemy's projectiles whistle over their heads. The *Sampaio,* which had stopped for a minute-and-a-half alongside, at a distance of 150 yards, was hit in all by thirty-eight Nordenfelt 1-inch bullets; but though nearly forty men were on board her, only one was very slightly wounded. The vessel herself was not at all the worse. The hits were distributed along her whole length. Two bullets struck the drum of the search-light, and the guns and torpedo tubes were scored.

We may now turn to the account of the *Aquidaban's* officers. That vessel had been expecting a boat from Desterro, probably the launch which engaged the *Silvado,* and when she saw the *Sampaio,* mistook her for this friend. In consequence, it was a minute or more before it became evident that the approaching torpedo vessel was an enemy, and not till it was recognised as the *Sampaio,* was fire opened. The *Aquidaban* used all those of her 1-inch Nordenfelts and 5·7-inch breech-loaders which would bear, but not her heavy guns. When the torpedo exploded, the shock felt was terrific; and the officer of the watch was thrown from the bridge into the sea. No one was killed by it, as there was no one in the forward compartments, which were torn open,

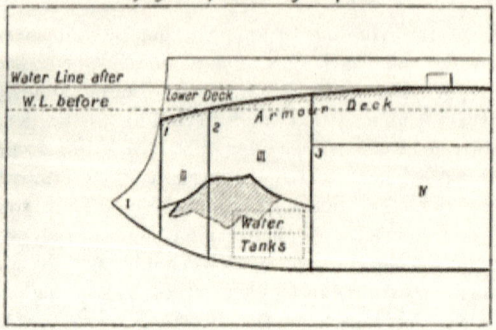

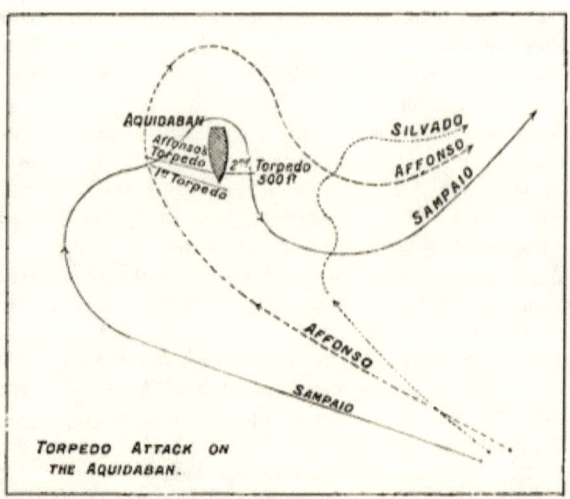

Map XXIII.

The water-tight doors were closed. There was great confusion after the ship had been struck, but in spite of this the ship's engines were started, and she moved a short distance to the north, where she slowly sank till she took the ground, in about twenty-two feet of water. As her draught was eighteen feet her deck and upper works remained above the water-level, and her crew were easily able to get ashore. On receiving promise of pardon one third of her men returned and surrendered.

The Brazilians took possession of her on the 16th. Making a careful examination, they discovered that the torpedo had struck close to the bulkhead, which separated compartments two and three, about thirty-five feet aft from the ram. It had blown a hole nineteen or twenty feet long, and six feet six inches broad, and at either end the steel skin of the ship had been torn.* The tear extended aft six feet longitudinally, past the bulkhead dividing the third and fourth compartments. The water-tight doors in this bulkhead had been loosened by the shock, so that the fourth compartment, which was a very large one, had filled as well as the first three. Inside everything was smashed beyond recognition. The armoured deck had been driven out a little, just over the place where the explosion occurred, and numerous rivets in the skin had been loosened. In all, it was calculated, she had taken on board 500 tons of water, and only the shoals saved her from foundering. On the starboard side forward, forty feet from the ram, was another hole, but much smaller than that to port, as it was only three feet in diameter. This may have been caused by the head of the first torpedo being driven right through the ship by the force of the explosion, or again it may have been made by the *Affonso's* torpedo. There is some doubt whether this hit, as no one appears to have heard or seen a second explosion; it is possible that it struck the ground, and, if it exploded, did not exert its full force on the

---

* *See* Diagram.

ship's bottom. This, however, is only a guess. The torpedoes used were of Schwarzkopf make, containing 57 kilogrammes (125lbs.) of gun cotton.

In this torpedo action there are one or two points of interest. Firstly, the *Aquidaban* betrayed herself; had her watch been a little less smart she might have escaped attack. The difficulty of discovering the vessel selected for destruction must always be great on a dark night when there is a wide sheet of water to be searched. A fleet of course would be easier to find, yet, on any but bright nights, even considerable squadrons should be safe from discovery on the open sea, provided no lights are carried. Secondly, having discovered that a vessel was approaching her, and knowing that hostile torpedo-craft were in the vicinity from the unsuccessful attempt of the previous evening, a mistake prevented her from opening fire till the enemy was too close. Thirdly, though the *Sampaio* lay almost motionless, not four hundred feet off her, the ironclad's gunners did not succeed in sinking the small torpedo gunboat. Probably the *Sampaio* was for nearly half-a-minute at very close quarters. The *Aquidaban** lacked medium quick-firers; between her 5·7-inch 70-pounder and her 1-inch Nordenfelts, she had nothing. All the evidence which we possess points to the utter futility of the machine-gun as a means of stopping torpedo craft of any size. A shell, with plenty of penetration, and a good bursting charge, is essential so that it, or its fragments, may rake the boat and open up its compartments. In fact, the larger the quick-firer the better, but for work against torpedo-boats the size of the gun must be conditioned by

---

* In the Table, which is based upon the report of Lieutenant Rogers (U.S.A.), in the U.S. Naval Intelligence publication, she is credited with six 4·7-inch quick-firers. There is no mention of these either in Mr. Laird Clowes' account of the war (Naval Annual, 1894), or in Lieutenant's Verlynde's, (R.F.N.) detailed description of the affair in the "Revue Maritime," March, 1895. We may, then, feel doubtful whether she mounted them. Of course, if she did, the remarks upon the necessity of heavy quick-firers lose their point, and the case against the large ship becomes very black.

the rapidity of its fire. The 12-pounder weapon would seem to be about the right weight, as it can deliver a good number of rounds in the minute, and its shells are large enough to do plenty of damage. One-inch Nordenfelts and Gatlings are useless for this special purpose. Fourthly, the *Aquidaban* had no nets out and was at anchor. No vessel has as yet been destroyed by the Whitehead, when in motion on the open sea, and it looks almost as if the conditions, which Mr. Laird Clowes has shown to apply to the use of the ram, apply also to the torpedo.* There were no vedette boats patrolling the bay, and there was no attempt to protect the ship by a boom. The mines, as is often the case, were a very delusory protection. Some of them were fished out of the water a few days later. Lastly, it is certain that her full complement was not on board, and that many of the men she carried were comparatively untrained. Admiral Mello is known to have suffered from shorthandedness, and he had lost some of his sailors in the land fighting.

From first to last the *Aquidaban* was the mainstay of the insurrection. She was able to take a certain amount of punishment from the forts, and though an antiquated vessel, secured for the insurgents the command of the sea, till the collapse came at Rio. It is not likely that Peixoto would have moved his squadron upon the harbour had she been present. The efforts made to effect her destruction showed that Peixoto did not feel himself safe whilst she was afloat.

The war cannot be said to have added greatly to our knowledge. It showed that an improvised fleet, without trained seamen, is a most untrustworthy instrument, but that is a self-evident fact. It showed that ships can pass forts with impunity, or something approaching impunity, if there is an unobstructed channel, but a long series of actions has already proved this.†

---

* As it is at present. *See* page 160.

† It is, however, possible that the pneumatic gun will modify this where the channel to be defended is narrow. This gun can be absolutely concealed, and is able to fire a shell a minute. The explosive charge of the shells is extremely heavy and one hit should disable any man-of-war afloat. *See* also p. 150.

Duckworth's brilliant run up the Dardanelles, Farragut's exploits on the Mississippi and, above all, at Mobile, had left no doubt. On the other hand, where there are obstructions in the channel, where there are mines or booms to hold the ships under fire, such proceedings become, not risky, but well-nigh impossible. It must be remembered that Marshal Peixoto's forts were not armed with heavy quick-firers, and that the powder used in the guns was very bad. The insurrection collapsed, not through any masterly activity on the part of the President, but rather through the incapacity of the Melloist leaders, and the fact that they could not collect an army. A fleet without an expeditionary force behind it is only valuable for defence, and lacks offensive power. Modern warships do not carry the crews of three-deckers, or even frigates, and have lost the power of landing a considerable body of men. The complements are barely sufficient to work the ships, and no one can be spared without risk.

Two months after she had been torpedoed the *Aquidaban* was patched and raised. She was then repaired at Rio, and her name changed to the 24 *de Maio*.

CHAPTER XIX.

THE STRUGGLE IN THE EAST, 1894-5.

Section 1.—The Combatants and their Ships.

FROM Brazil to the Yellow Sea is a far cry, but it is to this quarter that we must next turn our gaze. Hardly had the naval war in the West, a struggle so feebly and fatuously conducted that it seems almost absurd to call it a war, come to an abrupt conclusion, than the guns began to shoot in the East. Japan and China have been old enemies, and there is placed between them an apple of discord in the peninsula of Korea. In the spring of 1894 an insurrection broke out in the south of that kingdom. The King appealed to his nominal suzerain China for help, and help was granted. Two thousand Chinese soldiers were landed at the Korean port of Asan. Now by the treaty of April 18th, 1885, China was bound to inform the Japanese Government of the despatch of troops, and, as this had not been done, Japan promptly embarked a force of about 5000 men, and landed them at Chemulpho at the end of June. There followed the affair of the *Kowshing* and the declaration of war.

It seemed a most perilous adventure for Japan, a small and poor state with forty-one million inhabitants, single-handed to assail the colossal Chinese empire, with its four million square miles of territory and its three hundred millions of inhabitants. The task was not so formidable as it looked, for the colossus had feet of clay. The striking peculiarity of Japan is that there alone in Asia we find Western methods, Western organisation,

Western strategy, assimilated by an Oriental race. An observant eye might have discerned the prognostics of this astonishing phenomenon a generation back. At the close of the last century came a great revival of learning, bringing with it the study of earlier Japanese history, reminding a receptive and patriotic people that they possessed a great past. The revival of interest in their earlier history again contributed to the re-establishment of the Mikado's rule, in place of the Shogun, who had usurped much of his power. In 1853, when an American squadron threatened Yokohama, a historic debate took place at Yedo, in which the party who advised compliance with the American demands pointed out that, as they were, the Japanese must be beaten with no gain. The Kai-Koku, for so this party was called, went on to say, "Rather than allow this, as we are not the equals of foreigners in the mechanical arts, let us have intercourse with foreign countries, learn their drill and tactics, and, when we have made the nation as united as one family, we shall be able to go abroad and give lands in foreign countries to those who have distinguished themselves in battle." Japan was opened to the United States and later to England and Russia. In 1863 Japanese officers were sent to Holland to study naval war. In 1867-8 came the civil war, which ended in the triumph of the Mikado and the party of innovation. Japan steadily carried out the policy laid down by the Kai-Koku. Railways, telegraphs, roads were constructed. Schools were built and the population educated in a European fashion; a university at Tokio, as it was now called, was founded. First a deliberative assembly, and then, within recent years, a constitution was granted. The calendar was Europeanised, postage stamps, and an imperial post, were introduced; the criminal code was remodelled, and torture abolished. Newspapers were permitted to be published, and, in spite of a somewhat rigid censorship, there were 113 as early as 1882. Western manufactures made their appearance; cotton mills, paper mills, coal mines, and ironworks sprang up. At a

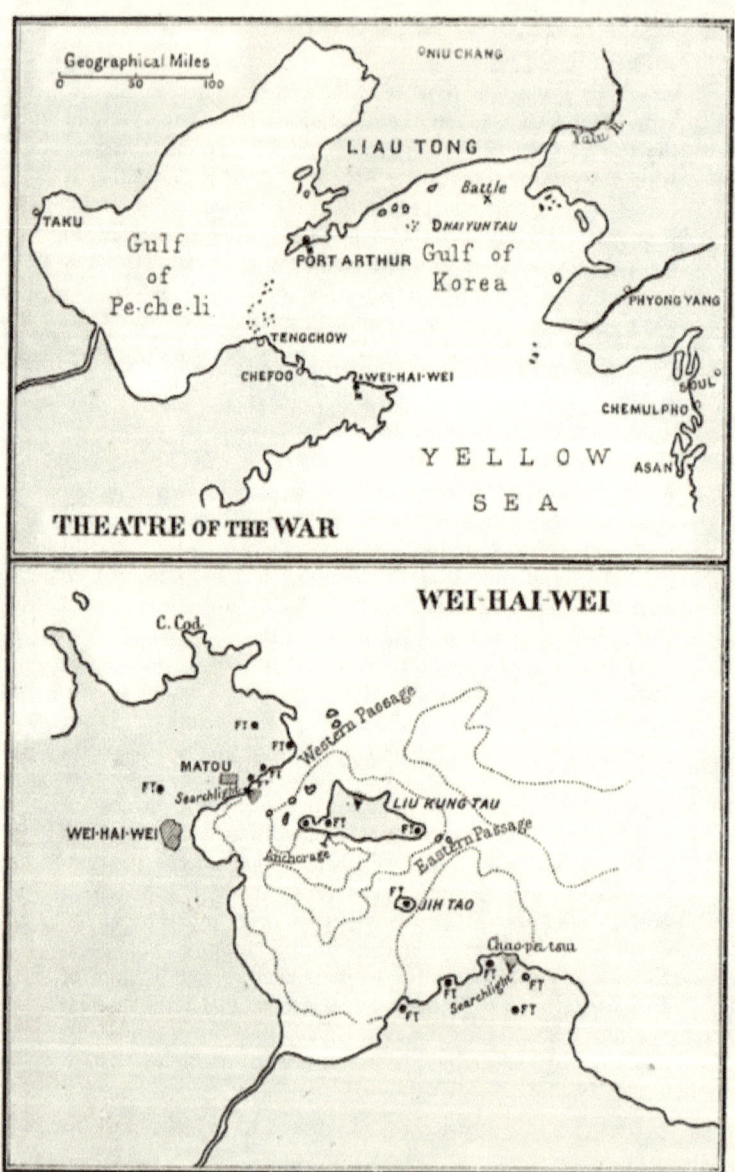

Map XXV.

bound the country passed from the twilight of barbarism into the daylight of full civilisation.

Japan, in its barbarous days, had always possessed a singularly high code of honour; its nobles had shown no want of self-sacrifice or devotion, whilst from end to end of the kingdom the story of their great past fired the Japanese, and made them ready to sacrifice their personal prepossessions for the welfare of Japan. The nation progressed eagerly upon the path of the Kai-Koku party. The army first received attention and was organised on the basis of universal service. That it is no paper force, but is not far from being the equal of any European force, is attested by its bravery and discipline in 1894-5, and by the unanimous evidence of observers. The Japanese fleet was similarly organised on the European model, England being chosen as the pattern; whilst English instructors were brought out to give practical and theoretical instruction. The islanders of the East take kindly to the sea. Captain Ingles, the ex-naval adviser to the Mikado's Government, states that they are just like Europeans—smart, constantly on the alert, cheerful, and patient. Their gunnery is excellent, though they are not so good with machine-worked as with man-handled weapons. The engineers are very good, keep the engines in capital order, and use them well. Thus Captain Ingles saw the *Naniwa* worked during the naval manœuvres at 100 revolutions, which was her maximum natural draught rate in England on her trial.*
Boilers and machinery are as efficient on the Japanese ships as they can be made. The discipline is comparable to that of an English squadron. The officers are hard-working and well up in the technical literature of their profession; in intelligence, capacity, and courage they are Europeans. Admiral Ito, the officer in command, is not a mere paper sailor, but has had training and experience in the annual

---

* On the other hand it is stated that the Japanese could never get anything like the trial speed out of their ships; and that the *Yoshino* in the action off Asan could not overtake the *Tsi Yuen*. *See* p. 69.

manœuvres. In short, the Japanese fleet is a war-force, and does not merely exist for show.

Some years were required to bring it to this pitch of efficiency, and as late as 1890 the Japanese Parliament refused to sanction a programme brought forward by Count Kabayama, which would have greatly increased the expenditure upon the navy, on the ground that the condition of the Japanese *personnel* was not satisfactory. It was thought that the naval officers had been selected by favouritism rather than merit, and there were objections to the pattern of the warships. But none the less a great deal was done between 1884 and 1894 towards providing modern and powerful vessels, and care was taken to procure the most modern guns and explosives. The naval officers in the war of 1894 proved that they were as good as their ships, and may be said to have surprised Japan no less than Europe.

On the other side of the China Sea is perhaps the most effete and barbarous state in the world. Whilst the national character of the Japanese stands high; whilst we admire them individually for their determined courage upon the battlefield, and for the intelligence and foresight which have won them victory, we can feel little but contempt for China. There is to be found an alien despotism, cruel and superstitious, ruling a vast congeries of human ants, nourished in filth, educated only in obsolete formulas and catechisms, taught to believe themselves infinitely superior to the "foreign devils" whom they so despise, and, if not without certain noble qualities, a certain passive stoicism, a remarkable faculty of application to work, yet ignorant, lethargic and bitterly opposed to Western innovation. The Government is as corrupt and treacherous as it is incapable. The rulers of provinces, the generals of armies, the admirals of fleets, are selected by an extraordinary system of examination, which would seem expressly adapted to choose the unfittest. Peculation is, with very rare exceptions, the primary object of everyone, from the exalted members of the Tsung-li-Yamen to the meanest mandarin.

Though time after time China had come into contact with Europe, and smarted sorely on each occasion, she had not learnt Japan's lesson. Indeed, we may doubt if she could have Europeanised herself so thoroughly and effectually, even had she followed hard in the footsteps of her rival, since there must have been a deep difference of character between the two races. China had no Daimios, no Mikado, no Kai-Koku; there was no one to hound her forward, and she remained true to Asia, true, that is to say, to hideous cruelty, to dirt, and to extortion; she retained an imposing exterior, but those who looked closely saw that it covered internal weakness and disorganisation. She deceived Europe, but she did not deceive Japan.

Needless to say, that a government which could do nothing but prevaricate, procrastinate, and peculate, would not be likely to bring into existence either a strong army or a strong fleet. Both might, indeed, exist on paper, for the benefit of the mandarins' private purse; neither would be found ready when the tug of war came. The army was a collection of dirty savages, whose tactics were grimaces and voluptuous music, whose arms were bows and arrows and unfamiliar rifles, to whom discipline was a word unknown, who fought to avoid the executioner's knife, not to defeat the enemy. The navy was more imposing. For a time China appeared to be following western models. Ironclads were bought in Germany, cruisers in England. A handful of naval instructors were enticed to China from Europe, and then insulted and thwarted till their forbearance was exhausted. Captain Lang, awhile an admiral of the Chinese navy, has, indeed, asserted that under Admiral Ting the Chinese navy was a splendid force. Against that we may put the evidence of the "Times" correspondent and Mr. Norman. Ting, it appears, was an ex-cavalry general, and is said to have been devoid of tactical and strategical knowledge, though he certainly did not lack courage. The discipline of the Chinese fleet may be judged from the fact tha th ewould play pitch and toss with the sentry at his cabin door, and, when he had won all the man's money, would order the paymaster to advance his subordinate more, that his game might continue. As to its efficiency, the ships were filthily dirty, which is, after all, only what we should expect; the water-tight doors were seldom closed or used, a fact which we must remember when we come to the Yalu; the guns were employed by the sailors as receptacles for pickles, rice, and chop-sticks; the heavy Krupps were kept in shocking order, and the rings on them were beginning to open out. As a foreign instructor said to Mr. Norman, so far from the Chinese squadrons being formidable, it was only a question who should get them as prizes. The officers were either inefficient nominees of the authorities, or more able but powerless. Quick-firers were not bought because there was little money to be "squeezed" out of them. One Chinese battleship is stated to have gone to the Yalu without one of her heavy guns, which her captain had pawned. There were shells loaded with charcoal; charges for heavy guns of stuff which would not burn, instead of cocoa powder; and there were docks silted up from neglect, or useless owing to the bad arrangement of their pumping machinery. In vain did the English and German advisers beseech the Government to add ships, to procure sailors, coal, stores, and oil. They pointed out that the Chinese engineers dared not use forced draught, and that the Chinese officers were so nervous when handling torpedoes that they fired them at 800 yards instead of 350. The Chinese seamen on occasions displayed both coolness and courage, though their gunnery left much to be desired; but they lacked that confidence in their leaders, which is, after all, an essential of success.

All this is of great importance as showing us what kind of a task the Japanese had before them, and how few deductions can be drawn from the way in which that task was performed. The Chinese fleet was not allowed to seek the Japanese. It was kept by orders out of the sight of Admiral Ito till compelled to fight. The strategy of the Tsung-li-Yamen may or may not have been sound, but it could have contributed to neither efficiency, discipline, nor *morale*. The officers were bad, the sailors certainly lacked training, and Ting, whatever his courage, was not a brilliant commander. Can any very serious tactical conclusions be drawn from the performances of such a force? We may, it is true, observe how the Chinese ships behaved under the Japanese fire, which will give us information as to construction. But can we get much more than this? Can we say that the Yalu proves line ahead to be the ideal formation, and not line abreast? We may be persuaded of it, but we shall do well not to rely overmuch upon this Eastern engagement. Had the Japanese made their onset in line abreast, it is quite probable that they would have won. In fact, the gulf between the two forces was immense. The only question from the beginning of the war was, how many of the Chinese ships would have to be sunk before the others could be captured and added to the Japanese fleet. The issue of an engagement with the Chinese was confidently predicted as a success for Japan by Mr. Norman, writing with full knowledge of both combatants, a month before the Yalu.

And now to turn to the Japanese fleet and naval resources. The Japanese sea-going ships fall into three classes: the first comprising three old and indifferent armoured vessels; the second, eight large protected cruisers; the third, eighteen gunboats and smaller vessels. Taking the first group, its members are the *Fusoo*, *Hiyei*, and *Kongo*. The *Fusoo* is an old central-battery ship, designed by Sir Edward Reed, and launched in 1877 at the Thames Iron Works. On her main deck in a citadel she carries four $14\frac{1}{2}$-ton guns, one at each corner of the citadel. On her upper deck are mounted two $5\frac{1}{2}$-ton guns. Her armour of wrought iron, 7 inches to 9 inches thick, is of very inferior quality and resisting power to the steel now used. Right ahead or astern she can fire two $14\frac{1}{2}$-ton, and two $5\frac{1}{2}$-ton guns; on the broadside two $14\frac{1}{2}$-ton and one $5\frac{1}{2}$-ton. Her speed at the beginning of the war did not probably exceed ten knots. She is very similar in type to the English *Iron Duke* class, but is smaller. The *Hiyei* and *Kongo* are sister ships of a very different type, protected by a 4½-inch iron belt, for one quarter of their length, amidships; they were both built in England in 1877-8, and carry on their main-deck six 3-ton guns and three 6-ton. Their ahead-fire is from two 6-ton guns, their broadside from two 6-ton and three 3-ton. There are no similar vessels in the British fleet. The *Riojo*, like the last two ships, has a wooden hull, and is an antiquated ironclad, built in England, carrying 4½-inch armour. She took no part in the fighting, and need not therefore be further considered.

In the second group, the *Chiyoda* alone has a chrome-steel armour-belt, 4½ inches in thickness, for two-thirds of her length. In addition, she has an inch protective deck from stem to stern with coal above it, and round her machinery compartments, coal and a belt of cellulose. She is divided into eighty-four water-tight compartments, and has a double bottom amidships. Her engines develop 5600 horse-power and give her a speed of nineteen-and-a-half knots an hour. She was built by Messrs. Thomson of Clydebank, and launched in 1891. Her armament includes ten 4·7-inch Armstrong quick-firers, and fourteen 47-millimètre Hotchkisses, with three Gatlings and three torpedo-tubes. Next come three larger vessels of 4240 tons, the *Hashidate*, *Itsukushima*, and *Matsushima*. They were designed by M. Bertin, and the last two were built at La Seyne in France, the first at Yokosuka in Japan. They carry an end-to-end steel turtle-back deck 2 inches thick amidships. Forward in the first two, aft in the last, is a barbette protected by 12-inch steel plates, but open at the top on which is mounted a 12 6-inch Canet 66-ton gun, the most powerful weapon of its size in the world, built to fire cordite, and loaded and manœuvred by hydraulic power. The loading machinery is sheltered by the thick armour, but the barbette is open underneath, having only a small armoured ammunition shaft for the passage of projectiles and charges from the magazine to the breech of the gun.

PLATE XXII.

THE JAPANESE CRUISER ITSUKUSHIMA.

A bullet-proof shield protects the gunner sighting the weapon. Besides this immense gun, which can at 2000 yards perforate any armour afloat, each vessel carries eleven Armstrong 4·7-inch quick-firers, five mounted on each broadside, behind bullet-proof shields, and one placed at the opposite end of the ship to the heavy gun.* The *Hashidate* further has six 6-pounder quick-firers, and six machine-guns, whilst the other two have each five 6-pounders, eleven 3-pounders, and six machine-guns. Each ship has four torpedo-tubes. Their speed is from sixteen to seventeen-and-a-half knots, and they were launched between 1889 and 1891. Their enormously heavy gun makes them something more than mere cruisers—indeed, in general feature they approximate most closely, on a small scale of course, to the huge Italian cruiser-battleships. Even more powerful than these is the *Yoshino*, an Armstrong-built cruiser of 4150 tons, launched at Elswick in 1892. From stem to stern she has an armoured deck 2 inches thick, but on its slopes amidships the thickness is $4\frac{1}{2}$ inches. The hull is minutely sub-divided, and there is a double bottom amidships. On the measured mile her speed reached the very high figure of 23·03 knots, and at the date of her trial she was the fastest cruiser in the world. Her armament is extremely strong. Forward she carries three 6-inch Armstrong quick-firers, mounted separately behind stout steel shields, whilst astern is a fourth. Two of the four are on the keel-line, the other two on sponsons forward. Amidships, on her upper deck, are eight 4·7-inch quick-firers, protected, like the 6-inch guns, by shields. Two fire right astern. She thus brings to bear, ahead three 6-inch, astern one 6-inch and two 4·7-inch, and on either broadside three 6-inch and four 4·7-inch guns, all quick-firers of the latest pattern. Besides this, her main battery, she carries twenty-two 3-pounders and five torpedo-tubes. Her supply of coal is 1000 tons, and the bunkers are so disposed as to protect the engine-rooms, boilers, and vitals.

* In the *Matsushima* two 4·7 inch quick-firers are placed forward, making her total of these guns twelve.

She has two funnels and two military masts, each with one top; on the forecastle is an armoured conning-tower. The *Akitsushima* is a Yokosuka-built cruiser, similar in design to the *Yoshino*, but carries two less 4·7-inch guns. She has a deck of steel, 1½ inches thick on the flat, 2½ inches on the slopes. Her speed is nineteen knots; her armament consists of four 6-inch quick-firers, placed on sponsons, two forward and two astern, with six 4·7-inch quick-firers mounted amidships. Her bow and stern fire is delivered from two 6-inch guns, but on the broadside she brings to bear two 6-inch and three 4·7-inch guns. She has fourteen 3-pounders and machine-guns, and carries four torpedo-tubes. By an error she is described in the older "Naval Annuals" as armed with a 66-ton Canet gun. Last come the *Naniwa* and *Takachiho*, two Elswick-built cruisers, launched in 1885. They are of interest as being practically sister-ships of the celebrated *Esmeralda*, which, designed by Mr. Rendel for the Chilian Government, was the prototype of the fast, heavily-armed, unarmoured ship. Their tonnage is 3600 to 3750. They have an end-to-end steel deck 3 inches thick on the slopes, 2 inches on the flat amidships and at the ends; above this deck are coal-bunkers. They have also cork-packed compartments running nearly round the ship on the water-line. Forward and aft, fifteen feet above the water, are placed two 28-ton, 10-inch guns, on central-pivot carriages, with hydraulic turning and loading gear. A steel screen revolves with the gun and protects the gunners, whilst an armoured loading-station is provided to the rear of each gun. Amidships are six 6-inch Armstrong slow-fire breech-loaders, mounted three on each beam, on sponsons. Besides these, twelve smaller weapons and four torpedo-tubes are carried. The trial speed was 18·7 knots, and there is bunker space for 800 tons of coal. Altogether, these are fine ships, if a trifle out of date.

The smaller Japanese ships do not merit any detailed description, as none of them, with the exception of the *Akagi*, a small gunboat, were at any time engaged. The torpedo-boats are forty-one in number, and of these, one, the *Kotaka*, has 1-inch armour: though an old boat, built in 1886, she did good service.

Behind her war-fleet, Japan had a very considerable mercantile marine, in which were included in 1894, 288 steamers of 174,000 tons. The Nippon Yusen Kaisha is one of the important shipowning companies of the world; to it belong fifty steamers, of which four can steam thirteen knots, whilst two, the *Saikio Maru* and *Kobe Maru*, of 2900 tons, steam fourteen knots. The government subventions this company to the extent of one-and-a-half million dollars yearly. Just before the war, in 1894, the Japanese Government purchased a large number of steamers at Shanghai and Hongkong, so that it wanted neither auxiliary cruisers nor transports. With the exception of the 66-ton gun, everything required by the Japanese ships is produced in Japan; machinery, castings, armour, guns, projectiles, melinite, powder, all are manufactured in the country.* At Tokio is the arsenal, employing in peace over 1000 men; whilst the dockyards are three in number. Onohama is the least important, building gunboats and torpedo-boats, and employing less than 1000 men. Yokosuka is more important; it builds cruisers, and has three docks, of which the following are the dimensions:

	Length.	Breadth.	Depth on Sill.
No. 1	392 feet	82 feet	22 feet
No. 2	502 ,,	94 ,,	28 ,,
No. 3	308 ,,	45 ,,	17 ,,

At Tokio, there is one dock 300 feet long, fifty-two feet broad, and fourteen deep; at Nagasaki, one 400 feet long; and at Osaka, where there are important ironworks, a dock 250 long. In addition, there are three slipways, one capable of lifting a 2000-ton ship At Kuré, on the Inland sea, is a new dockyard recently established. The Japanese are clever workmen, and were able to effect repairs with great rapidity.

* Japan has not, however, as yet been able to build large warships.

Whilst the Japanese fleet thus included a large number of thoroughly modern vessels, equipped with quick-firers, and capable of fast steaming, whilst these ships were manned by well-trained sailors, and officers who had studied strategy and tactics, the Chinese fleet had remained almost stationary since the close of the war with France in 1885. To begin with, it could scarcely be described as a fleet, being a local rather than an imperial force. In peace, it was organised, so far as it had any organisation and was not merely a jumble of badly kept ships, in four squadrons, the North Coast, the Foochow, the Shanghai, and the Canton. The first-named was the largest and strongest, though the Foochow was not much inferior. The total included two armoured second-class battleships, three small and indifferent armourclads, eleven old but heavily armed cruisers, ranging from 2200 to 2500 tons, nine smaller cruisers of over 1000 tons, thirty small vessels and gunboats, and forty-three torpedo-boats. In numbers, China, then, had a very decided superiority; but whilst the two large armoured ships were very much better than anything the Japanese possessed, the Chinese cruisers were old or in bad repair. The Chinese artillery was of a much older pattern than the Japanese, and there is reason to believe that very few quick-firers were included in it.

The two largest vessels which flew the Chinese flag, were the sisters *Chen Yuen* and *Ting Yuen,*, of about 7500 tons displacement, built by the Vulcan Company at Stettin in 1881-2. They carried 14 inches of compound armour upon a citadel which occupied about half their length; the other half was unarmoured. Thus forward and aft their ends were quite unprotected externally, but internally there was a 3-inch horizontal deck, a minute cellular sub-division, and a large number of cork-packed compartments. The extreme speed at their trials was fourteen-and-a-half knots, but their boilers were in a bad condition in 1894, and it is doubtful if they could steam much over ten knots in the hour. The heavy armament was placed at the forward end of

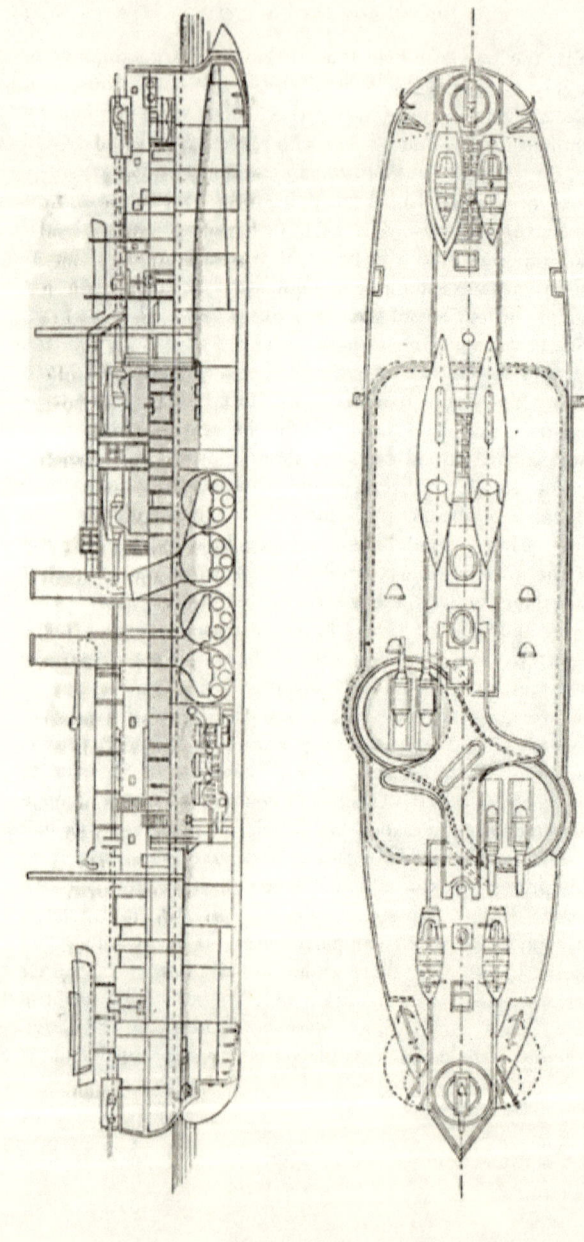

PLATE XXIII. ELEVATION AND DECK PLAN OF THE TING AND CHEN YUEN.

the citadel, in two barbettes, protected by 12-inch compound armour, and disposed *en échelon*. The barbettes were open at the top, but a bullet-proof shield protected the gunners.* Hydraulic power was used to work the four 12-inch 35-ton Krupps, placed two in each barbette. At each extremity of the ship, one 6-inch Krupp breech-loader was mounted in a small, lightly-armoured turret. In addition, there were twelve smaller guns, and two torpedo-tubes. The general design was similar to that of the *Ajax* and *Inflexible*—developing bow-fire at the expense of the broadside. As this feature may have exercised some tactical influence upon the battle of the Yalu, we may be permitted to dwell upon it.

In theory, each of these ships could fire ahead or astern the four heavy Krupps and one 6-inch gun; in practice, the blast of the inner gun in each barbette, when both were trained right ahead, would prevent the 6-inch gun from being worked, and might even demolish its turret; the same would be the case when they were trained aft. With an antagonist exactly abreast, all four guns would bear on the broadside, but yet on either beam there was a wide angle which was not covered by the fire of more than two heavy guns, since the barbettes, diagonally placed, obstructed each other. The funnels prevented the after barbette guns from bearing upon the port quarter; on the starboard quarter, the fore barbette guns could not fire, because of the after barbette; and six points before the beam on the port quarter, the after barbette guns could not bear for the same reason. The best fighting positions for these ships would then be to have their enemy right ahead or right astern—exactly abreast of them on the port (left) beam, or rather, ahead of them on the starboard (right) beam. In those positions, only would four heavy guns bear; in any other position, half their heavy armament would be useless. These vessels represented a tactical theory already obsolete in 1894, as they

---

* These shields were removed in the course of the war.

were built to attack in line abreast. At the date of their construction, the heavy gun was the dominant factor in naval tactics, coupled, it is true, with the ram. Since, 1882, however, had come the large calibre quick-firer, compelling a return to the fore and aft disposition of the guns, through the necessity of finding room for such useful weapons on the broadside. There followed in consequence a general return to line ahead as the battle formation, since the greatest weight of metal, with ships carrying their guns fore and aft, is discharged on the broadside, which line ahead leaves free.

The *King Yuen* and *Lai Yuen* were, again, sister barbette-ships of 2,850 tons, built by the Vulcan Company of Stettin, and launched in 1887. They had a very short armour-belt, $9\frac{1}{2}$ to $5\frac{1}{4}$-inches thick, over their engines and boilers, but the top of the belt lay flush with the water-line, and thus it was not of much service to them. Over the belt was a deck of steel $1\frac{1}{2}$-inches thick, which at the unprotected ends was twice as thick. Like the larger battleships, they had minute sub-division and cork packing outside their armour. Amidships only had they a double bottom. The nominal speed was sixteen-and-a-half knots with forced draught, but in 1894 they could not do much more than twelve or thirteen. A conning-tower protected by 6-inch armour was placed forward; and the armament consisted of two $8\frac{1}{4}$-inch Krupps mounted in a barbette forward, on which was compound armour 8 inches thick. The gunners had the shelter of a stout steel shield. Sponsoned out on either beam was one 5·9-inch breechloader, and thirteen smaller guns were also carried. The *Ping Yuen* of the same displacement was built at Foochow, and had a complete water-line belt of 8-inch armour, a deck 2 inches thick, and a barbette forward, protected by 5-inch plating and a shield. In this barbette she carried one 22-ton Krupp instead of the two 8-inch weapons; otherwise her armament was the same. Her speed was only ten and a half knots on trial, as money ran short when she was on the stocks, and her length was cut down, ruining her lines. The *Tsi Yuen* was a Stettin-built cruiser, having a 2 to 3-inch deck and a barbette forward, armoured with 10 inches of plating, in which were mounted two 8¼-inch Krupps. Aft she carried one 5·9-inch Krupp gun mounted in a steel turret. Her speed was nominally fifteen knots. The *Chih Yuen* and *Ching Yuen* were Armstrong cruisers, launched in 1886, and, like all that firm's ships, extremely powerful for their size: their armoured deck was 4 inches thick on the slopes, and 2 inches on the ends and centre; two 8-inch guns were mounted forward behind a shield and one aft, whilst one 6-inch gun was sponsoned out on either beam. The speed was at their trial eighteen-and-a-half knots. The *Tshao Yong* and *Yang Wei* again were of Armstrong construction, but older and smaller. Their speed had been sixteen knots, but their boilers were completely worn out and good for very little. They carried fore and aft one 10-inch Armstrong gun, on a central-pivot mounting, and amidships four 5·1-inch slow-firers and seven machine guns. All these vessels took part in the Yalu, with, in addition, two small cruisers of about 1000 tons, whose names and armaments are uncertain and unimportant.* Of the vessels which did not take part in that action the most important were the *Foo Ching, Ye Sing, Foo Sing, Kai Chi, Nan Shuin, Nan Ting,* and *Yang Pao*, all very similar to the *Chih Yuen*, carrying 8-inch or 8¼-inch, and 4·7-inch guns. The gunboats were mostly of the Rendel type, carrying one 38-ton or 35-ton muzzle-loader forward.

In her merchant marine, China was far behind Japan, having only thirty-five steamers of 44,000 tons. In docks she was better off. At Foochow, the most important Government yard, was one dock 390 feet long; at Amoy, one of 310 feet and two smaller; at Shanghai, one of 500 feet and four of over 300 feet, whilst one other was out of use. At Taku was

* *Kuang Chia* or *Kwang Kai* and *Kwang Ping*. The first is described by Captain McGiffin as of 1030 tons, armed with three 4·7-inch quickfirers; the second as of 1300 tons armed with three 6-inch, four 5-inch, and eight small guns. Details of the ships engaged at the Yalu are given in Tables XVIII.-XIX.

one of 340 feet and another of smaller dimensions; at
Whampoa, two of over 400 feet ; at Port Arthur, one 400 feet
long and one smaller dock. On the Gulf of Pe-che-li China
possessed two excellent harbours in Port Arthur and Wei-
hai-wei, each of which was defended on the seaward face by
strong works mounting very heavy guns, Armstrongs and
Krupps. But the management of the dockyards was not
good, though there seems to have been little fault to find with
the Chinese workmen.

Serving in the Chinese fleet were eight or nine Europeans,
of whom Major Von Hanneken, a German, acted as
strategical adviser to Admiral Ting. Five were Englishmen
with some knowledge of naval matters, one or two having
passed through the British Navy. The Japanese depended
entirely upon themselves, and had no Europeans.

CHAPTER XX.

THE ACTION OFF ASAN AND THE SINKING OF THE *KOWSHING*, July 25th, 1894.

ON July 25th the Japanese Flying Squadron under Rear Admiral Tsuboi, consisting of the ships *Yoshino, Naniwa, Takachiho, Akitsusu*,* all fast cruisers, was off Asan, a port in the Gulf of Korea, then in the possession of the Chinese. At anchor off Asan were three Chinese warships, the *Tsi Yuen* and two other vessels whose names are given as the *Kuwan-shi* and *Tsao Kiang*. These two are identified by Mr. Laird Clowes with the *Kwang Yi* and the despatch-boat *Tsan Chieng*, which had been purchased some time previously and was only a mercantile steamer, lightly armed. The Chinese ships weighed anchor, and putting to sea, perhaps to cover the transports which were coming up from Taku, fell in with the *Yoshino, Naniwa,* and *Akitsusu* in the gulf. According to the Japanese account the Chinese failed to salute Admiral Tsuboi's flag, as international etiquette requires, but were cleared for battle and gave various indications of a hostile purpose. Seeing this, the account proceeds, the Japanese stood out to sea to get out of the narrow waters in which they were manœuvring; but the *Naniwa* was so closely followed by the *Tsi Yuen* that she turned and headed for the Chinese ship, which was commanded by Captain Fong, one of the ringleaders in the intrigue to drive Captain Lang out of the Chinese navy. The *Tsi Yuen* being now in turn closely pressed by the *Naniwa*, though no shots had as yet been fired

* Or Akitsushima. Other accounts do not mention the *Takachiho*

on either side, hoisted the white flag above the Chinese naval
ensign. Under cover of this she approached the Japanese
ship, and, at a distance of only 300 yards from her stern,
treacherously discharged a torpedo at her. The torpedo
missing, the *Naniwa* opened at once on the three Chinese
ships, and was supported by her two consorts. This happened
about nine o'clock in the morning. The Chinese version is
very different, and is, for once, the more probable. The three
Japanese ships fell in with the Chinese, and without any notice
or warning opened upon them. The Chinese were taken by
surprise, their ships were not cleared for action, and there
was some delay before they could reply. At the first Japanese
discharge several shots struck the conning-tower of the *Tsi
Yuen*, piercing it, and blowing to pieces the first lieutenant
and a sub-lieutenant. The head of the first lieutenant was
left hanging on to one of the voice-pipes, whilst steering gear,
engine-room telegraphs, and voice-pipes were completely
wrecked. Captain Fong, who was inside, was not, however,
wounded, and at once gave orders to clear ship for action,
and went below. Before anything could be done a second
broadside struck the ship, doing great damage. A shell,
glancing from the armour-deck below, flew up under the
forward turret, passing through the armour of the turret. It
forced up the deck and damaged the shot-hoist, disabling one
of the two 8-inch guns. A few minutes later a second shell
perforated the turret, and bursting inside killed the gunnery
officer and six men,* but did no damage to either of the two
guns. Not a man was left on deck, so searching and deadly
was the fire; those of the crew who were not killed had fled
below the armour-deck, nor could they be driven to the guns
till the officers had drawn their revolvers, and threatened to
use them with effect. A large number of shells entered
between decks: one wrecked the officers' cabins and tore a huge
hole in the side; another struck the base of the funnel and

* This shot also wounded fourteen men. It was a shell with base fuse, and
struck the turret to the rear.

PLATE XXIV.

THE JAPANESE CRUISER YOSHINO.

See p. 59.

burst in it, killing several of the stokers. The boats were repeatedly hit and set on fire, whilst the military mast was struck more than once. Huge holes were blown in the side of the ship by the Japanese shells. On the armour-deck lay six torpedoes, which had not been sent below owing to the suddenness of the attack. One of these was discharged by the torpedo lieutenant without purpose or without aim, anxious only to get rid of a petard which might readily hoist his own ship. Strange to relate, not one of the other torpedoes was hit, though shells were bursting dangerously near them.

As soon as her hand-steering gear could be got to work, the *Tsi Yuen* turned from her enemies and headed for Wei-hai-wei. The ship was already in a dreadful plight, and with her stern to the Japanese she would not have appeared to have any chance of surviving a running fight, since the only gun which could fire astern was a 5·9-inch Krupp, mounted in a thinly-armoured turret aft. Behind her was the *Yoshino*, at least four knots faster, and bringing to bear upon her three 6-inch quick-firers, which were equivalent from the rapidity of their fire to six at least of the *Tsi Yuen's* guns. The awning aft, with the stanchions which supported it, had not been removed in the confusion, and hampered the single Chinese gun. Fortunately for the *Tsi Yuen* a shot from her one gun struck the *Yoshino's* bridge, whilst a second hit the Japanese conning-tower, and wrecked the chart-house. The Chinese also state that they knocked one of the hostile cruiser's guns overboard. In any case the *Yoshino* abandoned the pursuit —whether because of the damage to her bridge and chart-house, with possibly some injury to her steering gear, or because her machinery broke down at the critical moment, it is impossible to say.* The *Tsi Yuen* proceeded on her way, and reached Wei-hai-wei without further adventure. She had been terribly knocked about, losing sixteen killed, of whom three were officers, and twenty-five wounded. She had been

---

* It is now, however, stated that she was not injured, but that she could not overtake the *Tsi Yuen*; if so something must have broken down in her engines.

struck a great number of times by 6-inch, 4·7-inch, and smaller projectiles, but the damage was not quite so serious as might have been expected, owing to the fact that a great many of the Japanese shells failed to explode. A European officer who saw her on her arrival thus describes her appearance: "The vessel presented the appearance of an old wreck. The mast was shot through half-way up, the gear was torn in pieces, ropes hung loose and tattered. On deck the sight was cruel, and beggars description. Woodwork, cordage, bits of iron, and dead bodies, all lay in confusion. Between decks matters were as bad." And an English officer adds*: "The slaughter has been awful, blood and human remains being scattered over the decks and guns. Three of the five men working the 4-ton gun in the after-turret were blown to pieces by a 6-inch shell from one of the *Naniwa's* (?) quick-firing (?) guns, and a fourth was shot down while attempting to leave the turret. The remaining gunner stuck to his post, and managed to load and fire three rounds at the *Naniwa*, and, one shell entering her engine-room and another blowing her fore-bridge away, she hauled off. The Chinese admiral awarded the plucky gunner 1000 taels. One shell struck the *Chen Yuen's* [sic] steel deck, and, glancing, passed up through the conning-tower and exploded, blowing the gunnery lieutenant to pieces, and leaving his head hanging on to one of the voice-pipes. Huge fragments of armour and backing had been torn from their fastenings and carried inboard, crushing a number of poor wretches into shapeless masses, even the upper part of the funnels being splashed with blood. An engineer-officer (European) was sent for to repair the steam-pipe of the steering-engine, and tried to grope his way through the smoke of bursting shells and heaps of killed and wounded lying on the deck, when a shot struck his assistant and disembowelled him, covering the engineer with blood.

* In the earlier accounts the vessel inspected was described as the *Chen Yuen*, which, as now appears, was not engaged. The *Tsi Yuen* is evidently the ship referred to.

Conning Tower of the Tsi Yuen.

Plate XXV.

He nevertheless managed to reach the steering-engine, and repaired the pipe, for which he received a rather handsome reward from the admiral. The engagement lasted about one-and-a-quarter hours, when the Japanese hauled off, and the *Chen Yuen* made the best of her way back to Wei-hai-wei, their naval station, where she arrived the next day in just the same condition as she left the scene of action, no attempt having been made to wash away the blood or remove the dead bodies."

Meantime the *Kuwan-shi*, or *Kwang Yi*, had fought first the *Naniwa* and then both the *Naniwa* and *Akitsusu*, with gallantry, though only a small vessel and lightly armed. She was heavily hulled, and lost no fewer than thirty-seven men killed, before her ammunition ran short. In a leaking and sinking condition, her captain ran her inshore, beached her, and got the remnant of his crew away in safety. The Japanese left her for the time, but, returning later, fired thirteen rounds into her, one of which exploded a torpedo in her after torpedo-room, and blew her stern clean away. The courage of her captain and crew is evinced by the fact that only eighteen men, most of them badly wounded, escaped. The despatch-boat *Tsan Chieng* was not chased till after the sinking of the *Kowshing*, when she was quickly overpowered and taken. Captain Fong, of the *Tsi Yuen*, was, for deserting these two vessels, condemned to death on his arrival at Wei-hai-wei, but was subsequently given a chance of redeeming his character at the Yalu. It is hard to see what good he could have done by remaining to fight the Japanese; the odds were heavily against him, and his ship could only have fallen a prize to Admiral Tsuboi, or have been disabled. Fong behaved very badly at the Yalu, and for this was executed after that battle; on this occasion he does not seem to have been at fault.*

* Fault has been found with the Japanese, and with good reason, for their failure to capture the *Tsi Yuen*. They had four good cruisers, all much faster than the Chinese ship, and yet they let her get away, how or why, it is difficult to understand.

We must now go back a day or two to follow the fortunes of the *Kowshing*. Towards the end of July the Chinese decided to despatch troops to Korea by sea. Accordingly three British steamers were chartered, amongst which was the *Kowshing*, an iron screw-propelled vessel of 1355 tons, built in England, owned by Messrs. Jardine and Matheson, sailing under the British flag, and carrying a British captain and officers. On July 23rd she left Taku, having on board 1100 Chinese infantry, two Chinese generals, Major von Hanneken, and twelve field-guns, besides a large quantity of ammunition. Early on the morning of July 25th she sighted the islands in the Gulf of Korea, and about the same time noticed a large warship, which resembled the *Chen Yuen*, and appeared to have been in action, steaming westwards. This vessel was upon her port side, and must have been the *Tsi Yuen* in retreat before the Japanese. The *Tsi Yuen* might have signalled to the *Kowshing* what had happened, and so have prevented the catastrophe which was to follow, but no signals were made, either because Fong was not aware that the *Kowshing* was in the Chinese service, or more probably because all his concern was for the safety of his own ship. Some minutes later a vessel was seen under sail, upon a course which would cross the *Kowshing's* bows. This was the *Tsan Chieng*. An hour later still, at eight o'clock, a large warship came into sight from behind the island of Hsutan, and following her were three others. All appeared to the *Kowshing's* officers to be ironclads. At nine o'clock it could be seen that the nearest vessel flew the Japanese flag. She approached rapidly, saluted the *Kowshing*, and passed her. The four Japanese ships were now in line abreast and heading west; they appeared to be chasing the *Tsan Chieng*, and to intend no harm to the *Kowshing*. Presently, however, the ship which had saluted the *Kowshing* signalled to the English vessel to anchor, at the same time firing two blank shots. The order was obeyed, and was followed by a second, " Remain where you are or take the consequences," after which the Japanese ship was seen to circle and signal to her consorts. A few minutes passed, and the warship once more headed towards the *Kowshing;* as she drew near it was observed that her crew were at quarters and her guns trained on the *Kowshing*. A boat was lowered and a boarding party sent off to the English vessel, when Major von Hanneken and the English officers learnt that the ship observing them was the Elswick cruiser *Naniwa*, Captain Togo. The Chinese soldiers and generals were greatly excited, and when Von Hanneken and the English officers tried to persuade them to surrender, asserted that they would die rather than yield, and that if the Englishmen attempted to leave the ship they should be killed. Between the Chinese and the Japanese the Europeans were in no enviable plight. Meantime, several Japanese officers came on board and inspected the ship's papers. They were told by Captain Galsworthy that the *Kowshing* was a British ship, with the British consul's clearance, and flying the British ensign, and that she had sailed in peace. After some hesitation and argument she was ordered by the Japanese to follow the *Naniwa*. Whilst this short conference was proceeding the excitement on deck was growing, and the Chinese had set a guard upon the anchor. When the Japanese left, the Chinese absolutely refused to allow compliance with the *Naniwa's* demands. As argument was useless, Von Hanneken had the *Naniwa's* boat recalled. He explained to the Japanese that the position on board the *Kowshing* rendered obedience to their orders impossible, and asked that, as the ship had sailed in peace, she should be allowed to return to Taku. The Japanese officers understood, and promised to report to their captain.

Once more the boat left the transport, and some minutes of suspense followed. It reached the *Naniwa*, and the next thing was an imperious signal from her, "Quit ship as soon as possible." It was addressed to the Europeans, but they were helpless. Finally came the order, "Weigh, cut, or slip; wait for nothing." To attempt obedience in the face of a thousand armed Chinamen was hopeless. Captain Galsworthy replied, "We cannot," and his signal was acknowledged. Immediately the *Naniwa* began to move; she blew a loud blast on her steam siren and hoisted a red flag; then when she was broadside on to the *Kowshing*, abreast with her, and at a distance of 500 to 300 yards, fired a torpedo. At that moment, all the Europeans on board the doomed vessel mustered on deck, in obedience to Captain Galsworthy's orders. Whether the torpedo struck or not is doubtful, since, almost at the same instant as the torpedo left the *Naniwa*, that vessel fired with a terrific crash a broadside from her five guns, two of 28-tons, and three 6-inch. According to Von Hanneken the torpedo struck a coal bunker amidships. "The day became night: pieces of coal, splinters, and water filled the air, and then all of us leapt overboard and swam." According to the other survivors, the torpedo missed, and the damage was done by a 500lb. shell from one of the 28-ton guns, which exploded the boilers. The transport listed heavily to starboard, whilst the pitiless Japanese fire searched her vitals. From the *Naniwa's* tops, where were mounted Gatlings, and from Nordenfelts and small quick-firers on her upper deck, came a hail of small projectiles, tearing through the dense mass of Chinamen on the *Kowshing's* deck. The Chinese replied in a futile, though gallant manner, by discharging their rifles at the enemy. The result could not be long in doubt. The heel of the *Kowshing* grew greater, and she sank lower and lower in the water, till about 2 o'clock, an hour from the firing of the first shot, her deck was submerged. All this time the Europeans, and many of the Chinese who had leapt overboard, were in the water, exposed to stray projectiles from the Japanese, and deliberately fired upon by the Chinese who still were left on board the sinking ship. "Bullets began to strike the water on all sides of us," says Mr. Tamplin, the *Kowshing's* first officer, who had jumped overboard after the explosion, " and, turning to see

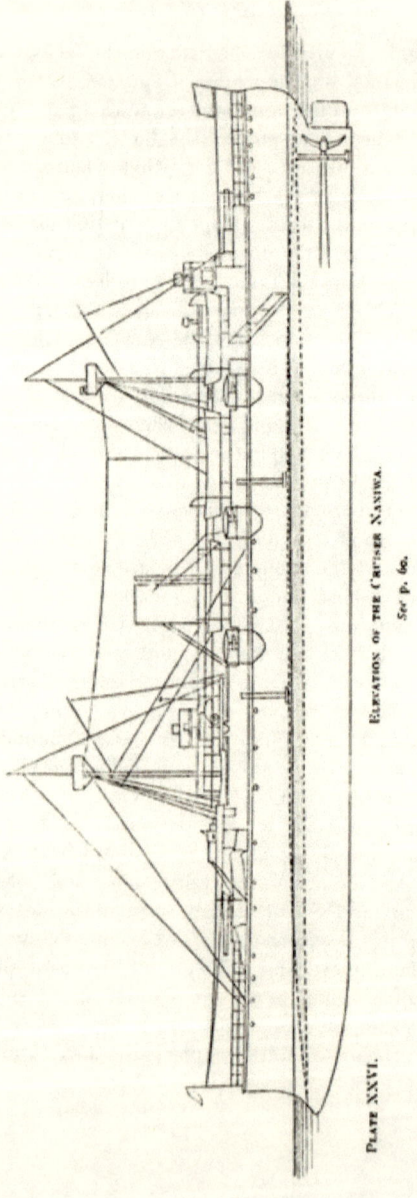

PLATE XXVI.

ELEVATION OF THE CRUISER NANIWA.

See p. 60.

whence they came, I saw that the Chinese, herding round the only part of the *Kowshing* that was then out of water, were firing at us. I swam straight to the *Naniwa*. I had been in the water nearly an hour when I was picked up by one of the *Naniwa's* boats." On telling the Japanese officer in charge that Captain Galsworthy was swimming for his life, Mr. Tamplin heard that he was already being looked after. The water was alive with Chinese soldiers, and two lifeboats had put off from the transport, crowded with Chinamen. What followed now was the most disgraceful feature of the day's proceedings. The Japanese made not the smallest attempt to rescue their drowning enemies; they did, indeed, look after the Europeans, but they left the Chinese to their fate or worse. For, when Mr. Tamplin was on board the *Naniwa's* boat, the Japanese officer told him that he had orders to sink the Chinese lifeboats, and in spite of remonstrances proceeded to do so. Two volleys were fired, and the Chinese boats were sunk. This atrocious act has been denied by the Japanese, but the evidence for it appears incontrovertible. Some of the Chinese succeeded in swimming to the island of Shopaul, whither Von Hanneken had escaped, after being hours in the water. The French gunboat, *Lion*, and the German warship, *Iltis*, saved, between them, three hundred, many of whom were wounded. Some of these had been in the boats and corroborated Mr. Tamplin's assertion that the Japanese fired on them. In one boat all were killed or wounded. Having completed her bloody work, the *Naniwa* steamed backwards and forwards till eight o'clock that evening. The Europeans were shown a shell in one of the officer's cabins, which it was stated, had been treacherously fired into the *Naniwa* by the *Tsi Yuen*. Next day they were transferred to the *Yaeyama*, which conveyed them to Japan, where they were set at liberty.

The Japanese had thus committed three questionable acts. 1. They had attacked the *Tsi Yuen* in peace and before a declaration of war. 2. They had followed this up by destroying a neutral ship, which had sailed before the action off Asan, and could not therefore know that a state of war existed. 3. They had fired upon the Chinese in the water.

As regards the first head, if we follow the Chinese account, the Japanese committed an act of violence—almost of unprovoked aggression, by attacking the *Tsi Yuen* and her consorts. At the same time, there was on the 25th every probability of war between China and Japan, and the *Tsi Yuen* had no business to be taken off her guard. It is the commonest thing possible for states to go to war without any declaration, or for acts of hostility to precede a declaration. Indeed, the vulgar idea that formal notice is necessary is probably due to the fact that histories use the phrase "war was declared" as a convenient expression for "hostilities commenced." That body of precedents which goes by the name of International Law, gives ample justification for such a course. Colonel Maurice has shown that between 1700 and 1870 there were only ten instances in which formal notice preceded acts of hostility. We cannot, therefore, condemn the Japanese for acting as they did: they were only once more copying the West. And if their story is true, which it does not appear to be, they themselves were the injured and the Chinese the aggressors.

We come now to their attack on the *Kowshing*. Here things are complicated by the presence of the neutral flag. Had they any right to treat her as they did, accepting their own version of the *Tsi Yuen's* behaviour? Before a declaration of war there is no contraband, and there are no neutrals, since all states are assumed to be friends in peace. There is then no obligation upon the neutral to avoid the conveyance of contraband, or the performance of non-neutral acts, unless, and it is an important proviso, it is notorious that a state of war exists. There is no right of search, no power to visit neutral vessels or examine their papers, resident with either of the two parties to the quarrel, till the quarrel has become war.

The whole question, then, turns upon this: Did a state of war notoriously exist on July 23rd? Apparently the English Government was satisfied of this, since we do not know that any demand for reparation has been made. And yet we question whether it could be said with truth that on the 23rd war was inevitable, or a state of war a notorious fact. The *Kowshing* cleared on that day from a Chinese port, and could not very well have received later intelligence. Therefore, if hostilities had commenced on the 24th she might yet with reason have been spared. She was carrying Chinese troops to Korea, it is true, but this was permitted by the treaty of 1885 between Japan and China. The neutral may carry whom he likes and carry him where he likes till war has been declared. Whilst the *Kowshing* was at sea the first act of hostility was committed—if by the Chinese, it does not follow that the neutral in the Chinese service should be injured, before he has had time to dissociate himself from the aggressor —if by the Japanese, still less. Nor can the neutral be converted into a belligerent by an attack at sea. He may, indeed, be requested to return to the port from which he had sailed, but for some reason or other, no such request was made in the case of the *Kowshing*, though the Chinese generals were ready to allow the English vessel to go back. Obviously, the English captain could not then and there discharge his living cargo into the sea. On the other hand, the Japanese captain saw before him a cargo of troops in a neutral ship, and these troops might be used against his country. He had, by his own account, received considerable provocation from the Chinese. If he ordered the *Kowshing* to proceed to a Chinese port, she might double when out of his sight, and return; to place a prize-crew on board her in the midst of a thousand or more armed Chinamen, was impossible; to escort her with his own ship, might well have been inconvenient. He therefore attacked her, but his attack was illegal, and constitutes a dangerous precedent. For it cannot be tolerated that neutrals should be treated with severity for breach of obligations which do not come into force till war begins.

On the second head, the *Naniwa* took the extreme course of sinking the ship, but only after the Japanese captain had requested the *Kowshing* to " weigh, cut, or slip," and the Chinese soldiers had refused to allow her to do this. He had further done his best to save the European officers. Having once decided to make her his prize, it is hard to see what else he could have done. As has been said, a prize-crew was out of the question. But his own act, which was illegal in the first instance, led to a grievous loss of human life in the second.

For the firing on the men in the water, there is no justification; it was an act at once barbarous and cruel. The principle which governs war is to avoid inflicting unnecessary suffering, and very few commanders have ever gone so far as to slaughter their enemies when they were helpless. Even the ancient Egyptians are depicted on their monuments as rescuing their drowning foes. It was an act comparable to the slaughter of the wounded after battle, and Japan with her fine professions should have shrunk from it. The Japanese commander might, indeed, allege that if he had rescued his enemies, they would have been an element of great danger on board his ship, for the Chinese are an ignorant, treacherous, and cruel race, who could not be expected to obey the rules of war. But when once in the water they were helpless, and care might have been taken to disarm them when they were got on board the *Naniwa's* boats. Again, he might have urged that he was punishing them for the *Tsi Yuen's* act. But one act of barbarism does not justify another, especially as Japan was contending with a barbarous power, herself a civilised state. It is the clear duty of the surviving combatant, after an action, to do his utmost to rescue his enemies.

About this same date, or perhaps a day or two later, the *Chen Yuen* is said to have encountered and driven off the

Takachiho and *Hiyei*. Details of this engagement, if it really happened, and the *Chen Yuen* is not an alias for the *Tsi Yuen* are still to seek. The *Hiyei* is said to have been very severely handled, and to have been left a cripple. But Captain McGiffin, the commander of the *Chen Yuen*, in a letter of August 2nd, makes no allusion to it, and indeed it is hard to see how the ironclad could have been at sea eight days after a pitched battle.

At the time of the action between the *Tsi Yuen* and the Japanese, the heavy Chinese ironclads were at sea, under Admiral Ting Ju Chang, the ex-cavalry officer who had been appointed commander of the Northern Squadron by Li Hung Chang. Captain Lang and numerous European officers have spoken well of him, and his acts testify to some obstinacy and no lack of personal courage. Though he inspired confidence in his foreign subordinates, he was great neither as a tactician nor as a strategist. "He knows nothing at all about naval matters; he is just the mandarin put on board by Li," said a foreign instructor in the Chinese navy to Mr. Norman. Perhaps the instructor somewhat exaggerated Ting's incapacity, as the admiral had held command in 1884 during the war between France and China, and must have picked up some fragments of knowledge from the various very capable foreigners in the service of China. He was not, however, a Nelson or a Tegetthof, if he never sank so low as a Persano, and he displayed the usual Chinese cruelty in orders that no quarter was to be given, at the same time encouraging a belief amongst his sailors that the Japanese would give no quarter. Of the Chinese preparations Captain McGiffin, who was present on board the *Chen Yuen*, gives us some details in a letter: "We are reinforcing our turrets on all the ironclads by bags of coal piled round them eight feet to ten feet thick. That is my own idea. Don't believe the sneers you may see at the Chinese sailors. They are plucky, well-trained, full of zeal, and will fight better against the Japs, their lifelong enemies, than anyone."

On the arrival of the *Tsi Yuen* at Wei-hai-wei, after her action with the Japanese squadron, six Chinese ships went out to attack the Japanese. Captain McGiffin, in a letter dated August 2nd, states: "We are now on our way . . . to meet the enemy, and I hope we will sink the dogs. We have been expecting war for days, but China has kept peaceable, and therefore Japan deliberately picked the fight. Admiral Ting and I wished to go to Chemulpho, and open fire on the Jap fleet, but at the last moment we got a direct cable from [the] Tsung-li-Yamen not to do so. It would have been splendid, for we would have destroyed their navy almost, I think. Our crews are full of enthusiasm. It is very pleasing to see them. We have had several alarms at night and by day from strange vessels, and the way we go into action is splendid. We are all clear for action, everything that could possibly cause splinters left ashore or thrown overboard. We have left all our boats behind. We will not need them, for if we sink the Japs will give no quarter, and we shall give them none either. The admiral is on the ironclad [the *Ting Yuen*]. . . . He made two signals to-day at noon. One, 'If the enemy shows the white flag, or hoists the Chinese ensign, give no quarter, but continue firing at her until she is sunk.' The other, 'Each officer and man do his best for his country to-morrow.'"

The expected battle never came off. For three days Ting hunted for the Japanese, but either could not find them or did not want to find them.* The Japanese were presumably engaged in convoying transports, and were quite content to be let alone; possibly some of their vessels had received serious injury off Asan, and were undergoing repair. In any case they do not seem to have paid much attention to the Chinese, who returned to Wei-hai-wei and remained strictly

* Ting was anxious to fight, but was dissuaded by his flag-captain, according to "Blackwood's" Correspondent in China. This writer's articles did not appear in time to be used in the text. They represent Ting in a very favourable light. Corrections from them are given at the end of subsequent chapters.

on the defensive, in obedience to orders from Li Hung Chang; to the effect that they were not to cruise to the east of a line drawn from Wei-hai-wei to the mouth of the Yalu. The Japanese appear to have known of this order, which practically effaced the Chinese fleet. Meanwhile from time to time, they reconnoitred Wei-hai-wei, and even on August 10th made some pretence of bombarding it.

CHAPTER XXI.

THE YALU* AND ITS LESSONS.

September 17th, 1894.

THE Chinese fleet, though on the whole superior, had thus abdicated the command of the sea, and lay, through no fault of its gallant commander, inactive at Wei-hai-wei. The Japanese were straining every nerve to convey as large a force as possible to Korea, and merely observed Admiral Ting, if, indeed, they gave themselves this trouble. Such was the situation through the last weeks of August and the first days of September. But early in September the Tsung-li-Yamen awoke from its lethargy, finding that China could not move men with sufficient rapidity into Korea by land. It decided to send a force by sea, and Admiral Ting was informed of the decision.

There were two courses open to Ting and his foreign advisers. They might collect every available warship, steam in search of the Japanese, and, having found them, fight a decisive action. If the Japanese were beaten in this the mastery of the sea would be to Ting. If the Chinese were beaten they would at least be unencumbered by transports, and would not waste men unnecessarily. Or Ting might convoy a flotilla of transports, holding his ships ready to protect them. His own inclinations appear to have been towards the former alternative. He was anxious to search

---

* I have retained the popular name, though the official Japanese accounts call it the battle of Haiyang from the island of Haiyang-tao, near which it was fought.

ADMIRAL TING.

ADMIRAL ITO.

for the Japanese and fight before he took the transports forward. But the defeat of the Chinese land-forces at Phyong-Yang forced his hand and compelled him to use all possible expedition. He was driven to convoy the transports with his fleet, whilst the Japanese were still at sea unbeaten, and whilst the command of the sea was in dispute.

The transports, five in number, had left Taku and taken on board four or five thousand men at Talien Bay. Here Ting joined them. He flew his flag on board the battleship *Ting Yuen*, and with him were the *Chen Yuen*, her sister, the three small ironclads, *King Yuen*, *Ping Yuen* and, *Lai Yuen*; five cruisers, *Ching Yuen*, *Chih Yuen*, *Tsi Yuen*, *Tshao Yong*, and *Yang Wei*; two revenue cruisers of the Canton flotilla, lightly armed and ill-protected, *Kwang Kai*, and *Kwang Ping*; and at least two torpedo-boats. Chinese accounts mention four Rendel gunboats and four torpedo-boats as present with the fleet. If the gunboats were there, they took little or no part in the fighting. The Chinese torpedo-boats were in bad order, having been used for scouting and despatch carrying. Their boilers were burnt out, and their machinery in bad condition. Two, however, a Yarrow and a Schichau boat, played some part at the Yalu.

Ting might with wisdom have detached his fast cruisers—the *Chih Yuen* and the *Ching Yuen*, which were still perhaps capable of making fifteen knots—to scout, as obviously he would be at an enormous disadvantage if the Japanese suddenly came down upon him. He did not do this, probably because he was afraid of dissipating his strength, and preferred to risk a surprise. On Sunday, September 16th, at one o'clock in the morning, he left Talien Bay, the convoy keeping inshore, whilst the fleet steamed a parallel course in the offing, drawn up in line ahead. That same day he reached the mouth of the Yalu, and the transports, with the *Ping Yuen*, the *Kwang Ping*, and the Schichau and Yarrow torpedo-boats, entered the river. Ting anchored his squadron twelve miles off the coast, which is a difficult one to approach, owing to banks and shallows. The night of September 16th—17th passed without event.

In the Gulf of Korea was a large Japanese fleet, having its headquarters at an island in the Gulf, where were facilities for coaling, a mine-field protecting the anchorage, a soft bottom in shallow water for running disabled ships aground, and a torpedo station. In command was Vice-Admiral Ito, an officer who had distinguished himself at various times in the Japanese naval manœuvres with the cruisers *Matsushima*, *Itsukushima*, *Hashidate*, and *Chiyoda*, all modern and fast, the old ironclads *Fusoo* and *Hiyei*, and the despatch gunboat *Akagi*. His second-in-command, Rear-Admiral Tsuboi, had under him the Flying Squadron, which had already been engaged with the *Tsi Yuen*, and which comprised the splendid Elswick cruisers *Yoshino*, *Naniwa*, and *Takachiho* with the *Akitsusu* or *Akitsushima*. In addition the *Saikio*, an improvised cruiser taken over from the Nippon Yusen Kaisha, was present, having on board Rear-Admiral Kabayama, chief of the naval staff, who was on a tour of inspection. The second Flying Squadron, including the old vessels *Tsukushi*, *Chokai*, *Maya*, and *Banjo*, and the third Flying Squadron composed of the *Kongo*, *Takao*, *Yamato*, *Muzashi*, *Katsuragi*, and *Tenrio*, were engaged in co-operating with the land forces, and took no part in the battle. The two more powerful Japanese squadrons had been convoying troops up to September 14th, and had not paid the least attention to the Chinese, unless, indeed, a telegram which reached Ting from Wei-hai-wei on September 14th, to the effect that there were two large Japanese vessels off that port, was correct. In that case these ships must have been detached on the 12th or 13th. Admiral Ito, leaving the convoy, anchored off Cape Shoppek, where he remained till the afternoon of the 16th. Thence he proceeded to the island of Haiyang-tao, off which some Japanese torpedo-boats were cruising. He reached the island at half-past six on the morning of the 17th. It appears that he expected a Chinese fleet, having perhaps been informed of

Ting's intentions by spies. At Haiyang-tao no Chinese fleet was to be seen. He then steamed east-north-east to Yalu Island, and at half-past eleven saw smoke on the horizon. From its volume he judged that there was the Chinese fleet, and steered towards it at a low speed.

Meanwhile, the Chinese had observed a dense cloud of smoke far away to the south-west, about ten o'clock, or an hour-and-a-half before they were seen by the Japanese. The Chinese ships, it will be remembered, were at anchor, and therefore would not be likely to emit much smoke, whilst the Japanese used their native coal, which produces a great deal of smoke, and they were probably at a distance of nearly thirty miles when they were seen by Ting. Had the latter possessed fast ships he might, by approaching the enemy at full speed, have surprised him. The Chinese had their fires banked, but steam was at once raised and anchor weighed, when the whole fleet proceeded in a south-westerly direction at the rate of seven knots.

Some time before the action, according to Mr. Laird Clowes, Ting had issued three very important orders. 1. In action, sister ships, or groups of pairs of ships, were, to the best of their power, to keep together and support each other. 2. All vessels were, if possible, to fight bows on. 3. All ships were, as far as possible, to follow the movements of the admiral. These orders were issued because Ting's squadron was not homogeneous, being composed of many types; because the Chinese signalling staff was extremely weak, and it was apprehended that signals could neither be made nor received in the heat of action; and because it was felt that the course of battle could not be foreseen, but must be left to individual judgment.* The heaviest battleships of the squadron—indeed, the only two which could be called ironclads—were built for end-on work, and possibly Ting had thoughts of using the

---

* This was very much Persano's defence of his action at Lissa. He held that the details of the battle could not be foreknown, and that orders were useless. Von Hanneken was perhaps responsible for these tactics.

ram. None the less, these instructions had a disastrous effect. The group formation is, perhaps, a good one with perfectly trained officers and men, but it destroys the unity of a fleet. It is obvious throughout the battle which followed, that the Chinese were little better than an incoherent mass of ships, whilst the Japanese were an organised and compact force, striking together and acting together. There seems to have been no definite plan on the Chinese side; but every captain instead was to do that which seemed good in his own eyes. For fighting in line abreast Ting cannot be blamed; it was the designers of his ships who had forced him to this tactic. But his dispositions were not good even for line abreast, and there was no preparation for the maintenance of that order in the face of a turning movement. The Chinese left their anchorage in what is described as "sectional line abreast," or columns of divisions line abreast; that is to say, the ships were in two lines, one behind the other, the ships of the second in rear of the gaps between the ships of the first. There does not seem, however, to have been overmuch order, and one very fatal mistake was made. The heaviest and most powerful ships were placed in the centre instead of on the wings, thus violating the tactical axiom that the extremities of a line should be strong. Had Ting placed the *Ting Yuen* on one flank, and the *Chen Yuen* on the other, some, at least, of the Chinese disasters might have been averted.*

On the Chinese ships, what were the preparations for action? The *Tsi Yuen's* fight off Asan had given Ting and his advisers some idea of the precautions necessary. The barbettes of the ships had been, as we have seen, protected by sacks of coal. Sandbags were used to shelter the lighter guns, and mantlets of rope were disposed in suitable places to catch splinters. The tops of some at least of the conning-towers had been removed to allow the gases and fragments of bursting shells a free escape, and to diminish the size of the

* The *Tshao Yong* and *Yang Wei* were slow in weighing and were left behind at the outset.

target. The shields on the barbettes in the ironclads had also been left on shore, and thin armour had been generally dispensed with, on the principle that no protection is better than a weak one. All the boats had been left behind except one gig for each ship. The decks of the *Chen Yuen* were well-drenched with water, a precaution which does not appear to have been taken on board other ships. And there was one very obvious provision which was neglected by the Chinese—to give their men a meal before action. The Japanese, with greater wisdom, had been piped to dinner on sighting the Chinese fleet. A full stomach is an important element in the battle.

Both sides were now approaching each other, cautiously, and at a low speed, neither wishing to run the risk of a violent shock, and each, perhaps, desirous to see what his opponent was going to do. It was the first time that fleets, equipped with the modern engines of destruction, monster guns, torpedoes, quick-firers, were going into action. With no knowledge of *personnel* it would have been hard to say which of the two was the stronger. The Chinese were infinitely the worse armed, but then they had two well-armoured battle-ships, a type which the Japanese did not possess, and the strength of their defence may have compensated for their offensive weakness. Ting had also a considerable number of Europeans to advise him, and to stiffen the resistance of his crews. On the *Ting Yuen* was Major Von Hanneken, Ting's chief of the staff, with Messrs. Tyler, Nichols, and Albrecht. On the *Chen Yuen* were Captain McGiffin and Herr Heckmann; on the *Chih Yuen*, Mr. Purvis, and on the *Tsi Yuen*, Herr Hoffman.

At about five minutes past twelve the Japanese could clearly make out their opponents, distinguish the types of vessels, and see what lay before them. Admiral Ito hoisted a large ensign and ordered his ships to clear for action, whilst the *Saikio* and *Akagi* were directed to move from the line of battle and take up a position on the port side of it. The fleet was now in line ahead, ranged thus: First, Admiral Tsuboi led the Flying Squadron, including the *Yoshino*, which carried his flag, *Takachiho*, *Akitsusu*, and *Naniwa*, a homogeneous force, steaming over seventeen knots and heavily armed. Then followed Admiral Ito with the Main Squadron, comprising the flagship *Matsushima*, with the *Chiyoda*, *Itsukushima*, *Hashidate*, *Fusoo*, and *Hiyei*. Last was Admiral Kabayama, with the *Saikio* and *Akagi*, to some extent covered by the rear vessels of the Main Squadron.* When the two fleets were separated by an interval of about five miles, the Japanese, according to Captain McGiffin, passed into line abreast, and in that formation advanced for some minutes before passing back again into line ahead.† There is no mention of this in Admiral Ito's account, and if it took place it is difficult to understand what the object was, unless to puzzle the Chinese. There was now a fresh east wind blowing, the sky was grey, and the sea rough.

Soon after twelve Ito had signalled brief instructions to his captains. His fleet was to attack the Chinese right and fight at ranges of from 2000 to 3000 yards, circling round the Chinese vessels. The Chinese had now combined their two lines into a single line abreast, facing almost south-west, but the wing ships had been slow in getting to their stations. In consequence the Chinese line approximated to a crescent, the horns of which were away from the Japanese. From right (starboard) to left (port) the ships were placed thus: *Yang Wei*, *Tshao Yong*, *Ching Yuen*, *Lai Yuen*, *Chen Yuen*, *Ting Yuen*, *King Yuen*, *Chih Yuen*, *Kwang Kai*, and *Tsi Yuen*.‡ The *Tsi Yuen* was some distance behind the other vessels as she had had trouble with her engines. On the other horn the *Yang Wei* and *Tshao Yong* were also to the rear, whilst the *Ping Yuen*, *Kwang Ping*, and the two torpedo-boats

---

* See Plan I.

† In the earlier versions which appeared in the *Pall Mall Gazette*. In the *Century* he states that they kept to line ahead.

See Plan I.

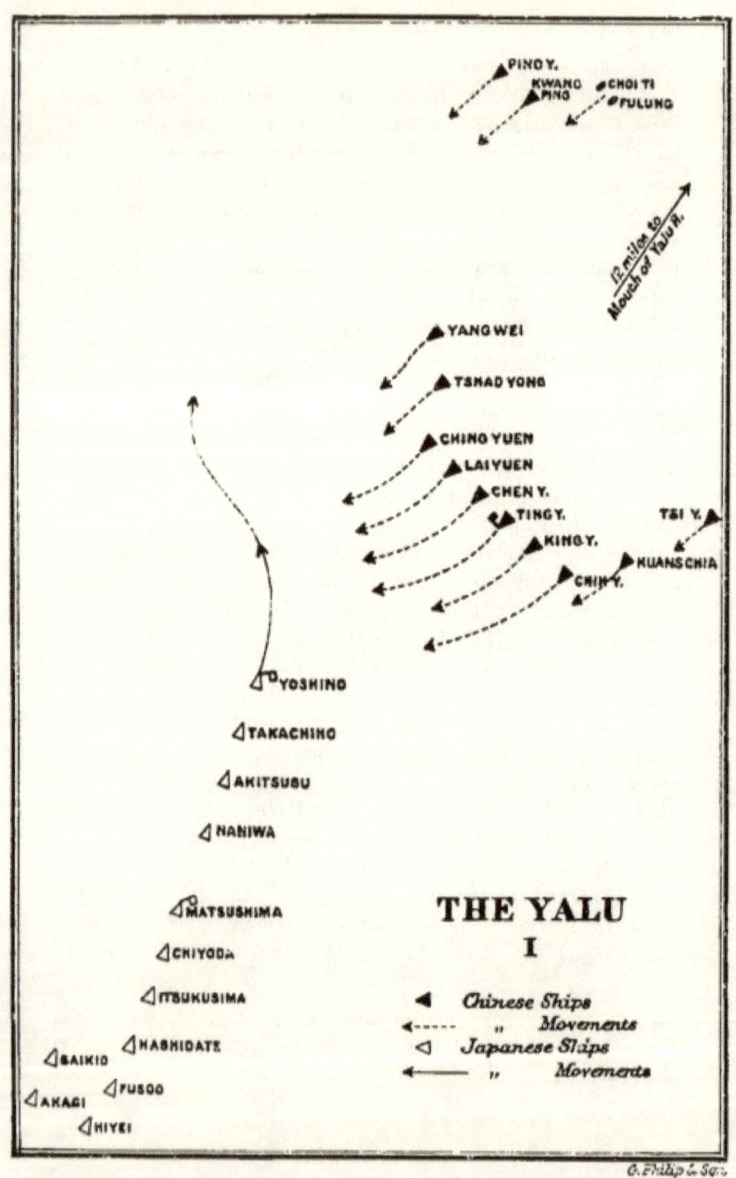

Map XXVI.

were just coming out of the harbour of Takushan, and were widely separated from the bulk of the fleet.

The Flying Squadron steadily approached the enemy, at first heading to starboard, and then keeping away to port, when Admiral Tsuboi had got within range. The speed was ten knots at 12.30, but at 12.45 was raised to fourteen. The first shots were fired from the *Ting Yuen's* 12-inch guns some time between 12.20 and 12.50. The concussion of the discharge was so great, or the Chinese officers were so unaccustomed to target practice with their heavy guns, that all who were on the bridge, which runs just above the barbettes, were thrown down, and Admiral Ting was so much shaken, that he was taken below. At a range of 6000 yards, the Chinese shots fell short. The Chinese ships were painted grey, and the Japanese white, so that mistakes could not very well occur. As the Japanese came on, the east wind blew their dense coal-smoke down, hiding them for minutes from the view of the Chinese. Their masts, however, were visible always, and enabled the Chinese gunners to lay the guns. This same wind carried away the smoke from the guns of the Chinese, so that it did not embarrass them much at this stage of the proceedings.

Ship after ship of the Chinese fleet opened fire, and the roll of heavy guns became continuous. The Japanese had not yet replied, but stood straight on till they were within 3000 yards of the centre of the Chinese line, when they turned in succession eight points to port, at the same time opening with their broadsides. Their 6 and 4·7-inch quick-firers poured in a hail of steel, which descended upon the Chinese, riddling the upper portions of the ships' superstructures, and filling the air with bursting shells. The water was lashed into foam by the shells, which, ricochetting, inflicted most of the hits. The sandbags piled up inside the vessels, prevented much damage being done as yet, and the Chinese gunners were kept lying down as far as possible, so that losses were small. As the Japanese came on, the two large Chinese battleships left the line and steamed forward as if to break through the enemy's line, or ram. The Japanese fired three or four times as fast as their enemies, sweeping their decks, cutting down masts, and riddling funnels. The Chinese appeared to fire slowly and at random; their shots went wide of the target. Admiral Tsuboi now raised the pace of his Flying Squadron, and rapidly neared the Chinese right flank. The Chinese had already lost what little order they possessed at the start, and were becoming a mob of ships, some of which masked the others. The left was practically out of the battle, whilst half-a-dozen vessels on the centre and right were bearing the brunt of the engagement.

The details of this engagement are hard to follow. No full and official account has appeared on either side, and, therefore, the story of it must necessarily be pieced together from various and often contradictory statements. The general features are, however, well ascertained, and we shall put these before our readers concisely, before proceeding to chronicle the fortunes of the various ships.

The Flying Squadron passed rapidly along the Chinese front, and when it reached the starboard flank, attacked vigorously the *Yang Wei* and *Tshao Yong*, small ill-protected vessels, and furnished with the most indifferent ammunition. Approaching till it was within 1700 yards range, it directed a tremendous fire upon them. In a few minutes the projectiles began to tell; the *Tshao Yong* was seen ablaze, and listing heavily to starboard. The *Yang Wei* was also in difficulties. On the other hand, the angular formation of the Chinese ships prevented half their vessels from bringing their guns to bear, and as each ship in the line turned, facing the enemy, they masked one another's fire. At this point, however, Rear-Admiral Tsuboi was recalled by a signal from his chief, and, circling to port, returned to assist the slower vessels of the Main Squadron.

We have seen that as the Main Squadron defiled past the Chinese front, the heavy Chinese battleships moved forward,

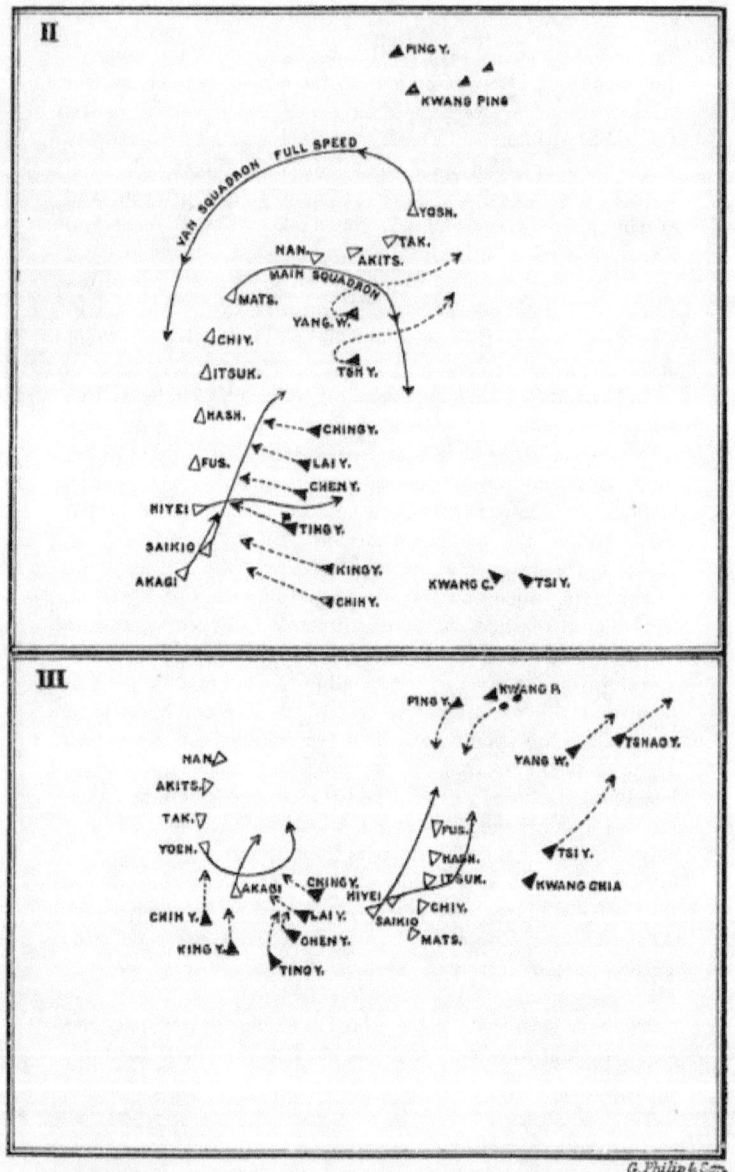

Map XXVII.

supported by the *Lai Yuen, King Yuen,* and *Chih Yuen.*
The faster Japanese ships drew ahead of the slower vessels
to the rear of their line, the *Fusoo, Hiyei, Saikio,* and *Akagi.*
On them fell the brunt of the Chinese attack. The *Fusoo*
cleared the advancing ironclads, but the *Hiyei* was compelled
to turn sharply to avoid the enemy's rams. With extraordinary audacity, she headed for the *Ting Yuen* and passed
between the flagship and the *Chen Yuen* at a range of 700
yards.* Fortunately for her, the Chinese ships could not
pour in their fire on her without risk of injuring one another,
but none the less she was hit several times. Two torpedoes
were discharged at her, but missed and passed astern. The
Akagi was near her and suffered more severely. Indeed, it
was only the arrival of the Flying Squadron which saved her
from destruction at the hands of the *Lai Yuen, Chih Yuen,*
and *Kwang Kai*. The *Saikio* kept further away to port, and
passed along the front of the Chinese line, receiving a very
heavy fire. On rounding the Chinese right, she found herself
confronted by the *Kwang Ping,* whilst the *Chih Yuen* was
coming up astern. Her steering gear was disabled, and to
add to her perplexities, the Chinese torpedo-boats assailed
her. Three torpedoes were fired at her, and as usual missed.
She was compelled, with the *Akagi,* to retire from the
engagement.

The position of the Chinese was now as follows: The line
abreast had become a disorderly conglomeration of ships fighting anyhow. On one side of them was the Flying Squadron,
on the other the Main Squadron of the Japanese,† which had
turned the right flank of the Chinese, and completed the
discomfiture of the unfortunate *Yang Wei,* which ran from
the scene of action ablaze. On turning the Chinese right,
the Main Squadron had a short but sharp engagement with
the *Ping Yuen* and the Chinese torpedo-boats, in which the

* See Plan II.

† It is said that this was due to a mistake and that Admiral Ito had intended to keep his ships together.

latter were driven off. The Chinese ships in the line were thus left to themselves, and had an enemy in front and an enemy behind. The Flying Squadron engaged the cruiser *Chih Yuen*, which broke from the line, and endeavoured to ram the *Yoshino*. The *Yoshino's* quick-firers, using cordite, covered her with bursting shells. At 3.30 she went to the bottom, having been sunk by gun-fire alone. Since the two Japanese squadrons were in danger of hitting each other, as they steamed backwards and forwards to the front and rear of the Chinese, they went further apart, always, however, drawing in when they were opposite the smaller Chinese vessels. The hail of projectiles upon the *Ting Yuen* and *Chen Yuen* was extraordinarily fierce, but there was no sign of their yielding. The *Tsi Yuen* and *Kwang Kai*, however, took to flight, and in her flight the former collided with the unhappy *Yang Wei*, damaging her severely. The *Ching Yuen* had also retired, as she was on fire. The *Lai Yuen* could be seen ablaze. To her and her sister, the *King Yuen*, the Japanese Flying Squadron turned its attention. At 3.52, the *Takachiho*, 3300 yards from the *King Yuen*, opened upon her. The *Yoshino*, with her 6-inch quick-firers, joined in at 2500 yards. At 4.48, the Chinese ship had a list to port, and could be seen ablaze. The bottom of the vessel showed; the rudder was useless; she veered to and fro in the smoke and uproar. The hail of shells tore her open; her stern dipped; and with a violent explosion she went to the bottom. On their side the Japanese lost the *Matsushima* which at 3.30 or 3.40 was fearfully injured by the Chinese. She steamed out of action and Ito transferred his flag to the *Hashidate*.

The loss of the Chinese so far had been heavy. Sunk were the *Chih Yuen* and *King Yuen*; disabled and sinking the *Tshao Yong* and *Yang Wei*; ashore the *Kwang Kai*, which had taken the ground in her desperate efforts to escape; in the offing the *Tsi Yuen* retiring to Port Arthur; and closer at hand the *Ching Yuen* and *Lai Yuen*, endeavouring to put out fires. The *Ping Yuen* and *Kwang Ping* were holding

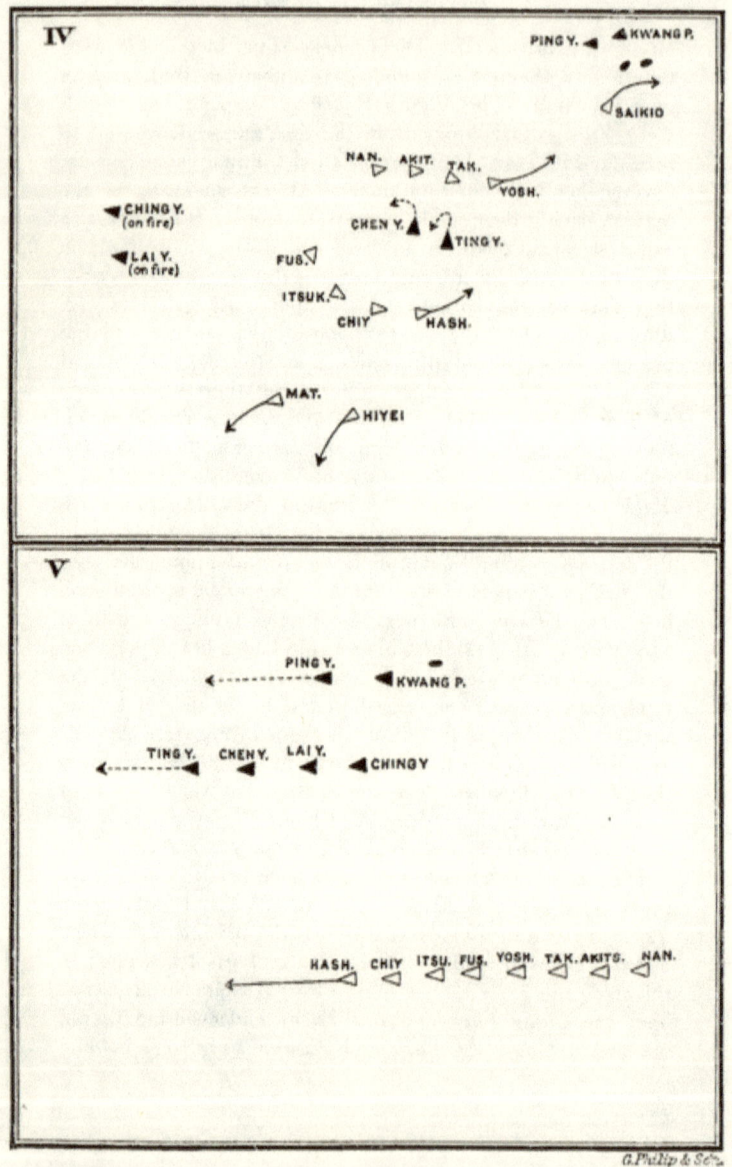

Map XXVIII.

discreetly aloof. In line there remained only the *Chen Yuen* and the *Ting Yuen*, both of which were on fire. The Japanese had lost the *Matsushima, Hiyei*, and *Saikio*, but the *Akagi* was already preparing to return to the battle. The Flying Squadron proceeded to chase the retiring Chinese ships, whilst the Main Squadron, after a brief respite, once more pounded the *Chen Yuen* and *Ting Yuen*, though the two combatants were now beginning to draw slowly apart, probably because ammunition on either side was running low.

The day was now verging upon evening. The Flying Squadron was recalled from its pursuit by signal, and the *Matsushima, Hiyei*, and *Saikio*, were sent off to the port of Kuré for repairs. Whilst the Japanese concentrated, the Chinese did the same, and the two heavy battleships—still unsubdued and still battle-worthy—collected round them the *Lai Yuen*, the *Ching Yuen*, the *Ping Yuen*, two Rendel gunboats and the two torpedo-boats. The Japanese, with wearied crews and exhausted ammunition, did not care to risk a night action, in which the torpedo-boats would have had an opportunity to show their efficacy. They followed, at a distance, towards Wei-hai-wei, but when day dawned the Chinese were not in sight. On this, they returned to the scene of action. The *Yang Wei* was destroyed by a spar-torpedo, used against her by the *Chiyoda's* launch. After coaling and taking on board ammunition at their base, the Japanese prepared for a second engagement. The *Naniwa* and *Akitsusu* reconnoitred the Chinese ports, whilst the rest of the squadron cruised in the gulf. In Talien Bay two Chinese ships were sighted. One, perhaps the *Tsi Yuen*, ran on seeing the Japanese, and succeeded in effecting her escape. The second turned out to be the *Kwang Kai*, hard and fast aground. She was destroyed, either by the Japanese or by her crew.

To turn now to the individual fortunes of the ships most hotly engaged. The *Matsushima* did not suffer heavily till she fought the *Ping Yuen*. With her she opened fire at

3000 yards, and continued firing till she was within 1300 yards. At 2.34 she was hulled by a 10·2-inch shell from the Chinese vessel, which killed four men at the port torpedo-tube aft, and striking the barbette was shattered against it. This shell passed very close to a loaded Whitehead torpedo, which would assuredly have destroyed the ship had it exploded It then coursed through store-rooms and an oil-tank, but proved, when it went to pieces against the barbette, to have been loaded with cement. In reply the Japanese disabled the *Ping Yuen's* big gun. Later in the afternoon, when facing the *Chen Yuen*, the *Matsushima* was far more severely handled. A 12-inch shell, from the Chinese ironclad, entered her battery, hurled the fourth 4·7-inch gun from its mount, and exploding, fired a great heap of ammunition. Two 4·7-inch guns were disabled, ninety officers and men killed or wounded. The gunnery officer, who was standing close to the ammunition, was blown to pieces, only his cap being afterwards found. The ship listed and a fire broke out. The crew, with unabated gallantry and courage, divided their attention between the fire and the enemy. The bandsmen went to the guns, and though the position of the ship was critical, and her loss appalling, there was no panic. The fire was on the lower deck, just above the magazine. In charge of the magazine were a gunner's mate and a seaman. The shell had, apparently, dented the plating over the powder, and the red glow through the crevices showed the danger. But these brave men did not abandon their posts. Stripping off their clothes they crammed them into the cracks and saved the *Matsushima*; though nearly a third of the men above the water-line had been put out of action, the remnant got the fire under. With fifty-seven officers and men dead, and nearly as many wounded, the ship steamed out of action.

The *Hiyei*, on escaping the torpedoes fired at her, had withstood the attack of the *Chen Yuen*. One 12-inch shell struck her in the neighbourhood of the ward-room, which was being used as a temporary hospital; bursting, it killed instantly her

PLATE XXVIII. OFFICERS OF THE MATSUSHIMA.

chief surgeon and paymaster, with a large number of wounded. The mizzenmast fell and the ship took fire. A second heavy shell, bursting on her upper deck, killed many of her gunners.

The *Akagi* had to face the Chinese ships of the left unsupported. At 850 yards she engaged them fiercely, clearing the *Lai Yuen's* deck with her starboard battery. At 1.20 a shot struck her bridge and killed her captain, Commander Sakamoto, with two gunners. The command passed to her navigating lieutenant, and a few minutes later she was hulled repeatedly about the level of her lower deck, losing four firemen. Her steam-pipe was shattered, and the scalding steam by its escape cut off the supply of ammunition, just when it was most wanted. The shot and powder had to be sent up by a ventilating shaft, thus incommoding the engine-room complement. Three gunners were killed at this time on the upper deck. The *Lai Yuen, Chih Yuen,* and *K'wang Kai* were coming up astern, and her position was one of extreme danger. Turning to port, she for a time eluded her enemies whilst repairs were made. Once more the Chinese neared her, and once more she fought them, now heading south. Her mainmast was shot away, and at a distance of only 330 yards from the *Lai Yuen*, her bridge was struck, and her new commander wounded. A lieutenant took his place, whilst those guns, which would bear astern, fired steadily at the *Lai Yuen*. At 2.20 a shell set the *Lai Yuen's* deck on fire, and the other Chinese vessels slowed down to give her assistance. The Flying Squadron, too, was coming up astern of them, and beginning to engross their attention. Thus the *Akagi* was enabled to steam out of range, when her steam-pipe was repaired and her crew given a much-needed rest. At 5.50, three hours after her withdrawal, she again joined the Main Squadron.*

The *Saikio* had an even more wonderful escape than these two ships. Passing along the Chinese line, she was struck in

---

* Admiral Ito imagined that she had been sunk.

quick succession by four enormous shells from the *Ting Yuen*.
Two went clean through her, doing no damage. Two common
shells, however, burst in the upper-deck saloon, shattered the
woodwork, and disabled the steering-gear. Signalling that
she could not steer, and manœuvring with her twin screws, she
passed through the Flying Squadron, between the *Naniwa*
and *Akitsusu*. Relieving tackle was fixed, and at 2.20 or
thereabouts she was again under control, and now found the
Ping Yuen before her. With this ship, the *Kwang Ping*,
which had already been in action with her, and the two
torpedo-boats, she began an engagement about 2.50. Opening at 3300 yards, she continued till she was only 500 yards off
the Chinese. One torpedo-boat was driven off by her fire,
but the second discharged three torpedoes at her. The first
crossed her bows, the second ran along her starboard side,
the third dived. Though she kept up a vigorous fire from
her machine-guns upon the boat, it escaped quite unharmed.
At 3.30 the *Saikio*, on fire astern, abandoned the engagement.
She had received a large number of projectiles, some of which
narrowly escaped her engine-room, but her damage was
inconsiderable and her loss of life nil. Nothing so surely
demonstrates the incapacity of the Chinese gunners as their
failure to sink this ship. A weak merchant steamer, she had
faced the Chinese line, and after this ordeal had engaged two
Chinese warships, one of which had considerable armour protection.

Next let us pass to the Chinese ships. The *Ting Yuen*
fought stoutly all through the day. At the beginning of the
battle a heavy shell, probably a *ricochet*, struck her fore military mast, killing seven men in the top and bringing it down.
As we have seen, Admiral Ting and Major von Hanneken
were injured by the concussion of the guns at the first discharge. Later in the day a very serious fire broke out forward. The smoke from it completely shrouded the barbettes,
and for some time the only gun which could fight was the
6-inch Krupp aft. The fire was got under through the exer-

tions of Herr Albrecht, after it had gravely imperilled the ship's safety. On board the *Ting Yuen* was killed Mr. Nichols, an ex-petty officer of the British Navy, who displayed great gallantry throughout the battle. The Chinese on board showed no great spirit; unlike the *Matsushima's* men, when they saw the ship ablaze they bolted, and did not even think of fighting the ship. Albrecht it was who saved her, standing to his hose amidst fire and exploding shells. It was found impossible to signal when the foremast had gone.

Like the *Ting Yuen* the *Chen Yuen* suffered much from fire. She was ablaze no less than eight times, but, mainly through the Europeans, each time the fire was got under. A European who was on board her records his experiences as follows:*
"In helping to put out one of these fires, I was wounded. The fire was forward, on the forecastle, and there was such a fierce fire sweeping the deck between it and the fore-barbette, that the officer, whom I ordered to go and put it out, declared it to be impossible to get there alive; so I had to go myself. I called for volunteers, and got several splendid fellows—some of our best men, unhappily, for nearly all were killed, but we got the fire under. The fire was on the port side, and as the starboard fore-barbette gun was firing across it, I sent orders that it was only to fire on the starboard side, but, as bad luck would have it, the man who received the order, the Number One of the gun, had his head shot off just after I had gone forward, and his successor did not know of it. As I stooped to pick up the hose, a shell, or a fragment, passed between my wrists, grazing each. Shortly afterwards, I heard a loud explosion, and saw a brilliant light behind me, was knocked down, and lay unconscious for a while—how long I do not know. I believe it was the flame from the gun which I had ordered to fire only on the starboard side, but it may have been a shell exploding, though, if so, I ought to have been blown to pieces.

* Captain McGiffin in the *Pall Mall Gazette*.

Anyhow, I was pretty badly burnt, and when I came to, I sat up leaning on my elbow, and found myself looking almost down the tube of the great gun, pointing straight at me. I saw the end move a little to one side, then to the other, up a little, then down; and I waited for years—a fraction of a second no doubt—for the gun to fire, for I knew that the gunner had taken aim. Then it suddenly occurred to me to make an effort. I rolled over on my side, and by great good fortune, down a hatchway some eight feet or so, on to a heap of rubbish, which broke my fall; as I fell I heard the roar of the big gun." Her Chinese crew behaved better than the *Ting Yuen's* men. The discipline was excellent, the guns were fairly handled, and she was manœuvred with some skill. With the *Ting Yuen*, she bore the brunt of the Japanese attack, and with the *Ting Yuen*, circled slowly round, attempting to keep end-on to the two hostile squadrons. She was much shattered in her upper works, but not seriously injured. The spindle of the port gun's hydraulic gear was struck, and the port gun put out of action for a time. The bow 6-inch gun was disabled by an accident. The foretop was hit twice, and six officers and men in it killed.

The *Chih Yuen*. under a brave and determined captain, Tang, had advanced from the line at the outset of the battle, and was hotly engaged with the Flying Squadron on its return. She was hit repeatedly, amidst loud cheers from the Japanese, and began to list to starboard. Her captain made a futile effort to ram, but the Japanese quick-firers were too strong for him. As he closed with the *Yoshino*, the list increased, the screws showed above water, racing in the air, and the ship went down with all hands. As she sank, a violent explosion was observed. She is said to have been finished off by a 12·6-inch shell from one of the big Canet guns. As in the case of the *King Yuen*, the appearance of an explosion may have been due to the bursting of her boilers, or the detonation of a torpedo in one of her above-water tubes. Her European engineer, Purvis, went down with her.

The *King Yuen* was badly on fire about the time the *Chih Yuen* sank, and dense volumes of smoke were seen pouring from her. She moved forward upon the *Yoshino*, and received a tremendous hail of shells. She was seen rolling very heavily, now to port, now to starboard, and as one after another the Flying Squadron plied her with their projectiles, she lost all power of steering and described wild circles. Her end is veiled in mystery. All the Chinese, who saw her go down, attributed her loss to a torpedo, but the Japanese fired none. There was a very thick cloud of smoke and an explosion just before she vanished, like the *Victoria* capsizing and showing her bottom. Of a crew of 270 men only seven escaped.

The *Lai Yuen* again was put out of action by fire. For an hour-and-a-half she was seen ablaze. A shell struck her deck, and though a bucket of water would, at the start, have put out the flames, with Chinese apathy she was allowed to burn, till the fire, having consumed almost everything above the water-line, burned itself out. She was left a mere shell, terribly damaged by fire and shell, yet, strange to say, her fighting and manœuvring qualities were little affected. She was brought safe to Port Arthur. Her deck, of 2-inch teak, and the large amount of paint and varnish lavished upon her woodwork, made her a ready prey to any shell.

The *Ching Yuen* was three times on fire. She retired to extinguish one of these fires, and, therefore, took little part in the battle.

The *Tshao Yong* had her steering-gear disabled and was seen ablaze. The *Yang Wei* ran aground, and was on fire when she was rammed by the *Tsi Yuen*. The *Kwang Kai*, *Kwang Ping*, and *Ping Yuen* took little part in the fighting, and were very slightly damaged.

The *Tsi Yuen* was commanded by Captain Fong, whose acquaintance we have made already. He is said to have run in a cowardly manner before his vessel had received serious injury, and it is certain that he managed his ship very badly,

but, in justice to him, we must record the statement of Herr
Hoffman, who was on board, and who gives us an interesting
picture of the battle. "We accomplished the journey to
Tatungkow in safety, landed the troops, and about 11 o'clock,
on the 17th ultimo, the whole fleet got up anchor, and prepared
to return to China. A short distance outside the mouth of the
river we met the Japanese fleet, and a battle followed, which
lasted till 5.30 in the evening. It was the most tremendous
fight I had ever dreamt about. Captain Fong fought the *Tsi
Yuen* with courage and ability. We had seven or eight men
killed on board, and continued firing away as fast as we could
until between 2 and 3 o'clock in the afternoon, by which time
we were terribly damaged, and had to leave the scene of
action. Our large gun aft, 16-centimètres, [15-centimètre?]
Krupp, was disabled, and the two forward guns had their gear
destroyed, so that they could not be used, and to all intents
and purposes the ship was useless,* so Captain Fong decided
to get out of the action and make the best of his way to
Port Arthur, to refit. The smoke was so dense that no one
could see very much of what was going on from the deck,
but from time to time we heard that this, that, or the other
ship was gone. Having left the fight in the *Tsi Yuen*, I know
nothing of what subsequently happened. We arrived at Port
Arthur five or six hours before the remainder of the fleet,
which came in about 8 o'clock. On the way in we had a
collision with another vessel [the *Yang Wei*] which sunk.
From the injuries to the *Tsi Yuen*, which were all abaft the
stern, I should say the other ship rammed us. The water
poured into the *Tsi Yuen* in a regular torrent, but we closed
the water-tight doors forward, and went on safely. I don't
think that the charges of cowardice, which have been brought
against Captain Fong, can be supported for a moment; he
fought his ship until it was no longer serviceable. The smoke

* Captain McGiffin, however, states that on examination at Port Arthur all
her guns were found in good order, except the 15-centimètre stern-chaser, which
must have been disabled during her retreat.

THE CHEN YUEN IN ACTION.
By J. T. Jane.

PLATE XXIX.

was so thick that one only had a chance of knowing what was going on in his own ship."

There is one point unexplained in the battle. The Chinese assert that the *Chih Yuen* was successful in her attempt to ram; the Japanese that she failed. Some foreign observers on the Chinese ships saw a vessel sinking with revolving screws, and thought that it was the *Chih Yuen's* victim. More probably it was the *Chih Yuen* herself, or the *King Yuen*. The torpedo-boats of the Chinese found it difficult to distinguish the combatants in the smoke, though there was the wide difference of colour, between white and grey. The speed of the torpedo-boats was found to be only fourteen or fifteen knots, instead of the trial figure of twenty. The boats were seen at once, and fired at by the Japanese, long before they got within range.

The five transports up the Yalu river received a message from Ting at the close of the engagement, ordering them to join him. They were too far up the river to do so at once, and their crews were prostrate with fear. Not till four days after the battle did they leave, and then they had the good fortune to return to Taku unmolested, in the face of the victorious Japanese fleet.

In its general features, there is a singular resemblance between this battle and Lissa, with this broad difference, that the Yalu was a long range fight, and Lissa a *mêlée*.* The Chinese, like the Italians, fought as a mob of ships, without orders, without plans, and without commander. Each vessel had to do what it could, as there were no signals after the *Ting Yuen's* foremast had been shot away. The Japanese, like the Austrians, knew what they meant to do, but they had this immense further advantage, that throughout they manœuvred and fought by signal. A signal at a decisive moment brought back Tsuboi with the Flying Squadron to support the *Hiyei* and *Akagi*; a signal recalled him a second

* There was, of course, some fighting at close quarters at the Yalu, but generally speaking the engagement was a long range one.

time at the close of the day, when pursuit might have been dangerous. Like the Austrians, the Japanese had won a distinct advantage, and yet failed to annihilate their enemy. Still directly, or indirectly, they struck off his list of ships five vessels, and so injured one that it was henceforward useless.*
Tegetthoff, on his part, sent two Italian ships to the bottom, and disabled a third. And like Tegetthof, if he did not exactly wipe out the Chinese, Ito left them with little stomach for future fighting. At the Yalu, China lost all chance of commanding the sea, and by losing it, brought on herself defeat in the war.

In another respect, the Yalu resembled Lissa. In each case the assailed fleet was engaged in covering a landing force. But whereas the Italians were encumbered with troops and transports, and whereas they were caught in the midst of disembarkation, after they had unsuccessfully cannonaded Lissa, the Chinese were taken at no such disadvantage. The earlier reports of the battle represented it as having taken place close inshore. As a matter of fact, the Chinese had no want of sea-room, nor were their movements in any way hindered by their convoy, which had ascended the river, and was therefore safe from attack, since the Japanese ships of heavy draught could not follow it. There seems no reason then to attribute the Chinese defeat to the presence of the convoy. Except that it brought the Chinese fleet to sea, it cannot be said to have had any effect upon the issue of the day. Had the Chinese warships been cruising alone, the result must have been the same. The bearing of this, and, indeed, of all the moves of both Chinese and Japanese during the earlier period of the war, upon the doctrine of "the fleet in being" is self-evident. It has been held that the presence of an inferior squadron at sea will prevent even a superior squadron from attempting to convoy or disembark troops. Yet by either side, up to the battle of the Yalu, this rule was

* The ships sunk were the *Chih Yuen, King Yuen, Tshao Yong, Yang Wei,* and *K'wang K'ai.* The *Lai Yuen* was disabled.

disregarded, though the two squadrons were approximately equal.*

As Tegetthof drew off at Lissa when apparently he had the Italian fleet in his grasp, so Ito retired at the Yalu. In either case it was probably the want of ammunition which led to the withdrawal.† With heavy guns and limited displacement the supply carried cannot be inexhaustible, and we may anticipate in future actions a similar indecisiveness at the finish, if from first to last the battle should be fought out a long ranges. That the Japanese did not close may seem strange, but in the first place they probably wished to capture the two big ironclads instead of destroying them. Captain Ingles in a telegram at the outbreak of war had recommended this course. Ulterior motives thus supervened to protect the Chinese, for if the Japanese had closed they must have employed the ram or the torpedo, their ammunition being exhausted, and either ram or torpedo would have sunk the battleships. Again, the Japanese with their unarmoured ships could not have thus closed without heavy loss. A cruiser is never very strong in the bows, and, disregarding the Chinese fire, the injury inflicted upon the "rammer" must have been very considerable. The Japanese, too, could not be certain that the Chinese, who fired very much more slowly, had used almost all their ammunition. There would have been risk of the two big battleships, protected as they were by armour, sinking, in a close action, the Japanese unarmoured cruisers.

The Chinese are stated to have made desperate efforts to come to close quarters with the Main Squadron, and to have been foiled by the superior speed of the Japanese. If the efforts were serious they should have succeeded. The *Fusoo*,

---

* The Japanese may, however, have known of the orders of the Tsung-li-Yamen to Ting. Still the Japanese fleet was hampered by no such orders. Captain Mahan's pronouncement on the "fleet in being," that such a fleet would not constitute such a deterrent force upon the movements of a resolute man, as had been supposed, seems to be fully justified by facts.

† *See* i. 240.

which kept her station in the Japanese line, was slower than one, if not both, of the Chinese battleships. The "Times" correspondent, however, states that the commanders of the ironclads gave orders to go at full speed, but that the Chinese lieutenants at the engine-room telegraphs, fearing for their own skins in a close action, did not transmit the orders correctly. The battleships are said to have begun the action with a very limited supply of common shell. Of this species of projectile they had but fifteen rounds per gun for their heavy ordnance, whilst the rest of their ammunition was armour-piercing shot. If this is true it was a piece of gross mismanagement.

On their return to Port Arthur, like the Italians, the Chinese claimed the victory. They represented that at least three Japanese vessels had been sunk, and probably still believe it to this day.

The first important battle since Lissa—the second since the introduction of the ironclad—had been fought and won, and the elaborate contrivances which had replaced the line-of-battleship had at last been tested at sea in a general action. It had been expected that the losses in such a battle would be very heavy, yet it cannot be said that this expectation was altogether justified. The Japanese lost, in killed, ten officers and eighty men, in wounded, sixteen officers and 188 men. This gives a grand total of 294. Their total force of sailors engaged in the battle could not have been much less than 3000, and may have been a little more. They lost then ten per cent. of their force.* The heaviest loss fell upon the *Matsushima*, where the killed numbered fifty-seven and the wounded fifty-four. As the flagship she would naturally be singled out by the Chinese, and would receive a heavy fire. She also went dangerously near the two large Chinese battleships. She carried no vertical armour, except on her heavy gun. Second came the *Hiyei*, with nineteen killed and thirty-

---

* See Table XXI.

seven wounded. She had no armour except a very short belt on the water-line, and, like the *Matsushima*, she was at very close quarters with the *Chen Yuen* and *Ting Yuen*. The *Itsukushima*, with thirteen killed and eighteen wounded, was third: she, too, was unarmoured. The *Akagi* lost eleven killed and seventeen wounded, but the *Saikio*, wonderful to relate, only eleven wounded. Bad gunnery on the part of the Chinese can alone explain this astonishing fact. The *Akitsusu* had fifteen killed or wounded, the *Fusoo* fourteen, the *Yoshino* and the *Hashidate* twelve each, the *Takachiho* three, the *Naniwa* one, and the *Chiyoda* none. The escape of the *Chiyoda* is extraordinary, when it is remembered that she fought in the line between the *Matsushima* and *Itsukushima*, both of which suffered severely.

The Chinese loss on board the ships which survived the encounter was not so heavy as that of the Japanese, but a very large number of men were killed, wounded, or drowned in the action on board the ships which went down. We shall not, perhaps, be exaggerating when we place the number of lives thus lost at from 600 to 800. In addition, thirty-six were killed and eighty-eight wounded on board the seven vessels which survived. The *Ting Yuen* lost fourteen killed and twenty-five wounded; the *Lai Yuen*, ten killed and twenty wounded; the *Chen Yuen*, seven killed and fifteen wounded; the *Ching Yuen*, two killed and fourteen wounded; the *Tsi Yuen*, three killed; the *Ping Yuen*, twelve wounded; and the *Kwang Ping*, three wounded. The Chinese may have had 3000 men present at the action: in that case, they lost from twenty to thirty per cent. of their force. It is interesting to notice how the armour of the *Ting Yuen* and *Chen Yuen* diminished their losses. They were the most hotly engaged of the Chinese ships, and had for hours to serve as the targets of five Japanese vessels. Yet the losses of these two ships, added together, are just half of those incurred on the Japanese flagship alone. And it must be remembered that the practice on the Japanese side was probably better than on the Chinese.

If it did nothing else, then, armour saved a very large number of human lives.

Comparing the losses at the Yalu with those at Lissa, and earlier battles, in the days of wooden ships, we get these results:

Battle.	Nation.	Total of Men Engaged.	Killed and Drowned.	Wounded.	Total Killed and Wounded.	Percentage of Casualties to Force.
Yalu	Chinese	3000 ?	600—800 ?	68	Say 700	23) Average
	Japanese	3000 ?	90	204	294	10 } of both 16½
Lissa	Italians	10,880	600 ?	39	639	6 } " 4
	Austrians	7831	38	138	176	2 }
Trafalgar	Allies	21,580	No returns accessible			
	English	16,820	449	1241	1690	10
Nile	French	9820	Estimated at	678	3000	29 } " 20
	English	7080	218		896	11 }
Camperdown	Dutch	3150	540	620	1100	36 } " 13
	English	8220	203	622	825	10 }
First of June	French	19,700	—	—	5000	25 (" 15½
	English	17,240	290	858	1148	6 (

Figures for earlier battles from Hodge: "Losses in Naval War."—Journal Statistical Society, vol. xviii. Average for thirteen battles 1782-1811, 11½ per cent.

It would thus appear that the loss of life in the engagement we are considering was rather heavier in proportion than it was before the days of ironclads. There is this also to be taken into account: on board the old line-of-battle ship very few men were below the water-line, whereas on modern ironclads and cruisers, quite a considerable proportion of the crew is busy in the engine-room or stokehold, below the armoured deck, and therefore out of the reach of shot, exposed only to death by drowning if the ship sinks. When this is remembered, the proportion of men killed and disabled amongst those exposed mounts considerably; especially on the Japanese side, where no ships were sunk, must this be recollected and the percentage adjusted. At least, five hundred men on the twelve Japanese ships must have been engaged in tending boilers or machinery, in addition to those at work at the ammunition hoists. If 2500 men only were exposed to fire, the loss would amount to twelve not ten per cent. It would not then seem that, with the modern engines of destruction, war has become less bloody, but, rather, that the risk of the individual serving in the modern fleet has been slightly increased. At the same time, the loss of life might be greatly diminished if special vessels were at hand to rescue the drowning in the water. Both at Lissa and at the Yalu, the greater proportion of deaths on the beaten side were due to drowning. At Lissa, Tegetthoff, as we have seen, endeavoured to give aid to the Italians in the water, but was prevented from doing so by the onset of the Italians.* At the Yalu, the Japanese saw the men crowding in the *Tshao Yong's* tops, after she had collided with the *Tsi Yuen* and sunk. They pitied, but as in Tegetthof's case, the stress of the battle would not suffer them to aid. The Chinese torpedo-boats, however, took off a large number of men, who must otherwise have fallen victims in the struggle, and their success in this mission of mercy suggests the question, whether some international agreement not to fire upon ships or vessels, engaged in saving the drowning, should not be arrived at. In the days of Nelson, chivalry and self-interest forbade the line-of-battle ship to fire upon frigates thus engaged. It would go without saying that these special vessels would not be armed, and would be distinguished in some way—by colour or build—from the combatants. The first principle of warfare is to crush the enemy; the second, which has only been recognised in modern times, is to inflict no unnecessary suffering in crushing him. When the enemy's sailors are in the water, they are as helpless as the wounded, and as ambulances are not fired upon, why should not ambulance-ships be given all possible immunity? The importance of this point was fully understood by Tegetthof, who was anxious after Lissa, that a European conference should be convened to deal with it. Unfortunately, this step has never been taken.

There seems little doubt that the stiffness of the Chinese resistance was greatly increased by their belief that no quarter would be given. With the exception of the *Tsi Yuen* and

* Vol. i. 237-8.

Kwang Kai, no ships in the Chinese line showed cowardice. Incapacity and blundering there was in plenty, but having their backs to the wall, as they supposed, the seamen fought courageously. On land, the Chinese invariably fled before their better-trained, and better-armed, opponents; here alone they stood up to them, and by doing so, once more proved that courage alone, without skill, will not win battles. The crews of the ships were one half composed of raw recruits, owing to the peculation of admirals and captains, who had maintained in peace a shadow of the nominal effective. Corrupt administration is a most fatal failing when there is fighting toward.

The gunnery of the Chinese was extremely indifferent, but their ships and guns were in part to blame for this. They had eight 12-inch Krupps and five 10-inch weapons by various makers, twelve guns of 8-inches, fifteen of 6-inches, and twelve of 4·7-inch, besides 130 machine-guns or Nordenfelts. Their heavy guns were of somewhat antiquated pattern, and they carried no large quick-firers. Thus they lacked the very weapons which would have been most effective against unarmoured ships. They were further very ill-provided with guns of moderate size. The Japanese had of weapons ranging from the 6·8-inch Krupp to the 4·7-inch Armstrong, no less than ninety-four, and of these sixty-six were quick-firers. The Chinese had of corresponding calibres only twenty-seven, with scarcely an exception, slow-firers. In auxiliary armament the Chinese were thus immeasurably behind their enemy, and the numerical difference is increased by the fact that, as a quick-firer will discharge in a given time from three to six times as many rounds as a slow-firer, each Japanese quick-firer was worth three Chinese guns. As the Japanese fought at a long range, and moved with fair rapidity, the Chinese gunners did little but miss them with the heavy guns, which have, indeed, a very long range, but are slow and awkward to lay upon the target. We find, therefore that, as we should have expected, the great guns made few hits. Five of their shots struck the *Saikio*, three or four the *Matsushima*, two the *Hiyei*, one the *Naniwa*, and possibly some the other ships. This would give a total of twelve to fifteen. Now the *Chen Yuen* and *Ting Yuen* alone fired between them 197 12-inch projectiles, to say nothing of 268 of 6-inch calibre. The other ships must have discharged fully as many 10·2 and 8·2-inch shells. Thus a very rough estimate of heavy shots fired will give at the least 400, of which not twenty, or four per cent., struck the target. This is curiously in accord with the experience of the *Shah*. Of the Chinese heavy projectiles which did strike, a considerable proportion again appear to have been armour-piercing shot, which would do little damage. It was estimated by a Japanese officer that ten per cent. of the Chinese shot, and fifteen per cent. of the Japanese scored hits.[*] The estimate appears a high one, when we consider the long ranges which prevailed, and if it is correct for the Chinese, the low percentage of hits made by their heavy guns must have been compensated by the high percentage of their medium and light guns. Applying this estimate to the figures given for purposes of comparison in Table XX., each Chinese ship was firing 32·8 shots a minute, with 3·28 hits, each Japanese 188·3 shots, with 28·24 hits.

The slow fire of the Chinese moderate-sized guns made strongly against good shooting, besides placing them far behind the Japanese in the weight of metal thrown in a given time. With a moving target, the shorter the period that elapses between the shots, the less the need for a fresh adjustment of sights, and the more likely the shot to hit the target. Not only is the 6-inch quick-firer a longer, a heavier, and a more powerful gun than the earlier 6-inch or 5·9-inch, but it is also more accurate. One man does the training and aiming, having the weapon completely under his control; and this, of course, applies with still more force to the 4·7-inch quick-firer. The rapidity of fire, which is so

[*] Captain McGiffin, however, gives the Chinese percentage as 20, and the Japanese as 12. The truth lies perhaps between the two.

startling a feature in these weapons, is not, perhaps, of so much importance as the excellent mounting of the gun. Yet, of course, the power to discharge a great number of shells at a critical moment may be of immense value. A rough calculation shows that the whole Chinese fleet could fire on the broadside, in a period of ten minutes, 58,620lbs. weight of projectiles, whilst the ships which fought in the line, as opposed to the inshore squadron, could only fire 53,100lbs. The weight of metal discharged in the same period by the Japanese was, on the other hand, 119,700lbs., so that their artillery preponderance may be expressed as 119 to 58, or as two to one.*

The structural damage inflicted by the Chinese fire may now be considered. Two 4·7-inch guns appear to have been the only Japanese weapons hopelessly disabled. The *Matsushima*, though so severely hulled, was not externally much the worse. Indeed, even the *Hiyei* and *Akagi*, which were in the very hottest of the fray, carried little trace of the Chinese handiwork. The *Akagi* had several small holes in her starboard side, and her funnel was riddled. The *Hiyei* had a large hole in the stern, and several smaller ones in her sides. The *Naniwa* had a shell in her coal bunkers on the level of the water-line. The *Itsukushima* had one shell in her torpedo-room, another in her engine-room, and a third some way up the mast. The *Hashidate's* barbette was struck by a 5·9-inch-shell. The *Saikio* had the following hits distributed about her hull, boats, and funnel: 12-inch shell four, 8·2-inch one, 5·9-inch two, 4·7-inch four, 6-pounder, or smaller, ten. It is an extraordinary fact that such a vessel could take so much punishment. The 8·2-inch shell, however, was within ten feet of the engine-room, and, had it struck, it must have disabled the ship.

Unlike the Chinese, the Japanese had obtained the very best guns, and mounted them on their ships. They had three very

---

* *Vide* Table XX.

SHOT HOLES IN THE CHEN YUEN AFTER THE YALU.

PLATE XXX.

heavy Canets of 12-inch calibre, but these guns were unnecessarily large for the work to be done. On the proving ground they had indicated a perforation of 44-inches of wrought iron; we must, therefore, feel exceedingly doubtful whether any of their armour-piercing projectiles struck the Chinese battleships.* Three dents, three inches deep, are mentioned in the Chinese armour; these must have been from their common shell, as armour, 12 to 14-inches thick, could scarcely resist the 66-ton armour-piercing shot, even at 2000 yards. It has been said that the Japanese were bound to mount heavy guns in their fleet, as otherwise the Chinese ironclads might, early in the action, have simulated disablement, and so have enticed the Japanese unarmoured cruisers to close quarters, to their destruction.† For a fleet, which might have to oppose ironclads, to be absolutely destitute of armour-piercing guns, could not be expedient. But, whilst heavy guns were a necessity, it is not certain that the 66-ton gun was not too heavy, and that a pair of 30-ton guns would not have been as efficient a deterrent to such tactics, whilst their more rapid fire would have rendered them better able to deal with unarmoured ships. A shell from one of the 66-ton guns is said to have struck the *King Yuen*, perforating her armoured deck, and wrecking her compartments so hopelessly that she foundered. Another may have hit the *Chih Yuen*, and produced the curious explosion noted by on-lookers. Yet, there seems no evidence of this. The *Ching Yuen* is stated to have had a 12-inch shell in her bunkers amidships, but not to have been injured.

The quick-firers were most efficacious and deadly, shattering the structure of the Chinese ships outside their armour, and

---

* Captain McGiffin in the *Century* states that the *Chen Yuen* was struck by the heavy Japanese projectiles. But on examination, after her capture, no trace of hits from the Canet guns could be found.

† This would, however, have exposed the Chinese to the risk of being rammed or torpedoed by the more numerous and better handled Japanese ships.

killing or wounding all on their decks. The *Yang Wei* was riddled by the 4·7-inch shells of the Japanese, whilst the two heavy battleships had each about 200 shot marks. The hail upon them is described as most terrible, and forward and aft they were full of holes, whilst their superstructures were reduced to a tangle of ironwork and splinters. Yet, in all, on the Chinese ships, which survived, only three guns were dismounted. The *Yoshino* fired cordite, which performed admirably, enabling her gunners to shoot with great accuracy, owing to the absence of smoke. It does not appear that this explosive was used on board the other ships—the practice of the Japanese was better than that of the Chinese; when we remember the far greater number of shot and shell which the former were projecting, we get at once one explanation of their success. The empty cartridge cases, from the Japanese quick-firers, were got rid of, by pitching them down the hatchways.

In the matter of size and speed the Japanese had the advantage. The Chinese had, indeed, two large battleships, but their other vessels were small and weak, with the exception of the Elswick cruisers. The slowest Chinese ship in line was the *Chen Yuen*, whose speed from the state of her boilers could not have exceeded twelve knots. The average tonnage of their twelve larger ships in company was 2950 tons. The Japanese had no vessels of the size of the *Chen Yuen* or *Ting Yuen*, but their ships were more nearly even in displacement, and, excluding the *Saikio* and the gunboat, had an average tonnage of 3575.* The slowest ship with their fleet was the *Hiyei*, which could not do twelve knots, and the *Fusoo*, *Saikio*, and *Akagi* were little better. The rest of the fleet was both fast and homogeneous, with an effective speed of about fifteen knots. It is doubtful whether anything whatever was gained by bringing the slow vessels into line. The Japanese cruisers could have manœuvred with more freedom,

* *Vide* Table XX.

and would have been stronger in fact, if not in appearance, had they eliminated their obsolete and improvised ships. It is interesting to notice that of the four slow ships, three were greatly damaged, and two lost very heavily. On the other hand the *Fusoo* kept her station with the Main Squadron, though at her best she was only a thirteen knot ship.

The number of fires which occurred on board the ships of both combatants is a striking feature of the battle. The *Lai Yuen* was so severely burnt that nothing but her ironwork remained above the waterline; the *King Yuen*, her sister ship, was seen to be blazing before she was sent to the bottom: the *Ting Yuen* was on fire three times, and it was only through the courage of her foreign officers that the fires were got under; the same might be said of the *Chen Yuen*, which had eight fires; the *Ching Yuen* was on fire three times, but owing to her powerful pumps and the discipline of her crew got the flames under; the *Yang Wei*, *Tshao Yong*, and *Kwang Kai*, were all on fire at least once. This makes a total of eight ships, and nineteen fires, out of ten ships seriously engaged. The Japanese, on their part, suffered somewhat from fire, though not so seriously as the Chinese. Doubtless their ships were in better order, and discipline on board them was more thoroughly maintained. It is also probable that less wood was used in the construction of their vessels. It does not appear that any high explosive was employed as a "burster" on either side, though it has been stated by Japanese officers that they had melinite, and used it in some of their shells. The fires seem to have been the effect of gunpowder alone.

It had been widely prophesied that machine-handled guns would break down under the stress of war. This forecast has not been altogether justified. In all, there were nineteen such guns mounted in the ships which were seriously engaged. One Canet gun of three was temporarily put out of action; one of the eight 12-inch Krupps on board the two Chinese battleships was not in working order at the close of the battle;

and four others were temporarily disabled through causes which we do not know. This, if not a good record in itself, is by no means so bad a one as the opponents of the heavy gun would have led us to anticipate.

From the tactical point of view, there are two very astonishing features in this battle. Neither the ram nor the torpedo scored a single success. The explanation of this, is that the Japanese, with their superior speed, and the mobility which obedience to signals conferred upon them, deliberately decided against both these arms—arms which are as deadly as they are uncertain. To use the ram, ships must come to very close quarters, and, as during the battle, with a few exceptions, the Japanese kept at a distance of over 2000 yards, it was obviously impossible for the Chinese to ram. The *Chih Yuen* made an attempt, and was thought to have succeeded; but it was the *King Yuen* which was seen sinking, and not a Japanese ship. The torpedo was still more useless. The Chinese ships engaged carried forty-four tubes; the Japanese, thirty-two. These, with their accompanying supply of torpedoes, were so much dead weight, conveyed to no purpose, except to endanger the ships which conveyed it. It would have been more profitable for both sides to have devoted the space and weight thus absorbed to guns or ammunition. The Chinese torpedoists were only too eager to get rid of their above-water torpedoes, when the hail of projectiles began to descend upon their ships. They had gone into action with torpedoes in the tubes, and others charged, ready on deck to reload. On the *Chen Yuen*, all these were hastily discharged when she came under the fire of the Japanese, and hardly had this been done, when the stern tube was struck by a shell.* On the *Ching Yuen*, the same course was followed, and it is probable that the other Chinese ships took steps to get rid of these truly remarkable weapons. Some were set to sink on

---

* Captain McGiffin, however, denies that on the *Chen Yuen* the torpedoes were discharged merely to get rid of them, and asserts that they were used against the Japanese. If so they must have been fired at very long ranges.

PLATE XXXI.

CHINESE CRUISER CHIH YUEN.

See p. 63.

being discharged, but not all; and thus there must have been a certain number of live torpedoes floating about in the water. The question suggests itself: Did the *Chih Yuen* strike one of these, or was a torpedo in one of her tubes exploded by a Japanese shell? In her case, it will be remembered, there was an explosion, as if of a torpedo, just before she sank. There are no means of answering the question, but there can be no doubt that in future, captains will be very careful what they do with their torpedoes. The torpedo-boats which were present with the Chinese, effected nothing during the action. They did not dash into the battle under cover of the smoke and uproar, and fall upon their enemies, as had been prophesied. The engines of one went wrong, and a second missed at the closest quarters, three times. Yet this boat, though some minutes under fire, received from the Japanese not the slightest damage. At the close of the day, however, the torpedo-boat did exercise some influence on the tactics of the Japanese: since the mere possibility of a night attack upon his worn and tired crews, decided Ito against a close pursuit. He stood badly in need of destroyers, or fast torpedo gunboats, to make an end of the hostile boats before dusk came on. Thus, if the torpedo proved ineffective in the battle, the influence of the torpedo menace after the engagement, must yet be acknowledged.

The collision of the *Tsi Yuen* and *Yang Wei*, recalls the fact that there were similar accidents at Lissa on the Italian side, though in no case was a ship sunk. Whenever the vessels of a fleet are fighting independently, it would seem that this is a real danger. The duration of the battle was four-and-a-half hours, from 12.30 to 5 p.m., but during the last hour-and-a-half the fire maintained on both sides was very desultory. Trafalgar and the First of June both lasted about five hours, and St. Vincent five hours-and-a-half. Lissa was over in less than three hours.

From the battle the most varied and contradictory deductions have been drawn. Indeed, each naval expert would seem to obtain from it confirmation of his own particular fancies. Let us, however, pass in review the different lessons and consider how far they are really supported by the evidence we possess. Full and accurate details of the injuries inflicted on the Japanese ships are not at present accessible, but doubtless when the official history of the war is published by the Imperial Government, much new light will be thrown upon the battle. And before we proceed to consider the lessons, let us ask how far any of them would be applicable to a Western engagement—a battle, say, between the French and English Mediterranean fleets. The first point to emphasise is that no large battleship of modern construction fought at the Yalu. Between the *Royal Sovereign*, of 14,150 tons, or the *Brennus*, of 11,000, and the *Chen Yuen*, of 7430, the gap is immense, both in defensive and offensive power. The *Chen Yuen* represented obsolete naval theories—the sacrifice of broadside to end-on fire, and the enormous preponderance of the heavy over the auxiliary armament. She was at least four knots slower than either of the two ships named. It is not fair, then, to regard her as the type of the modern battleship, which carries heavy guns firing with twice the rapidity of her 12-inch Krupps, and a large complement of the most powerful quick-firers into the bargain. Secondly, the average size of ships in the line-of-battle in a Western engagement would be much greater than it was at the Yalu. Taking ironclads and large cruisers, as opposed to scouts, the average displacement of the British Mediterranean fleet is 10,000 tons, the average of the French Evolutionary Squadron 9500. The British average is thus nearly three times that of the Japanese fleet, at the Yalu. With these increased displacements come increased sub-division, increased protection, increased armaments, increased steadiness, and therefore better shooting, whilst there is for practical purposes no difference in speed between the two rival fleets in the Mediterranean, England having, perhaps, a very slight advantage here over France. It is obvious that their greater size would render these ships less likely to be sunk by gun-fire, but it would at the same time conduce to a desire to employ the torpedo-boat, since the blow dealt to an opponent by the destruction of such large vessels would be a very serious one. Thirdly, the inequality in training, the lack of a courageous, well-disciplined, and thoroughly efficient *personnel*, which so handicapped the Chinese, would be found on neither side. Both England and France have reduced naval training to a fine art, and it is impossible to say to-day which has the better men. Probably both are equally good. Fourthly, an engagement would not be likely to be fought at long ranges with the gun only, without some attempt, and perhaps very determined attempts on the part of the weaker side to come to close quarters, and bring on a general *mêlée*, in which the weaker cannot lose. There would on neither side be any motive to spare the enemy's battleships. Each would endeavour to sink and spare not. Fifthly, without doing any injustice to either Japanese or Chinese, it may be prophesied that the gunnery in a Western engagement would be better, and that more hits at the same ranges and with the same number of projectiles would be inflicted. Thus the damage done would probably be far greater.

We will now review the deductions which have been made. In the first place, as regards strategy and tactics, we are told with varying truth, that, since the value of line ahead, and the weakness of line abreast, were foretold, naval tactics have been demonstrated to be an exact science; that steam has not vitally affected tactics nor strategy; that line ahead is the one formation for battle; that on sea, as on land, increased rapidity of fire has necessitated open order; and that it has been proved that the tumult and confusion of the encounter do not prevent fleets from acting coherently and obeying signal. First, as to line ahead; the only authority who has found fault with the disposition on the part of the Japanese, is the Italian Admiral di Amezega. His opinion is that the Japanese ideas of strategy and tactics were based on the days of sailing ships, and that their operations have taught us little. He holds that the group system is the ideal one, and that the Japanese should have linked their vessels in "homogeneous groups, which should have entered action in succession, pouring in their fire." The criticism is not very clearly expressed, and it is difficult to discover any particular point in it. Yet line abreast cannot be said to have been fairly tested at the Yalu; had Ting made his wings strong, and manœuvred his fleet together, keeping bows on—a difficult, but not impossible, evolution—the condemnation of this formation might not have been so vehement. It cannot be maintained that it lost him the battle; it was only one, and apparently a minor one, of numerous causes, amongst which were bad ships, bad guns, bad officers. His heavy ironclads more or less compelled him, by their structural peculiarities, to fight in line abreast. To blame them, is only to find fault with the past. And when line ahead is pronounced the only battle formation, and it is unhesitatingly laid down that for battle in line ahead alone must ships be constructed, we may be allowed a protest. For how do we know that the enemy will not decide to fight a stern battle, if we leave our bow-fire weak? In that case, of course, a chase formation, or line abreast becomes necessary. And the peril incurred by the rearward Japanese ships is noticeable.* The line ahead defiling past the line abreast, runs the risk of the wing of the line abreast being projected, so as to cut off the rearward ships. The Chinese movement was not well executed, but with capable gunners and good captains against them, the Japanese would indisputably have lost the *Hiyei*, *Saikio*, and *Akagi*. On the whole, however, we may endorse the value of line ahead. If it is not strong at all points, what formation is better? Only both van and rear should contain very powerful ships, a truth which the Japanese forgot.

* Captain Mahan in the *Century* criticises the Japanese severely for thus defiling past their enemy's front, instead of doubling upon the Chinese left.

The Japanese made full use of signals under fire. It is most important to know that signalling is possible in battle, but with good gunnery on both sides, we may be permitted to doubt whether much could remain of signalmen, who are generally very much exposed, or of signalling gear, after a very few minutes. The Japanese shot away halyards and masts, and if they did not suffer the loss of their own gear, this was due to their enemy's indifferent marksmanship. We should be very rash to conclude that it will be possible to communicate orders in a Western engagement, after the battle is fairly joined.

The strategy of both sides is certainly against the view of the *Jeune Ecole* that steam has changed conditions and replaced the warfare of squadrons by such old substitutes, under new-fangled names, as the *guerre des côtes*, and the *guerre de course*. Neither Chinese nor Japanese fleet attempted the one or the other. Ting might have given us a great deal of information had he been obliging enough to detach his only two fast ships, the *Chih Yuen* and the *Ching Yuen* to bombard Japanese ports, and interrupt the Japanese lines of communication. The Japanese, during this part of the war, obeyed the old principles of strategy, husbanded their resources, and only used their ships against ships.

Turning now to details of structure and tactics, it has been in one and the same breath maintained, that the Yalu has proved the necessity of vertical armour; that it has shown that the unarmoured ship can face and defeat the ironclad. The latter is the deduction of the *Jeune Ecole*, and will not be found to bear inspection. Doubtless the Japanese fleet was, for all practical purposes, an unarmoured one, whilst the Chinese included two well-armoured ships. But the real test, which enables us to discriminate between the resistance of armoured and unarmoured ships, is given by the behaviour of these two different classes in the Chinese fleet, where both cruiser and ironclad had to withstand a hail of 6-inch and 4·7-inch shells, with an occasional shot from a heavier

gun. The two Chinese ironclads came out of the
encounter much battered, but still battle-worthy. Their stout
plating stood them in good stead. They could still manœuvre
and fight their guns, whilst the loss of life on board them was
small considering the vehemence of the attack delivered upon
them. Far different was it with the unarmoured ships, in
which class virtually fall the *King Yuen* and her sister, the
Lai Yuen, with their very short belts below the water-line.
Of the eight in line, two fled before they had been punished;
one withdrew on fire; one was sunk by collision, and three
were sunk or hopelessly damaged by the Japanese fire. One
only fought through the battle and survived it without serious
injury. Both the *Chih Yuen* and *King Yuen* had end-to-end
decks which did not save them. It may have been that their
hatches and water-tight doors below were not closed, or, to
give yet another hypothesis for their loss, that the water poured
through the breaches in their sides and collected upon their
armour decks till they capsized, because, as they rolled, the
great weight of water rushed to one side and destroyed their
righting power. We know the sea was rough. Yet, though
all this damage was done, a good many of the Japanese shells
failed to explode, even when they struck armour. With high
explosives the ravages might have been much more frightful.
On the other hand, owing to the Chinese lack of common
shell, the Japanese ships were not fairly tested. Armour-
piercing projectiles would only pass through both sides of
unarmoured vessels without scattering fragments or splinters,
and it is the bursting of a shell, rather than the perforation
of a shot, which does the damage. The long range, at which
the action was fought must also be taken into account, and
the fact that at five in the evening, after four-and-a-half hours'
incessant fighting, the two ironclads were as formidable as
ever, whereas the *Matsushima* had suffered terribly and was
out of action. Three of the Japanese cruisers carried heavier
guns than are found on board any European vessel of their
type—approximating, indeed, to the great Italian battleships—

and they had armour on their heavy gun positions. They would, as we have seen, have done better with smaller guns, or perhaps with one smaller gun, and plating on their 4·7-inch battery.

So far then from demonstrating the superiority of the unarmoured ship, the Yalu has shown that armour is necessary for ships which are to lie in the line-of-battle. As at Alexandria, it was proved that under practical conditions the resistance of plating is far greater than would be imagined after experiments on the proving ground. A large extent of surface protected by a moderate thickness of steel will be best—such as we find in the design of the *Majestic*. It will save human life and it will make the ship very hard to sink. And it seems a matter of the utmost importance to protect the powerful guns with armour—whether on board cruiser or battleship. Indeed, one gun behind six inches of steel is worth two without any protection. Farragut's maxim that a powerful fire is the best protection is admirable, but can be pushed too far.* It is, of course, urged that the 6-inch guns on the *Chen Yuen* and *Ting Yuen* were not disabled, though only very weakly protected. They were, however, only four in number, and were placed at the extremities of the ships, so that that the Japanese gunners could not be likely to aim at them, since the range was great, but rather at the centre of the hull. The heavy guns would have been disabled, judging from the number of hits on the armour round them, had they been unprotected as on our cruisers. Water-line hits appear to have been very few, a fact which has an important bearing on the question of unarmoured ends. Both the Chinese ironclads had only a citadel amidships, which did not extend over half the ships' length. Forward their unprotected ends were exposed to the tornado of projectiles, and were hit without the ships suffering much. This may be held to point

---

* Farragut himself recognised the value of armour. He did not attack at Mobile till ironclads had been sent to him, and indeed delayed expressly to wait for them. i. 119-20.

to the uselessness of armour, but it must not be forgotten that the vitals were strongly protected, and that shot or shell could not reach the engine-room or boilers. Their athwartship bulkheads, too, prevented shells from raking them. Whilst the armoured citadel thus served a useful purpose, we cannot doubt that the complete belt is the best and most efficient means of water-line protection. It need not be broad forward and aft, but should just cover the space between wind and water in fine weather. Had the two fleets come to close quarters the damage to unarmoured ends might have been very great. There is no record of hits below the water-line.

The experience of the *Ting Yuen*, where all four heavy guns placed close together were temporarily disabled, is strongly against concentration of armaments. The more the guns are dispersed the better, and perhaps in time we may see introduced on English battleships the lozenge-wise disposition of heavy guns, so long adhered to and so late abandoned by the French. We have fully recognised the importance of dispersing the auxiliary armament, placing, as we now do, each gun in a separate casemate. Though the heavy guns on our ships have the advantage of thick armour, it might be wise to reduce the thickness on the barbettes, whilst increasing the number of barbettes. Another important point to note is the danger of placing torpedo-tubes above the water-line. With quick-firers the risk of this has become so great that henceforward all tubes in cruisers and battleships must be submerged. The sooner the above-water tubes on our completed vessels are removed the better; they are mere lumber, except in the few cases where they are protected by armour. No wise captain would employ them in action. Again, wood should never be employed above the water-line, owing to the risk of fire, which is great when gunpowder is used as a burster, and perhaps even greater with melinite. The substitutes are metal, papier-maché, or linoleum, the latter being very largely employed between decks on modern French vessels.

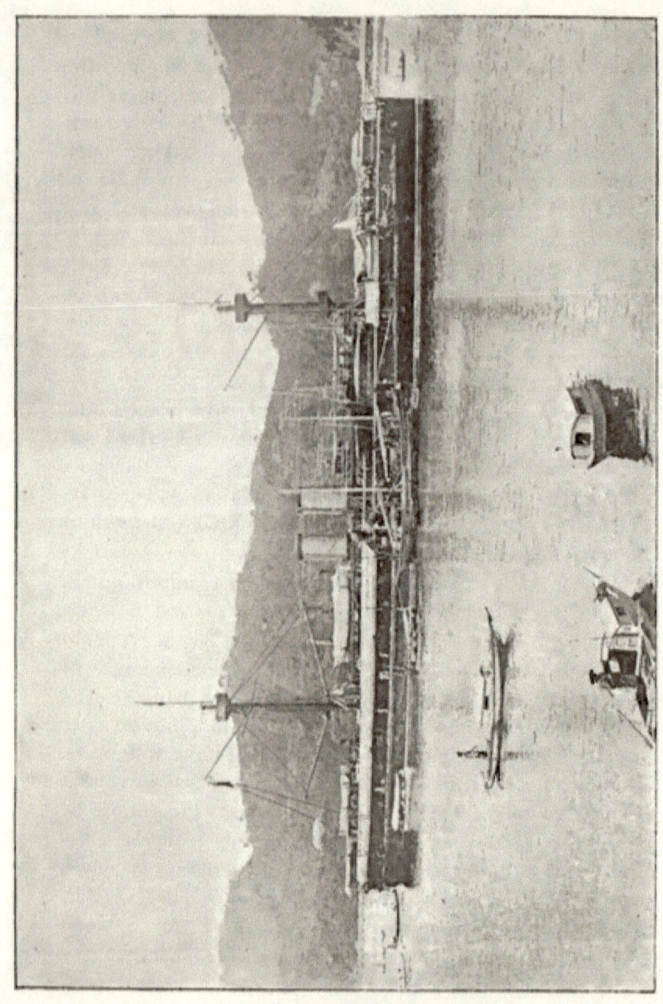

Plate XXXII.

THE CHINESE BATTLESHIP TING YUEN.

See p. 62

It is somewhat disquieting to notice that the weight of wood employed on board the *Majestic* is as great as that on board the earlier and older *Royal Sovereign*. Further, wooden companion ladders, and perhaps wooden boats, should be replaced by iron.

Tactically it is argued that the gun is now supreme in the contest between the gun, the ram, and the torpedo. We have already seen that there was a special circumstance which would hinder the Japanese from using either ram or torpedo in the later stage of the battle, this circumstance being the wish to capture the heavy battleships. Hasty deductions must therefore be avoided. The newest type of torpedo-boat was present with neither fleet, though, according to a European officer on one of the Chinese ships, there were whole minutes when he could see nothing owing to the smoke, and when a torpedo attack must have succeeded. The Japanese were undoubtedly in the same case. The torpedo-boat, from its small size, can steam under the shelter of the battle-ship, only emerging at the critical moment, and is particularly dangerous in battle. The gun has always been the most important weapon, and such it still remains, but it is not everything. There is the fullest evidence that very great attention to gunnery will be amply repaid on the scene of action, and that, with the possibility of an engagement fought at high speed, the gunner should be trained in peace by firing at targets from rapidly moving ships at sea. He must be taught to husband his ammunition, and not to waste a shot, although it is so very easy to load and fire the quick-firer. And great care should be taken, not only to have an ample supply of ammunition with the fleet, but also to see that every ship is well provided with the various descriptions of ammunition which may be needed—common shell, armour-piercing shot, and shrapnel.

As to speed, it has been concluded that the Japanese owed nothing of their success to it, and also that they entirely owed their success to it. Which view is right? Speed, we know, is a strategical factor of the utmost importance. Is it also a tactical factor? We may recall the fact that the slowest ship of Admiral Ito's squadron was slower than Admiral Ting's slowest, but the Flying Squadron, which fought and manœuvred quite independently of these slow ships, was able to use its speed. This speed enabled it to double, and arrive in time to support the hard pressed rear of the Main Squadron, thereby undoubtedly contributing to the Japanese success.* The Japanese slow ships, with the exception of the *Fusoo*, were out of action quite early, because they could not avoid coming to close quarters with the Chinese, when, being feeble ships at the best, they were overpowered. But for the Chinese to endeavour to ram or torpedo the Flying Squadron was hopeless, and in the Main Squadron, at the close of the battle, there was only one vessel which they could catch. And if their slow ships, without a round of shell, had endeavoured to fall upon the *Fusoo*, the speed of the Japanese would have enabled them to concentrate round these assailants, and, having more numerous and faster vessels, to ram or use the torpedo. For under no circumstances, given sea-room, can the fast ship be rammed by the slow, whilst the slow is always exposed to the ram of the fast. This may at the bottom have been the circumstance which deterred the Chinese from closing in the final phase of the battle. Therefore speed had a very real influence on the engagement, giving the Japanese an advantage comparable to the possession of the weather gauge.

As for guns, the heavy quick-firers of 6-inch and 4·7-inch calibre were irresistible when pitted against slow-firers of their own size, nor was any difficulty in supplying ammunition fast enough experienced. It has been concluded that an armament of numerous, moderate-sized guns is best, and that the heavy gun is played out. If the moderate-sized gun be of about 8-inch or 10-inch calibre, with good penetration, we

---

* The loss of life on board the ships composing the Flying Squadron was less than on those of the Main Squadron. The former steamed 14 knots, the latter 10. Undoubtedly the high speed of the Flying Squadron rendered it hard to hit, and acted as a protection.

may agree, but for the attack upon the newer type of battle-ships, which carry a considerable amount of armour sufficiently thick to resist the shells of the quick-firer, heavy guns are still necessary. The smaller quick-firers and machine-guns appear to have only inflicted trifling damage, though firing from the tops at close quarters they might be found useful. Rifle-fire is not mentioned as causing any loss, and we may conclude that the days of small-arms at sea are over.

Lastly, in the opinion of all, training and discipline are shown to be indispensable if victory is to be secured. However good are the guns and the ships, they will be well-nigh useless if the officers and men who are working them are not thoroughly acquainted with their business, if gunners cannot shoot straight, if officers do not handle their ships with skill. A well-trained *personnel* is the first requisite, if an efficient navy is to be created. Constant manœuvring and target practice at sea, expensive though they may be, are the one royal road to success.

Having reviewed the various deductions, we see that, with the exception of the danger of fire, the Yalu has done nothing but emphasise principles already known and understood. Its teaching is in no way revolutionary, but tends strongly to confirm the argument of those who hold that naval science is an exact science, and that its issues can be predicted. On the points where we most want practical information, such as the true place of torpedo-boats in a fleet action, or the possibilities of ramming, it yields no light.

NOTE.—According to "Blackwood's" correspondent, the *Ting Yuen's* flag-captain did not signal to the Chinese fleet the dispositions which Ting and Von Hanneken had ordered; and it was impossible at the last minute to change the formation. Of steel common shell the Chinese heavy guns had not fifteen rounds apiece, but only three between them. These were of 3'5 to 4 calibres long, with a bursting charge of 132lbs. of powder. All three were on board the *Chen Yuen*, and one was used with such terrible effect upon the *Matsushima*. The other shells supplied were 2'8 calibres long, carrying a charge of 25lbs. of powder, and, being made at Tientsin, fitted badly, and often were loaded with non-explosives. It thus appears that not the Chinese gunners, but the Chinese ammunition was at fault.

CHAPTER XXII.

NAVAL OPERATIONS AT PORT ARTHUR AND WEI-HAI-WEI.

October, 1894, to February, 1895.

DEFEATED at the Yalu, the Chinese fleet was seen no more at sea. It retired to Port Arthur much crippled and battered, and was refitted in a leisurely manner. The Japanese fleet kept the sea, and all the ships, except the *Matsushima*, *Hiyei*, *Akagi*, and *Saikio*, effected their repairs without having recourse to port. Admiral Ito, in the weeks which followed the battle, was chiefly engaged in convoying transports, paying no attention whatever to the Chinese. On October 24th a Japanese army disembarked near Port Arthur, and the attack on that place began. Admiral Ting had before this, in obedience to orders, withdrawn to Wei-hai-wei, where he remained till the last act of the war was concluded. As soon as Admiral Ito had accomplished his transport work, he steamed to Wei-hai-wei, and offered battle to the Chinese. Their fleet included the *Ting Yuen*, *Ching Yuen*, *Tsi Yuen*, and *Ping Yuen*, with the cruiser *Foo Sing*,* which was a powerfully armed ship, and the *Kwang Ting*, a smaller vessel. There were also in harbour: the *Chen Yuen*, which had run on a reef whilst entering the harbour, and was therefore temporarily disabled; the belted cruiser *Lai Yuen*, which had not been repaired since the Yalu; the *Kwang Tsi*, which was unarmed; six Rendel gunboats, and fourteen torpedo-boats. The Chinese showing no inclination to come out, Admiral Ito

---

* It now appears that this ship was not at Wei-hai-wei.

returned to Port Arthur, where he supported the land attacks of the army on November 20th and 21st.

On November 21st the Japanese ships shelled the forts at long range, doing very little damage, and the Chinese spasmodically replied, without, however, securing a single hit. At four in the afternoon the fleet was six miles from the forts on which the troops were making their assault, when a squall of rain came down. Under cover of this, ten torpedo-boats, led by the Yarrow-built *Kotaka*, which has 1-inch armour, dashed into the harbour, their rush being supported by two cruisers. The rest of the fleet assisted them by maintaining a long-range fire upon the forts. The Chinese soldiers were crowding down to the water, whilst those who were in the forts could be reached with ease from the interior of the harbour, as the works faced landward. The torpedo-boats opened with their machine guns, doing terrible execution, and completed the confusion of the Chinese, who, seeing themselves taken in rear, were seized with panic. Thanks to this audacious rush, the place was within half-an-hour in the hands of the Japanese. The Chinese had laid mines in the entrance, but these failed to explode. Altogether this was a brilliant performance on the part of the Japanese sailors, and inflicted no loss upon them. By the fall of Port Arthur they obtained an excellent naval base, with docks and workshops in the Gulf of Pe-che-li.

There still remained to the Celestials Wei-hai-wei, and on this place fell the next attack. On January 18th and 19th the fleet bombarded Teng-chow-foo, which lies eighty miles west of Wei-hai-wei, and on the 20th the army landed to the east of the naval port, and the investment began. The harbour is formed by two bays, off which lies the island of Leu-kung-tau. There are thus two entrances: one to the east of considerable width, with the island of Jih-tau almost in the very centre of the fairway, and one to the west, which is not half the width of the other, and is rendered difficult by reefs. The entrances to the harbour were protected by very strong forts and batteries mounting heavy breech-loaders and quick-firers, whilst the mountainous islands of Leu-kung-tau and Jih-tau were also well fortified. On Leu-kung-tau were the naval headquarters with a gunnery school, a naval school, and a coaling jetty. There was no dock, but only an anchorage, in some degree sheltered from a bombardment from the sea by the Island. Here the remnants of the Chinese navy had gathered, and as the *Chen Yuen* had now been repaired, they constituted a formidable force on paper, and were capable of giving a great deal of trouble, if used with vigour. Towards the end of January the Japanese fleet appeared off the place and watched both entrances. Admiral Ting was now caught in a trap. At sea were the victors of the Yalu, on land the conquerors of Port Arthur. He could only extricate himself by sacrificing his slow ships, and from making this sacrifice he shrank.

On January 30th the Japanese ships in concert with the army, opened a long range fire upon the forts. The *Naniwa*, *Akitsusu*, and *Katsuragi* assailed the works at Chao-pei-tsui, on the eastern entrance, which, after the explosion of a magazine, they silenced. Meantime the rest of Ito's squadron bombarded Leu-kung-tau. Most of the land forts had passed into the hands of the Japanese, when, on the night of the 30th-31st, a torpedo attack upon the Chinese fleet was decided upon. Both entrances to the harbour were closed by booms constructed of three steel hawsers, one to one-and-a-half inches in diameter, supported at intervals of 30 feet by wooden floats. The eastern boom was about 3300 yards long between Jih-Tao and the shore. There were two openings in it, a small one near the coast, which was obstructed by numerous rocks and therefore very dangerous, and a larger one 300 feet wide in the centre, to allow the Chinese ships to go in and out. Mines were plentifully sown but did not prove very effective. On blockading Wei-hai-wei the Japanese had attempted to clear the central passage but had not entirely succeeded. Their boats, therefore, moved to the attack by the landward passage where they would be covered by the forts which were now in the hands of their army. The boats were sixteen in number, and were formed in three divisions, thus constituted :

I. Division.—No. 23 (Division boat), *Kotaka*, Nos. 7, 11, 12, 13. Total, six boats.
II. Division.—No. 21 (Division boat), Nos. 8, 9, 14, 18, 19. Total, six boats.
III. Division.—No. 22 (Division boat), Nos. 5, 6, 10. Total, four boats.

The *Kotaka* had been built by Yarrow in 1886, and her trial speed was nineteen knots ; Number 21 was a Normand boat launched in 1891, when she steamed twenty-three knots ; Numbers 22, 23 were by Schichau, built in 1891, and of twenty-three knots extreme speed. These were all first-class boats. The others were of the second-class, built at Onahama, and probably did not steam more than 18-19 knots at their trial.

The boats proceeded under the cover of darkness towards the passage, and were nearing it when the land forts suddenly opened upon them and betrayed their approach. The Japanese gunners had taken them for Chinese boats, and there was nothing for it but retreat. They retired accordingly without suffering any loss.

Next day the weather was extremely bad, a violent wind blowing, and snow falling heavily. The larger ships of Admiral Ito's squadron retired for shelter to Teng-chow, but left a cruiser to observe the Chinese. On February 2nd they re-appeared. The sea was now calm, but the cold was intense. At a distance of 2400 yards they steamed rapidly past the forts on Leu-kung-tau, which now alone remained to the Chinese, and bombarded them, producing very little effect. On the other hand the Chinese gunners could not hit the Japanese moving at a rapid speed. On the night of February 2nd a second torpedo attack was attempted, but failed, as the Chinese discovered the boats and opened upon them. On the 3rd and 4th the bombardment of the island was vigorously carried on both from the sea and from the land. The Chinese forts and ironclads replied, but the latter were in difficulties as they had not much room for manœuvring. On the 8th twelve Chinese torpedo-boats made a desperate attempt to escape by the western entrance. The Japanese fleet opened upon them as they came out, and gave chase, capturing or sinking most of them. With engines and boilers in bad order they could not hope to elude fast cruisers of the *Yoshino* and *Akitsusu* type.

A third torpedo attack was planned for the night of February 4th. The first division was to head for the western passage and create a diversion by making a false attack. The real assault came from the east, and was delivered by the 2nd and 3rd divisions. About 3 a.m., when the moon went down, the ten boats composing these divisions steered for the central opening in the boom, using the masts of the *Ting Yuen* as their guide.* Numbers 8 and 21 were unlucky enough to touch rocks on their way in, and though they got off, could take no more part in the attack. The cold was intense— there were eighteen degrees of frost—and the spray froze on the boats, clogging their torpedo-tubes, as they travelled through the water. At four o'clock the attack was opened by Number 5 as she drew close to the Chinese, but firing too soon her torpedoes missed. Number 22 followed her; in silence she advanced, in silence fired three torpedoes, then turned and retired without anyone on board being able to say whether the Chinese ships had been hit. The orders were strict, forbidding the men to expose themselves. As she was backing to get away from the Chinese, who had now opened a heavy fire, she either struck a Chinese boat which had come up to drive her off, or, more probably, grated against a rock and lost her rudder. In another minute she ran with great violence upon the rocks, and as her plight was hopeless the crew decided to abandon her. They had only one boat, which would hold six men, whilst there were sixteen on board. The first boat-load got to the shore safely, but on the second trip the boat foundered close in shore, leaving on board one sub-lieutenant and six sailors. There they remained till daylight

* So the account in *Le Yacht*, April 22, 1895. I do not understand how they could be seen in the darkness.

in the bitter cold, and the sub-lieutenant, Suzuki, and one sailor, half frozen, fell overboard into the water and were drowned. With daylight the Chinese opened a sharp fire upon the five who were left, but being seen from the shore, a boat was sent to them, which brought the survivors off in an exhausted condition. Next came Number 10. Steaming at the rate of ten knots, she passed very close to a number of small vessels and Chinese torpedo-boats, moored to the west of Jih-tau. As she drew near the enemy's large ships, she collided with another torpedo-boat engaged in the attack, but suffered no harm. Approaching through a hail of Gatling bullets a great grey mass rose suddenly up before her. It was the *Ting Yuen*, and at it she fired her bow-tube. Owing to the ice, the torpedo did not leave the tube but stuck projecting from it, half in, half out. Her commander turned gently to port and fired his broadside tube. In spite, however, of the fact that the sights were most carefully laid, and the speed corrections accurately applied, the torpedo which had been pointed at the centre of the *Ting Yuen*, distant about 300 yards, only just caught her stern. A man looking out from the boat saw it explode. Number 10 at once circled under a heavy fire from the Chinese, and turning, touched, with the projecting torpedo in her bow-tube, Number 6. The two boats ran a terrible risk, for the trigger of the torpedo was actually smashed, without exploding the detonator. They separated, and Number 10 retired, whilst Number 6 went forward to continue the attack. When within range her bow-tube was fired, and once more the torpedo stuck. Circling, she brought her broadside tube to bear, but the torpedo broke in two on leaving the tube. A hail of 1-pounder shells from the ironclad's Hotchkisses was falling about her, and yet strange to say no harm was done her. One only struck her hull abreast of her engines, and stuck in her side without exploding. The screw of the fuse must have come loose in flight. Number 9 fired a torpedo at a despatch-boat, when she was herself pierced by a projectile which burst her boilers, wounding fatally two men and slightly two, whilst four were scalded to death. The boat, however, remained afloat. For some minutes she lay helpless under a heavy fire till Number 19 came to her aid and took her in tow, but she sank before she could be got out of the harbour. The attack was now over, and the Japanese boats retired. Numbers 8 and 18 had their rudders or screws injured by touching rocks or by contact with the boom. They were, however, towed off. Number 6 had been hit by forty-six rifle shots, and one Hotchkiss shell; Number 10 by two rifle shots. The loss of life was not, however, heavy. The damaged boats were either repaired on the spot or sent to Port Arthur.

The fourth and final attack was made on the night of February 5th. This time the first division was selected to do the serious work, whilst the remnants of the second and third watched the western entrance. The Chinese did not discover the boats till they were right in amongst them, and then made only a feeble resistance. Seven torpedoes were discharged by the *Kotaka*, and Numbers 11 and 23. The *Ting Yuen* seems to have received another; the *Wei Yuen* one, and the *Ching Yuen* one. The *Lai Yuen*, too, was hit on this occasion, and capsized, her bottom showing above water. Her crew were imprisoned alive in an iron tomb, and were heard knocking and shrieking for days. It was a work of great difficulty to cut through the bottom, and when at last this had been done, all were found dead. The *Ting Yuen* floated in spite of the torpedoes, but was seen to be slowly settling next day. Her water-tight doors were either closed before the explosion, or immediately after, and thus delayed her loss. The *Ching Yuen* was disabled, but not destroyed, and she could still fire her guns. Whilst the Japanese suffered trivial loss—twelve killed, and two torpedo-boats sunk—they had thus, in one way or another, reduced the Chinese fleet to the *Chen Yuen*, *Tsi Yuen*, *Ping Yuen*, and *Kwang Ting*.

On the 6th, fresh parties of sailors and marines were landed on the island to support a force which Admiral Ito had placed ashore previously, and a heavy fire was poured in upon the last remnants of the Chinese fleet, and upon the forts. On the 7th, the Japanese fleet was very hotly engaged with the Chinese works, and suffered considerably. The *Matsushima* was struck by a shell, which destroyed her bridge and wrecked her funnel, and almost immediately after, by a second, which passed through the engine-room and entered the torpedo magazine,* but, luckily, glanced up and exploded harmlessly above the armour deck. The *Yoshino* and *Naniwa* were also hit. On the other hand, a Chinese magazine was blown up. On the 8th, the island forts were attacked by storm, and all but one captured. The *Ching Yuen* was sunk on the 9th, just after she had delivered her broadside. A shell from a 9-inch gun in one of the land forts in the possession of the Japanese, struck her bow a little above the water-line, and sent her to the bottom. Yet Admiral Ting still held out, though his enemies were closing in upon him, and the western mine-field had been destroyed. This day the *Itsukushima* was hit on the water-line by a shell which failed to explode. The 10th and 11th, the bombardment went on, and of the Japanese ships, the *Katsuraki* and *Tenrio* were hit and damaged. On the 12th, Admiral Ting bowed his head to fate and surrendered. He did not outlive his defeat, choosing rather to die by his own hand.

The total losses of the Japanese fleet in these various actions, were two officers and twenty-seven sailors killed, whilst four officers and thirty-two sailors were wounded. Those Chinese ships which were afloat, including the *Chen Yuen*, were taken to Japan.

Thus ended, with the fall of Wei-hai-wei, the career of Admiral Ting. The ex-cavalry officer had shown patriotism and pluck, but perhaps he made a mistake in refusing at the beginning of the siege to put to sea and risk an engagement.

---

* *Pall Mall Gazette*, April 8th, 1895. This is suspiciously like the shot which struck her at the Yalu (*vide* page 94). It is possible that the correspondent misunderstood his informant.

If defeated, the result could not have been worse for him or his country than it was, and his two heavy battleships might have got safely away to Foochow, where they could have been reinforced by cruisers from the southern squadrons. The Japanese would have found it difficult to stop them. His slow and feeble torpedo-boats, his battered ships, his treacherous officers, his disheartened seamen, were not capable of conducting an active defence, or of harrying the blockaders on the dark nights, and his fleet played a purely passive *rôle*. It degenerated into a target for the projectiles of the land forts, for the guns which the Chinese themselves had mounted.

The combats of the Japanese fleet with the land forts, teach us little, yet that little confirms the lessons of the past. It was not the ships, but the heavy guns on land, which silenced the forts, and Admiral Ito had a narrow shave of losing his flagship. The torpedo attacks were well conceived and well conducted, but the demoralised condition of the Chinese must be taken into account. We see clearly that booms and mines are a very futile defence, if they are not covered by heavy guns, and if the openings in them are not closely watched by launches and torpedo-boats. It is increasingly evident that only in absolutely enclosed harbours can fleets rest absolutely secure. The fact has been already recognised in France, where, at great expense, sheets of water have been surrounded with breakwaters both at Cherbourg and Brest. It has also been recognised in England—witness the new works at Portland, Dover, and Gibraltar. Not that British fleets are likely to copy Chinese strategy, and lie in port whilst their enemy is sweeping the sea. But it is necessary to possess havens of refuge, where isolated battleships and cruisers, perhaps harassed by weeks of blockading, perhaps damaged in action, will be able to lie without needing even to keep a watch. The mere possibility of a torpedo attack, imposes a terrible strain upon officers and men.

The Japanese boats, when once they got in amongst the

Chinese, did not effect such wholesale destruction as we had been led to expect. They did not sink ships right and left. On the other hand, the losses both in men and boats were singularly small considering the results achieved. Five vessels are claimed to have been injured, representing a displacement of at least 14,000 tons, considerably larger than the displacement of the ships sunk or destroyed at the Yalu. Not one of the attacking boats was directly sunk by gun-fire, but then the Chinese ships were almost entirely devoid of the larger quick-firers—6-inch, 4·7-inch, 20-pounder, and 12-pounder, which would probably stop these small and delicate craft with a single hit; nor were they over well provided with 6-pounders and 3-pounders. Certainly the attack upon a *Royal Sovereign* or *Brennus* at anchor would be quite a different matter. The boats, too, had the support of the land works, which would not be the case if a European squadron of these vessels went to look for its enemy in harbour. This is the first occasion on which the torpedo-boat, pure and simple, has succeeded in sinking larger vessels with the Whitehead. Both the *Blanco Encalada* and the *Aquidaban* were torpedoed by torpedo-gunboats. The torpedoes used were of the Schwartzkopf type, fired by electricity, with a charge of 200 grammes (less than ½ lb.) of powder.

NOTE.—"Blackwood" gives the following account of the torpedoing of the *Ting Yuen*, from Commander Tyler's journal: "I saw a torpedo boat approaching us end on. When about 300 yards off she turned hard-a-port. Just then I saw one of our shot take effect, a cloud of steam rising from the boat. A few seconds after she turned her torpedo struck us. It was a loud dull thud and a heavy quivering shock, a column of water dashed over the decks, and a faint, sickly smell of explosives. . . . Within a minute of being hit I was down below. The water was bubbling up from one of the water-tight hatches, and there was about a foot of water in my cabin. . . . The water-tight doors . . . were in working order, and were kept clear. They were all leaking badly, however. The ship was beached; she did not fill and sink at once, though all her bulkheads leaked owing to the shock. Thus it appears that the 125lb. charge of gun cotton will not necessarily inflict a fatal wound."

CHAPTER XXIII.

THE NAVAL BATTLE OF TO-MORROW.

THE difficulty of forecasting the future is nowhere greater than where the mind has little material upon which to base its judgment, where, in other words, the instances are insufficient for an induction. On land there have been two great wars within the memory of the present generation, and yet even with this experience it is difficult to predict the details of a future land-battle; so considerable have been the changes of *matériel* in recent years. At sea, changes in *matériel* have been far greater, and have exercised an influence more profound upon the science of war. Monster guns, torpedoes, rams, are factors which no soldier has to consider. So rapid is the progress of invention, so swift the march towards perfection, that at sea what was yesterday the most formidable of fighting machines may be looked upon to-morrow as little better than lumber. On land it is men who fight, at sea, men and machines. And though we have no warrant for thinking the machines all-important, they must necessarily affect in some degree the issue of any war. Naval progress is a race to obtain the best machines, and the constant structural changes, made to obtain that best, exhibit a state of flux unparalleled in the past. But whilst the implements of war are in this transitional state there is no sign of a similar flux in principles. Strategy, it would seem, remains the same as in the past, and tactics have only altered in detail.

We may now sum up the world's experience since the introduction of the ironclad. There have been two pitched battles: Lissa, in 1866, and the Yalu, or Haiyang, in 1894. The former is not of much value, as it represents an action between ships as different from those of to-day as they themselves were different from the ships of Nelson. The Yalu, as the more recent, is also the more valuable. But here really modern battleships were absent on either side, and there were certain ulterior motives interfering with the conduct of the engagement. Moreover, both at Lissa and the Yalu one side was greatly inferior to the other in discipline and *morale*. We may say that there is no instance of fleets approximately equal in skill, discipline, and numbers encountering one another. Some such encounter is necessary to test our *a priori* conclusions concerning the value of particular classes of ships, of particular types of construction and armament, and of particular formations.

Actions of single ships are rather more numerous, and fairly numerous, too, are actions of ships with forts. The American Civil War abounds in this last type of engagement, and since then there has been the bombardment of Alexandria, when, however, the Egyptians, being Orientals, did not make the resistance which we should have to expect from Westerners; and the various actions at Rio. Of single-ship actions the most important are the fight between the *Merrimac* and *Monitor;* between the *Tennessee* and Farragut's fleet; between the *Alabama* and *Kearsarge;* between the *Shah* and *Huascar;* between the *Vesta* and *Assar-i-Chevket*; the two engagements in which the *Huascar* faced the Chilians; and the encounter of the *Tsi Yuen* with the Japanese Flying Squadron. Of torpedo actions, the French affairs with the Chinese on the Min and at Sheipoo, are of little value, as in this case the torpedo-boat encountered enemies who were careless to an extreme degree. The Russian attempts upon Turkish ironclads in the Black Sea, the sinking of the *Blanco Encalada* and the *Aquidaban*, and the repeated attacks of the Japanese at Wei-hai-wei are more instructive, but cannot be said to have definitely decided the powers and limitations of the torpedo-boat.

The remoter past is of importance as illustrating certain questions, such as the influence of dimension, the general requirements of battleships, the methods of attack, and the formation to be adopted. Though steam has changed a great deal it has not changed everything, and though French writers of the *Jeune École* often tell us that the abyss which separates us from the past is profound, we may obtain some profit by crossing it. Naval warfare is as much a matter of men as of ships, and even if ships have changed, men have not.

First as to the battle dispositions of a fleet. What ships are to be placed in the line, if line there is? Is the protected cruiser to figure in it, and if so, what class of protected cruiser? Can the older and smaller battleships take their place with the newer and heavier vessels? Where are the torpedo-boats to be stationed, and what is to be their business in the conflict? Is there to be a reserve, or is the whole of the fleet to come into action simultaneously?

The history of the past shows that "a special class of ships to fight in the line of battle" was necessary. In "Naval Warfare,"* Admiral Colomb has pointed out how at first a heterogeneous medley of vessels, with scarcely a break from the largest to the smallest size, lay in the line; but that by slow degrees, experience showed it was inexpedient to place small vessels side by side with large ones. Gradually the English line of battle tended to uniformity. The 120-gun ship was found too large, the fifty-six-gun and forty-four-gun ship too small. It was the mean which conquered in the shape of the seventy-four. The frigate, during the revolutionary war with France, took no place in the line. It did not venture to encounter the crushing broadside of the line-of-battle ship,† but rather acted as an auxiliary to the

---

* Page 80.

† With very rare exceptions, *e.g.*, *Melpomène* (French), engaged *Agamemnon* (English), and *Agamemnon* was a good deal injured. Three frigates, Nelson thought, had an advantage over one ship of the line (64 guns). Laughton, Nelson, 54.

combatants, saving men and towing disabled ships. At Camperdown, it is true, it played some part in a fleet action; but Camperdown was the exception rather than the rule. Still less do we find sloops or corvettes engaging side by side with heavier ships. The frigate, the corvette, and the sloop were built for one purpose, the battleship for another. Each stuck to its last.

The next question to be considered, is whether the protected cruiser of to-day stands to the battleship as did the frigate to the line of battle vessel. To decide, we must examine the defensive and offensive power of the two contrasted classes of ships in each period. As type of the frigate, we may take the thirty-eight-gun ship of 1805; as type of the battleship, the seventy-four-gun ship of the same date. The armaments, broadsides, and complements of the two classes are given by James as follows:—

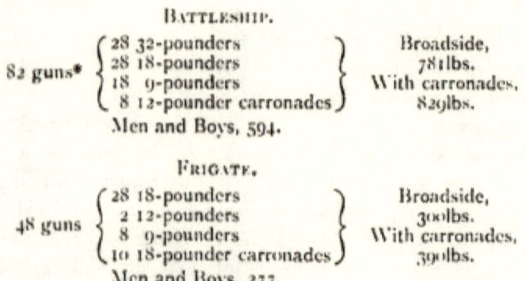

BATTLESHIP.

82 guns* { 28 32-pounders
28 18-pounders
18 9-pounders
8 12-pounder carronades } Broadside, 781lbs.
With carronades, 829lbs.
Men and Boys, 594.

FRIGATE.

48 guns { 28 18-pounders
2 12-pounders
8 9-pounders
10 18-pounder carronades } Broadside, 300lbs.
With carronades, 390lbs.
Men and Boys, 277.

In weight of metal discharged, and in the number of men carried, important for boarding, the seventy-four was to the frigate about as two to one. But in reviewing the gun-power of the two vessels, there is this to be considered, that in James' words " the destruction caused by discharges of cannon is in a great degree proportionate to the diameter and weight of

---

* Carronades were not taken into consideration in the official classification of ships, by the numbers of their guns. Hence the seventy-four was really an eighty-two.

shot." * Now the seventy-four carried the 32-pounder, which was nearly twice as powerful as the frigate's heaviest 18-pounder, when we look at the weight of metal thrown, and far more efficacious in battering the sides of the wooden ship. In defensive strength, the seventy-four-gun ship had stronger scantling and thicker sides, so that here again there was another point of superiority.

And now to pass to the modern cruiser and battleship. As type of the former the medium vessels of the *Eclipse* class may stand, and of the latter the medium battleship *Renown*, when we get these results:

BATTLESHIP.

Offensive. Guns 45 { IV 10-inch 29-ton guns, X 6-inch quick-firers, VIII 12-pounder „ „, XII 3-pounder „ „, IX Machine, &c. } Broadside, 2566lbs. 19 projectiles.

Torpedo tubes, seven. Cannot force a torpedo action. Ram.

Defensive. All guns over 12-pounder behind armour 6—10 inches thick.
Water line and side 8 to 6 inches armour. Coal.
Deck below water, 3 inches maximum thickness.
Bulkheads to prevent raking-fire.
Minute sub-division.

Crew. Probably 600.

CRUISER.

Offensive. Guns 25 { V 6-inch quick-firers „ „, VI 4·7-inch „ „, VIII 12-pounder „ „, I 3-pounder „ „, V Machine, &c. } Broadside, 486lbs. 11 projectiles.

Torpedo tubes, &c., four. Can force a torpedo action.

Defensive. Guns protected only by shields.
Coal protection.
Deck below water line 2½ inches thick.
No bulkheads.
Minute sub-division.

Crew. 430.

In weight of metal thrown, the battleship is to the cruiser as five-and-a-half to one, and all that has been said of the

* James, i., 44.

value of heavy artillery in the case of the seventy-four applies here with redoubled force, since the *Renown's* largest projectile is not twice, but five times the weight of the *Eclipse's* largest. The side of the *Renown* is impervious amidships to all the *Eclipse's* shells; the *Eclipse* from stem to stern is open, and exposed to the smallest projectiles. The *Eclipse* can be raked in the end-on position, the *Renown* cannot. The *Eclipse*, with her main armament ill-protected on deck, cannot hope to silence by gun-fire the *Renown's* well-protected weapons. Both ships are of the same date and designed by the same hand, yet it can scarcely be denied that the disparity between them is enormous. The sole advantage which the *Eclipse* possesses is that of forcing a torpedo action, and to do this she has to approach closely to the battleship, thereby giving the latter the opportunity of crushing her by gun-fire. We may conclude that this cruiser could not lie in the line of battle beside the battleship, as she exhibits a comparative inferiority very much greater than that of the frigate.*

It may be said, however, that whilst the medium and smaller cruisers are manifestly unable to enter the line, the larger and more powerful ships of the class, which are beginning to abound in our navy, could do so. To test the statement we will make one more comparison between the most powerful cruiser afloat, the *Terrible*, and the most powerful battleship, the *Majestic*. These are the figures:

BATTLESHIP.

Offensive. 52 guns
- VI 12-inch 46-ton guns
- XII 9-inch quick-firers
- XVI 12-pounder ,, ,,
- XII 3-pounder ,, ,,
- VIII Machine, &c.

Broadside, 4104lbs. 24 projectiles.

Torpedo tubes, four. Cannot force a torpedo action.
Ram.

* It is a common argument of some theorists, that the large battleship's equivalent weight in moderate cruisers would be more than a match for her. Three cruisers of the English *Astræa* class would slightly exceed the displacement of the *Renown*, and would bring to bear on the broadside between them, six 6-inch and twelve 4·7-inch quickfirers, with a broadside of 1140lbs.

BATTLESHIP—*contd.*

Defensive. All guns above 12-pounder behind armour 6—14 inches thick.
Water line and side 9 inches armour. Coal.
Deck below water line 4 inches maximum.
Bulkheads against raking fire.
Minute sub-division.

Crew. About 750.

CRUISER.

Offensive. 53 guns.
$\left\{\begin{array}{l}\text{II } 9\text{·}2\text{-inch 24-ton guns}\\ \text{XII } 6\text{-inch quick-firers}\\ \text{XVI } 12\text{-pounder ,, ,,}\\ \text{XII } 3\text{-pounder ,, ,,}\\ \text{XI Machine, &c.}\end{array}\right\}$ Broadside, 1474lbs. 22 projectiles.

Torpedo tubes, four. Can force a torpedo action.

Defensive. All guns above 12-pounder behind armour 6 inches thick.
Coal.
Deck below water line 4 inches maximum.
No bulkheads.
Minute sub-division.

Crew. About 850.

The inequality between these two is far less than that existing between the *Renown* and *Eclipse*, but it is still very great. The four heavy guns of the *Majestic* are the factors which give her her preponderance in broadside fire. Omitting these, and the *Terrible's* 9·2-inch guns, the two ships are almost identical in armament. But whereas the *Majestic's* gunners can fire with effect at every square yard of the cruiser's side, the casemates exposing a negligible surface of armour, the vitals of the *Majestic* are proof to all the cruiser's shot below 9·2-inch in calibre. The absence of bulkheads, as in the *Eclipse*, is a further handicap to the big cruiser. Still she is in a very different position from the small cruiser, as a very

against the *Renown's* 2500lbs. (excluding the smaller guns on either side). If the old seventy-four was considered a match for three frigates (Naval Chronicle, xxxix., 459), though these fired a greater weight of metal, the battleship may be considered more than a match for three cruisers, since, in addition to her advantage of concentrated size and power, she fires twice their weight of metal. The torpedo is the only factor which can affect her superiority.

considerable amount of cover is given to her guns, and her broadside is heavy.* It is difficult, then, arguing wholly on *a priori* grounds, to suppose that she cannot lie in the line, but if she does lie in the line it will be at no small risk to herself.

Armoured cruisers, if they have water-line protection, plated gun positions, and bulkheads, are better fitted for action in the line of battle. The *Dupuy-de-Lôme*, for instance, at long ranges, under service conditions, might be found proof to the shot of the 6-inch quick-firer, and could not be perforated by the shell of that gun. The *Impérieuse* verges very closely upon the second-class battleship of her date. The belted cruisers of the *Aurora* class are of an older epoch in design, and could not face battleships of their own date. Their armament is unprotected, their gunners exposed to every shell, and in a hot or close action their batteries could not be fought.

Cruisers thus fall into three classes. (1) The medium or small cruiser, unfit for the line of battle. (2) The very large cruiser which may fight in the line but at considerable risk. (3) The armoured cruiser with water-line protection as well as armour on her guns, fit for the line. The belted cruiser of *Aurora* type will fall somewhere between the first and second class, and is unfit for the line. In general it will be best to keep cruisers to their own proper duties as far as possible, but with a large cruiser squadron present on either side, the temptation to place them in line will necessarily be great.

Why should they not be placed in line? it may be asked once more. There will be vessels of the same class present on either side, perhaps cruisers in the opponent's line, and surely, even if they go to the bottom, it will not be till they have done very considerable mischief by their fire to the enemy's

---

* It is still, however, weaker in proportion to the battleship's than was that of the thirty-eight-gun frigate to the seventy-four's. But the armour in the *Terrible* protects the gun-crews well.

cruisers and the unarmoured surface of his battleships' sides. The battle will, as far as we can judge, be fought fleet to fleet in its earlier stages, and there will not be a number of actions between individual ships in which the cruiser will run the risk of having to encounter the battleship. In answer, we may say that the loss of ship after ship will be very discouraging to the crews of such vessels as survive, and that if one fleet of ten battleships and five cruisers, all of which are placed in line, assailing another of similar strength, in which the cruisers are held in reserve, succeeds in damaging severely one or two of the hostile battleships but only with the loss of three or four of its cruisers, the infliction of this damage will not compensate for the moral effect of the loss of the ships.* It is very evident that at the Yalu, neither Chinese nor Japanese gained anything by bringing the weaker ships into battle, whilst, though the *Saikio* and *Yang Wei* were poor and feeble vessels when contrasted with modern cruisers, they were not faced by any ship comparable in offensive power to the *Royal Sovereign* or *Renown*. It looks as though fleet to fleet actions made uniform battleships as necessary as in the past. Again, cruisers, being longer in proportion to their beam, are not generally so handy at a moderate speed as the shorter and broader battleship. Their inclusion in the line will thus reduce the manœuvring power of the battleship.

The best solution of the difficulty would seem to be the sharp separation of cruisers and battleships. As the Japanese placed their fast ships in one squadron, the slow in another, which each acted independently of the other, the same should be done with cruisers and battleships. The cruiser line may attack independently, seeking first the enemy's cruiser line,

* When the *Tecumseh* sank at Mobile (i. 124), the Confederate gunners, imagining that their fire had sent her to the bottom, at once redoubled their exertions, whilst the Federal fire grew perceptibly less vehement. It was only Farragut's dauntless handling of the *Hartford* that restored the Northern *élan*. So also at the Yalu, a foreigner on the *Tsi Yuen* states that the news "Another ship gone," greatly depressed the Chinese, as well it might.

or may face the hostile battleships, keeping at a great distance from them to neutralise the cruisers' want of armour, and to minimise the risk of hits from heavy shells. By pouring in a hail of 6 and 4.7-inch shells, the enemy's attention may be distracted from the more serious assault of the battleships, which will be simultaneously delivered. The ships of each class will then be together, and the principle of like with like carried out. The cruisers will be able to use their high speed if it is found desirable; the battleships will not be hindered by ships with large turning circles.

The line of battleships will be composed, naturally, of vessels similar to those figuring in the hostile fleet. At the outbreak of war, each side may be expected to employ its newest and best ships. Provided the smaller vessels are well armoured and armed, there is no reason why they should not lie in the line. The French *Jemmapes*, the Russian *Admiral Ortshakoff* are of the battleship type, though smaller and less powerful than armourclads such as the *Renown* or *Royal Sovereign*. To the latter, they are what the fifty-gun ship was to the seventy-four or 120-gun ship. It will be well if older vessels, whose speed is low, armour thin, and armament weak, are formed in yet a third division, should they be present.* Such vessels would be well adapted to act as a reserve; if they fight, it must be at long ranges, where their moderate armour will stand them in good stead. If introduced among more modern or first-class battleships, they will lower their speed, and in some cases reduce their manœuvring qualities. Let us by way of illustration take the squadrons of England and France at their present strength, and further suppose the English Mediterranean fleet to have been reinforced by the Channel Squadron. The

---

* A further argument for this is that more than eight or ten ships cannot, at intervals of two cables (400 yards), be handled in one line. If the interval is diminished, there will be the risk of collision between friends. In the grouping of ships, speed is especially to be considered, as if an eighteen-knot vessel is with a fourteen-knot ship, an element of tactical superiority is wasted.

following would be the division of the English force on this principle :

Main Squadron.* 16·7	Second or Reserve Squadron* 14	Second Cruiser Division. 17·8	Third Cruiser Division. 16·7
Royal Sovereign	*Collingwood*	Flora	Arethusa
Empress of India	*Camperdown*	Charybdis	Scout
Repulse	*Howe*	Cambrian	Surprise
Resolution	*Rodney*	Sybille	Fearless
Ramillies	*Rupert*	Barham	
Hood	*Polyphemus*	Bellona	
Nile			
Trafalgar	First Cruiser Division. 20	Torpedo Gun-boats. 17·7	
Barfleur	Hawke	Dryad	
	Blenheim	Gleaner	
	Endymion	Hebe	
		Skipjack	
		Sandfly	
		Sharpshooter	
		Speedy	

The French fleet, if the same principle were followed, would be drawn up thus :

Main Squadron.* 14·2	Reserve Squadron.* 13	Cruiser Divisions* 16·8	Torpedo Gun-boats. 18
Baudin	*Caiman*	*Dupuy de Lôme*	D'Iberville
Courbet	*Indomptable*	Tage	Bombe
Dévastation	*Terrible*	Sfax	Léger
Formidable	*Richelieu*	Suchet	Couleuvrine
Magenta		Condor	Flèche
Marceau		Cosmao	Lévrier
Neptune		Faucon	
Duperré		Lalande	
		Troude	
		Forbin	
		Milan	
		Vautour	
		Wattignies	

* Armoured Ships in *italics*. The figures give the trial speed of the slowest ship in each squadron, in knots.

Engaging vessels as little armoured as are the French first-class battleships represented in this list, there seems no

reason why the first-class cruisers should not lie in the line.*
The "Admirals" are almost too good for a reserve, and could
perfectly well lie in line; though lacking quick-firers, as they
do at present, and thin armour to keep out high explosives,
they would be distinctly inferior to the other nine. The
Polyphemus would be with the reserve squadron, as her time
does not come till the fleets close.

The precise position of the torpedo craft is also a matter of
dispute. The torpedo gunboat offers too large a target in
broad daylight, and is too vulnerable to attack at the commencement of the engagement. All the objections which
have been urged against the cruiser in the line of battle apply
with additional force to the torpedo gunboat. Two possibilities remain. The torpedo gunboat may be placed to leeward
of the battleship and emerge only to defend its larger mate
from the assault of the enemy's torpedo-boats.† This would
appear to be the original intention of the designers of such
craft. When the assailing flotilla arrives at a distance of
600 yards, just outside torpedo range, it will find itself faced
by the torpedo gunboats. The assailants would be under fire
from about 3000 yards up to 600 yards, for a distance of
2400 yards. The assailed torpedo gunboats would only be
under the enemy's heavy fire for the time occupied in steaming out 600 yards from the battleship, if that; for the battleships of the assailants would have to fire over their own boats,
which might prove a dangerous experiment. The torpedo
gunboat, thus placed, must to some extent hamper the movements of the battleship, if a sudden turn becomes necessary,
but it will be fairly sheltered and ready at hand when wanted.
Should it be formed up with others of its own class, upon
it will fall the duty of watching and combating the enemy's

---

* The *Baudin* and *Formidable*, broadside on, expose 400 square yards of unprotected target.

† In such a position, however, unless very close under the battleship, it will be exposed to the risk of hits from such projectiles as pass over the battleship, owing to too great elevation.

torpedo craft, whether gunboats or boats simply. It may also have to make a rush at the critical moment, when the enemy's quick-firers have been dismounted or silenced, and only the heavier guns have to be faced.

The torpedo-boat, whether of the sea-going or the still larger "destroyer" class, is nearly certain to be present in some force with either fleet. It lacks protection, as its only defence is its diminutive size and very high speed, and it is valueless for offence outside 500 yards. But it has this great advantage. The torpedo which it carries will, if it gets home, deal a crushing blow, and almost certainly disable, or there and then send to the bottom, any ship which it strikes. It is not likely that torpedo-boats will be sent against intact battleships, whose quick-firers are in good order and whose gunners are unshaken. The boats' time will come towards the close of the battle, when the fight has left great masses of iron wreckage; when the targets have lost their power of movement; when their crews are diminished in number and wearied by the intense strain of action. But even then it will not be as easy as it might appear to destroy the damaged battleships, since they, too, will have auxiliaries, who will be able either to meet the assault of the torpedo-boats or to destroy the opposing battleships. As it will be an anomalous position for the boats of each side to deal the final blow to the ships of the other side simultaneously, it seems probable that at the close of the action between the larger ships there will follow a fierce contest between the smaller craft.

The immense moral effect of dealing a heavy blow at the enemy when the battle begins, may, however, in defiance of prudence, lead to a rush of the boats of one side upon the ships of the other early in the engagement. To meet such a rush the assailed must have boats ready. They will steam forward, as we have said, to the limit of torpedo range, leaving the hostile boats as long as possible under the big ships' fire. The line of big ships will probably draw off, so as to prolong the duration of the assailant boats' approach. But supposing that the hostile boats steam straight upon the battleships, and the latter maintain a course at right angles to their approach, they will be under fire for the time taken to cover 2400 yards, which at a speed of twenty knots would be about three and a half minutes, during which time a 6-inch quick-firer would discharge from ten to fifteen shots, or a 12-pounder twenty to thirty. Though in the torpedo attacks of the Chilian and Brazilian civil wars, torpedo gunboats have come off without much more than a scratch, these attacks were made at night and upon ships which had not a powerful quick-firing armament or well-disciplined-crews. It will be a different matter attacking by day modern battleships, equipped with quick-firers, and using smokeless powder, though the torpedo-boat is never a target easy to hit. It remains possible that one or more boats may succeed in their onset, and that an odd battleship may fall victim, but the price paid will be a very heavy one.* The best plan would seem to be to hold in the boats at the beginning of the battle. For it may be better to throw away a battleship, than to abandon the chance of following up a victory, or striking a heavy blow later in the action, which intact torpedo-boats may give.

Of course, if there is anything like a *mêlée*, then comes the opportunity of the boats, but even then there may be danger to friends as well as foes. If it is true that the Chinese boats could not distinguish their enemies at the Yalu, it is a very noteworthy fact, for in that battle there was little that savoured of the *mêlée*, though there was more smoke than would be produced with cordite or amide powder. The sphere of action of the torpedo-boat upon the battlefield very closely

---

* To lose half-a-dozen Cushings would be disastrous to any fleet. The ideal torpedo-officer will be too rare and valuable a being to be risked for small gain, and if the torpedo-officer is not ideal in courage, coolness, and sagacity, his attack will miscarry. In manœuvres there is no ordeal of fire to sink and slay. Crews of boats under a terrific fire, to which they can make no reply, will need extraordinary steadiness and heroism. A torpedo flotilla once beaten off with any loss, will be good for little, owing to the bad moral effect of such a repulse on the men. *cf.* Cipriani, *Journal United Service Institution*, xxxviii., 763.

resembles that of cavalry upon land, and these craft should be used like cavalry. They act by surprise; they complete the ruin of the beaten.

The English fleet includes one vessel which is specially built for ramming—the *Polyphemus*; and the United States have in the *Katahdin* a similar craft. There are many who are in love with "the small swift ram," but it is doubtful how far such a ship is attainable, and how far she would be useful if the ideal could be obtained. Ability to ram depends upon speed and handiness in the assailant and the want of these qualities in the assailed. To obtain a high speed, not only upon the measured mile but in a sea-way, the boilers must be heavy and the engines powerful. This necessarily involves a high displacement, as the hull must be strong to withstand the jar of the machinery and the violent concussion of ramming. If the ram is given guns and armour, she becomes a battleship; if she is left without them, she is liable to be destroyed by gun-fire long before she can use her sole weapon. And that weapon is a most uncertain and two-edged one, as we shall presently see.

There is yet another species of ship which has appeared within the last decade—the ship armed with the Zalinski gun for projecting large charges of dynamite to a great distance. The Zalinski gun is at present in an undeveloped stage, but there is good reason to suppose, like many other inventions, it will be perfected in time. It offers a very large target to hostile quick-firers, and it is not strongly constructed, but it can project shells containing 200lbs. of the highest explosive, to a range of 2000 yards at the rate of one a minute.* The shell is a long time in the air, in some cases as much as twelve

---

* Firing at a target which represented the *Philadelphia*, at ranges of from 2000 to 1000 yards, 44 per cent. of hits were made. The projectile is practically a torpedo, with from two to five times the range of the Whitehead, and, it is probable, at least as great accuracy. The fuses of the projectiles and the valves of the gun have been vastly improved of late, but there seems to be some scepticism as to the value of the weapon. v. Schroeder, Proceedings, U.S.A. Naval Institute, xx., l. ff.

seconds. Like the torpedo it could not be used with much effect against a single ship, which could turn, or stop, or increase her speed, and thus avoid it; but it might be deadly against a squadron, where the individual ships cannot act with entire freedom, but are dependent upon their neighbours. Such a dynamite vessel might lie to leeward of the heavier ships, and throw her aerial torpedoes over them at the enemy. The effect of her projectiles exploding against a ship's side or deck would probably be most destructive. At the same time, the American *Vesuvius*, which has been built for this purpose, does not give entire satisfaction, and it has been proposed to take her pneumatic guns out of her.

The perfecting of the pneumatic gun would be the death-knell of the battleship in its present form, and it is hard to see what protection could be devised against its bolts. As the jar to the ship is very slight with air-impulse, it can readily be fitted upon merchant steamers, and was so employed upon the *Nictheroy* in the Brazilian Civil War.*

The position of the commander-in-chief in battle has been much canvassed. Persano chose to leave the line at Lissa when the Austrian attack was impending, and placed himself outside it, to be the better able to communicate his orders. As he had failed to acquaint his captains of the purposed change, the effect was most disastrous. It has been pointed out that for the admiral to withdraw to a light ship, or to place himself with the reserve, is very dangerous, as he thereby becomes a simple spectator, and cannot be at hand to change the formation of his fleet instantly, if this should be required.† Farragut at Port Hudson, Tegetthof at Lissa, Nelson at Trafalgar, are good precedents for the admiral's ship leading.‡ It is more difficult than it was in the past, when ships

---

* Page 40-41.   † Mahan, Influence of Sea Power, 353, ff.

‡ Nelson before Trafalgar, Farragut at Mobile and New Orleans, were each entreated by their captains not to lead, because of the danger. Nelson complied with the request, but only momentarily, and on second thoughts took his

were closer together, and when it was comparatively safe to stand on deck, for the leader, by his personal example, to encourage his fleet. But if his personal heroism cannot be witnessed, the behaviour of his ship can be seen by all. And if he takes his place at the head of the line, it will be possible to fight the battle without signals, upon the "follow-my-leader" plan, a plan which was constantly practised by the Mediterranean fleet under Admiral Tryon. In most English battleships and cruisers the signalmen are altogether unprotected. The masts are naturally liable to be shot away, and the semaphores are placed in a very exposed position upon the bridge, where the hail of shells, fired direct, or ricochetting from the water, would exert its most devastating effect. If a really simple arrangement can be devised whereby the signalmen can work the signals, whether at the mast-head, on Prince Louis of Battenberg's plan, or otherwise, from under cover, a great many of the difficulties of the commander-in-chief will vanish. It will still, however, be hard to communicate with those divisions of the fleet which are at any distance, and with a large fleet and open order the distances must be great. It is therefore absolutely necessary for the subordinate commanders to know before-hand what the chief is going to do, for them to discuss with him the various possibilities, and to be saturated with his ideas. Thus it was that Nelson's captains learnt what were their admiral's intentions, and were prepared at the moment of action to execute his wishes.* The want of such knowledge amongst the Italian commanders was one of the chief factors which produced the disasters of Lissa.

At the Yalu Admiral Ito led the Main Squadron, and Admiral Tsuboi the Flying Squadron, and then signalling was

---

position in the van (James, iii., 391). Farragut gave way on both occasions, but never ceased to regret his compliance at Mobile. There he resumed the lead at the critical moment of the attack (i., 125).

* "The necessity for many signals was thus entirely done away with." Laughton, Nelson, 215. For Persano at Lissa see i., 224, 232-5.

comparatively easy on board the Japanese Fleet. On board French ships, with their strong military masts, which contain a steel core inside the stairway, signalling would probably be easier in action than from the tops of English ships. It is possible that in one or other of the tops the admiral might find it advantageous to take up his position during action, as Farragut found it necessary to climb the rigging at Mobile. But special protection should be given to the admiral's battle-quarters, as his life is a valuable one, and the military masts are much exposed. It was a saying of Admiral Tryon that the captain must look forward, the admiral aft. From the forward conning-tower ships astern cannot be seen, and from the after conning-tower the same is the case with ships forward. From the upper tops a good all round view can be had, and there is not much to obstruct the field of vision. Two or three inches of Harveyed steel would, however, at that great height, diminish the ship's stability. It goes without saying, that if the top be selected by the commander as his battle position, communications with the captain must be fully assured.

We must now pass to the general formation of fleets for battle. There are five various dispositions which may be adopted: line ahead, line abreast, one or other of these two with the ships *en échelon*, triangular or group formation, and naval square. The tactics to be adopted and the structural peculiarities of the ships engaged, must to some extent dictate the nature of the formation adopted. What is suited to a *mêlée*, or to the use of the ram and torpedo, is not necessarily best adapted for the employment of the gun.

A single line, whether abreast or ahead, has these disadvantages: its extremities are weak; a great extent of water is covered; the force is not concentrated; and the attack of the enemy may be delivered upon a section of the fleet. In addition line abreast masks the broadside fire of the ships placed in it and compels an end-on attack, which can only result, if the enemy meet it by a similar disposition, in a confused *mêlée*, in which the fleets will break up into a number of isolated units and chance will decide the issue.* The end-on attack was favoured by our constructors between 1875 and 1885, when most of our "écheloned" turret-ships were designed. Vessels of the *Inflexible* and *Colossus* class are clearly meant to fight in line abreast, since the disposition of their turrets limits the arc of their broadside fire. But our more modern ships are marked by a return to the older and sounder conception of a powerful broadside,† and no admiral would be likely, of his own choice, to place such ships in an order which would reduce their efficiency, and prevent them from employing their numerous and formidable quick-firers. Any disposition adopted, however, must be to some extent guided by the action of the enemy, and if we look at the French battleships we shall see, between 1880 and 1890, the tendency, which has been observed in England, to secure powerful end-on fire, though stern fire is not neglected, as it has been neglected in England in the past. This looks rather as though the French meant to fight a stern battle in line abreast. In such an order ironclads like the *Magenta* and *Carnot* can bring to bear three heavy guns to our two in our more recent designs. But since France has followed us at last in the fore and aft disposition of the heavy guns, in her very latest ships—the *St. Louis* class—it is somewhat doubtful what she intends to do.

In a stern battle the French fleet would form in line abreast, and steam away from our ships. As we hold at present the weather gauge in speed, we could overtake the retreating fleet, but we should probably have to chase in line abreast. If our ships were in line ahead, the leaders would run the risk

---

* It can also be easily doubled upon, or turned on either wing, as the broadside fire of all the ships in it except one is masked.

† At the same time good end-on fire is a necessity, as the enemy may otherwise, with equal speed, fight a stern battle, and torpedo attacks are generally best delivered from ahead. *cf.* Sturdee. *Journal United Service Institution*, xxxviii., 1244-5.

of disablement from the enemy's concentrated fire. If we attacked in two columns in line ahead, one coming up on each quarter of the French line abreast, our two divisions would be widely separated, and the enemy might concentrate upon one or other, since steam lends itself to sudden and rapid movements. Our cruiser divisions, or at least our powerful first-class cruisers, would no doubt thus move on the French flank, striving to disable one ship, and so compel the others to delay in order to support it. The superior speed of vessels, such as the *Edgar* and *Powerful*, would enable them to avoid any sudden concentration by the enemy.

Line abreast can be readily converted into line ahead by a quarter turn. But for ordinary purposes it has all the disadvantages of line ahead and none of the advantages. The strong point in line ahead is that it leaves the broadside clear, and allows the ships to follow the movements of their leader. It is the most elastic and the most simple formation, and the target is small. For, as errors in shooting more generally arise from vertical than horizontal misdirection, and as more shots fly over the target than fly wide of it on either side, with a trifling error in elevation, projectiles will be more likely to drop on the deck of the ship end-on, when the target is 300 feet to 380 feet long, than upon her, broadside-on, when the target is only 60 feet to 80 feet. The armour is no doubt presented at a sharper angle end-on, and glances will be more frequent, but the gain in this direction does not compensate for the loss in other ways. The object in battle is not so much to elude the enemy's projectiles as to pour projectiles upon him.

A third formation is the bow-and-quarter line adopted by Tegetthof at Lissa. This leaves free both broadside and bow guns, but there is some risk of gunners in the uproar and excitement of the battle hitting friends. It is, too, a formation which lacks elasticity. A better disposition is a slightly indented line ahead, which permits the bow chasers to fire at the enemy, but does not avoid the risk of accidental injury to friends. A line ahead, re-inforced at both of its extremities is, perhaps, the best formation of those considered. The vessels supporting the leaders can be ships with a powerful bow fire, when the most will be made of them. In the same way, those supporting the rear should have a strong stern fire.

Groups of ships acting together, and supporting one another in a series of triangles, each composed of three ships, were in great favour as a battle formation some years ago, but now receive less enthusiastic support. The group-commander is a fresh intermediary between the admiral and his captains, and an unnecessary intermediary. It may be taken for granted that whatever order is adopted, each vessel in the line will support her neighbour, so that there is little gained by detailing B and C to cover A. Nor does there seem any reason why the group should consist of three ships and not of two or four, since, as has been said, there is no transcendental power in the number three. For the group it may be argued that, with such a formation, ships of similar type can best act together, and that often there are not more than two or three ships of identical type. But it will always be the case that ships in line will be similar, and small differences of construction do not necessitate a total change of formation. Where the differences are important, the ships of the different types will not be jumbled together, but placed apart. The group as a sub-division of the division is hardly necessary. Most group formations, too, offer a good target to hostile fire.

Naval squares, or the arrangement of ships in quincunces, again, are open to the same objection. They offer too good a target to the quick-firer. Such dispositions lack elasticity, and inevitably mask the fire of the ships in more than one direction; and, as in all complicated formations, there is risk of gunners hitting their friends. The formation of six ships in a triangle, recommended by MM. Montéchant and Z.,* is open to all these objections.

* Guerres Navales de Demain, p. 185.

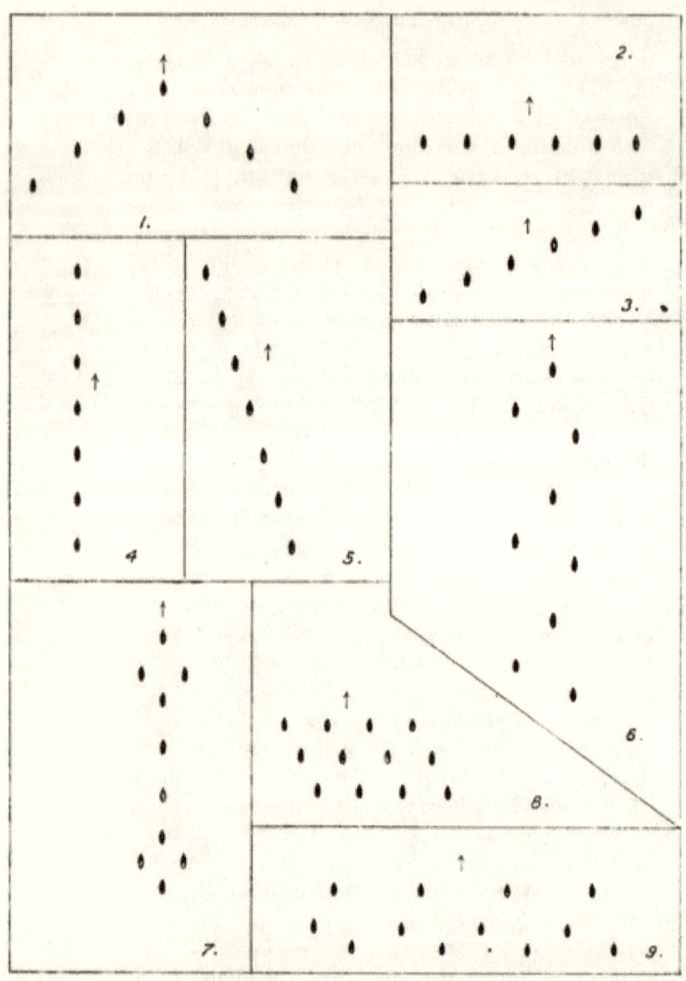

1. Bow & Quarter Lines 2. Line Abreast. 3. Indented Line Abreast.
4. Line Ahead. 5. Indented Line Ahead. 6. Line of Groups Ahead.
7. Line Ahead Re-inforced 8. Quincunx 9. Line of Groups Abreast

Naval Formations.

Map XXIX.

There is little doubt that line ahead is the best formation when the enemy does not run away. It is the normal formation for cruising, and thus its adoption for battle does not necessitate a sudden alteration at the last minute; it is easy to make changes of direction or formation with it; the broadside is left clear, and there is no danger of firing into friends; station can readily be kept; signals are almost unnecessary, or reduced to a minimum; and bow fire can be obtained by reinforcing the head with ships powerful in this direction, or by indenting the leading vessels. There is no other disposition which offers so many advantages and which has so few defects; moreover, there is this additional argument in its favour, that it was the formation adopted by the deliberate experience of the past. Though details may have changed principles have not, and if our ancestors feared to allow their ships to be raked, we shall be wise if we refuse to expose ours to such a risk.

Line ahead, then, will probably be the formation adopted by a judicious commander-in-chief, and the various classes of ships will be in separate divisions or lines. The main battleship squadron will engage closely with the enemy, at ranges not exceeding 3000 yards and not less than 1000. To leeward of it will lie the reserve battleship squadron and the torpedo divisions, whilst the first-class cruisers will attack at long range, not going inside 2000 yards, and the second-class cruisers will lie yet farther out. The cruisers will endeavour to divert the enemy's attention. They may steam round and round him if they are fast and he is slow, or they may steam alongside on the same course. It is certain that the battle will be a running battle, that both sides will be in motion, and perhaps in rapid motion. For either side to lie to whilst the other cannonades him would be suicidal. At the outset, ships will probably use their highest trustworthy speed, leaving a reserve of one or two knots for contingencies. As the action progresses the speed will diminish, when ships are more or less disabled. Vessels very severely injured will leave the line, and their place will be taken from the reserve. Should both fleets steam fast upon a parallel course in one direction, injured ships will be left behind, a prey to torpedo-boats. Over these there may be fierce engagements, as the one side will detach torpedo-boats to attack, and the other, boats to protect: or the two fleets may, like the *Alabama* and *Kearsarge*, circle on a common centre. Generally speaking, the battle will, in its earlier stage, be fought fleet to fleet, and individual ships will not fight isolated actions. The effort of each commander will be to concentrate upon a portion of his enemy's force his whole strength. This will be a vastly harder proceeding with alert opponents, whose ships are propelled by steam, and can move rapidly to any point, than it was when the line had only to be broken to secure the advantage, and when our enemies were hardly our match in skill, however fiery their courage may have been. The victory in war goes to the side which makes fewest mistakes, and no doubt mistakes will be made, otherwise it looks as though, with even forces, battles would be a matter of hard pounding, and therefore indecisive. The certainty and rapidity of movements with steam, however, render great combinations, sudden changes, rapid developments, possible. The fleet which is skilled and practised in steam tactics will threaten attack in one direction and in one formation, and then, perhaps, change to quite another with great speed. If the opponent is less skilled in station keeping or manœuvring, he will fall into disorder, his ships will mask one another's fire, and he will lose from the start.

A period of manœuvring may thus precede the opening of fire, as well as succeed it. Unless the ships are well handled and the formation simple, there may in this be some risk of collision between friends, as the fleet will be possibly under fire, and the tension and excitement tremendous. To get the weather gauge, so that the smoke from funnels and guns blows down upon the enemy, may be very advantageous;

advantageous, too, it may be to place the sun in the hostile gunners' eyes. If neither side forces a *mêlée*,* there will be a long range cannonade, during which considerable damage may be done on either side, but such a long range encounter cannot, it seems, be decisive.† Careful gunnery, strict fire discipline will be essential to prevent the waste of precious ammunition. The quick-firers and heavy guns will be alone employed, as small projectiles cannot do much harm at these considerable ranges, and their turn will come later.‡ The preliminary cannonade will continue till one side is getting decidedly the worst, or till both have expended a good proportion of their ammunition. In the former case, the weaker side may be desirous to close, to redeem if possible what has been lost in the chances of the *mêlée*. In the latter case, each side will be anxious to end the battle by bringing it to a decision. It is possible that the weaker may attempt to steam off; but if he does, he will abandon his disabled ships to destruction or capture.

So far, there is mention of neither ram nor torpedo as being employed by ships. The ram has been shown by repeated analyses to be a most difficult weapon to use.§ It involves actual contact with the enemy, and if the enemy has engines unimpaired, and sea room, the past proves that such actual contact is difficult to effect. The ram has hardly ever scored

---

* It will require great self-restraint and strict discipline to prevent a dash upon the enemy, when he comes into sight. There may be disobedient captains, like Tang of the *Chih Yuen*, who will be anxious to use the ram, and may for that object, leave the line. The strain upon the crews is, perhaps, less at close quarters, as the duration of the action must then be shorter.

† The Yalu was a striking demonstration of the resistance of ironclads to long range fire.

‡ As it will be useless to attempt to pierce armour at long ranges, common shell will be used. When the fleets close, it might be advantageous to use common-shell and armour-piercing shot alternately.

§ Mr. Laird Clowes shows that of seventy-four attempts to ram, in twenty cases the rammed ship was much damaged, disabled, or sunk. There is only one case in which serious damage was inflicted upon a ship under steam with sea room. *Journal United Service Institution*, xxxviii., 223.

against ships in motion. When the *Merrimac* rammed the *Cumberland*, the latter was at anchor; when the *Ferdinand Max* charged the *Re d'Italia*, the Italian was motionless. In the American Civil War, few attempts to ram were successful out of the great total. The use of the ram involves a *mêlée* which tacticians agree in regarding as detestable. And if vessels are charged whilst running at a high speed, the damage to the "rammer" may be great. The *Camperdown* was badly injured, though the *Victoria's* speed was only five knots; the *König Wilhelm* had her stem badly twisted by the *Grosser Kurfürst*, which was steaming at ten knots. The *Iron Duke* alone charged a vessel under way without being much the worse for it. The utmost skill will be necessary to deal a blow with a ram. At very close quarters a furious hail of projectiles will crash upon the conning-tower, and render the direction of the ship a matter of extreme difficulty. And there is the great risk of being rammed by the enemy if there is any miscalculation, or without miscalculation, of being torpedoed. Indeed, the torpedo may be said to have relegated the ram to the background. Yet, if the fleets charge one another end-on, there may be cases when the ram will be used, but there will be great danger then of end-to-end collisions should the commanders on either side be determined, and these will almost certainly result in the loss of both ships, unless, indeed, the bows of the ship on one side are so weak as to take the full force of the collision and to break it. More probably the less determined man will swerve at the last minute and expose his side, as did Buchanan at Mobile.

The torpedo has a limited range, though not so limited as the ram.* As long as the fleets fight at a distance it cannot be used, whilst even at close quarters it is somewhat uncertain.

---

* In Table XXV. will be found all the instances of the employment of the torpedo in war up to this time. So far as any result can be deduced it is that the torpedo is not successful against ships in motion. But, unlike the ram, it has not been often enough employed to give grounds for any induction, and it is moreover an essentially progressive weapon, improving every year.

It is a most deadly projectile when it strikes, but the difficulty is to ensure its striking. If the submerged tubes on shipboard are used whilst the ship is in rapid motion there is danger of the head of the torpedo being wrenched off as it leaves the side, or of the torpedo being deflected in quite a different direction to that intended. Instances have occurred in which it has fouled the screw of the vessel from which it was discharged; this was only in practice but had the head contained a charge, very unpleasant results might have followed.* With a line of ships astern the danger is greatly increased. Above water tubes, when they are not protected by armour, could hardly be used in action,† and even from them the deflection is just as great. Moreover, when the torpedo has safely left the discharger and is running straight, its course can be followed, and the ship at which it is aimed can elude it by a quick movement. Doubtless the Whitehead is improving day by day, and may ultimately be brought to the comparative perfection of the gun. But at present it is by no means perfect, and could not be employed by the large ship except in a *mêlée*, or at the close of the artillery duel. It is useful as a protection against the ram; otherwise it is best left to its special craft, the torpedo-boat.

There is one other kind of torpedo that might conceivably play some part in a fleet action, the dirigible torpedo of pattern similar to the Brennan and Nordenfelt. At present it is in an undeveloped stage, but there is no doubt that a torpedo which could be steered and directed from a distance,

---

* Some of these difficulties are now overcome, or are in a fair way to be so. The chief causes of deflection must be (1) The speed of the ship, which can be ascertained and allowed for. (2) The inclination of the ship, which cannot be ascertained, as it varies with the helm used, and the state of the sea. Torpedoes are fitted with an arrangement which prevents their explosion till they have run a safe distance. The chief arguments against the torpedo are these: (1) Its complication. (2) Its limited range. (3) The terrible effects of an accidental explosion. (4) Its many failures in peace when uncharged. Lloyd and Hadcock. Artillery, 261.

† At the Yalu the Chinese are said to have emptied their above-water tubes, though this is denied by Commander McGiffin.

would be a most formidable weapon. Still the difficulties are great. It is one thing to direct such an engine of destruction from a stationary ship at a stationary ship, and quite another to manage it when both ships are travelling at a high rate of speed through the water. One or two French cruisers have been fitted with such a dirigible torpedo, but its employment in battle would probably necessitate special craft.

At the close of the long range cannonade will come the close action. The range will be diminished to 600 yards or 700 yards, and the stronger side will steam in to assure its victory. This will be the most terrible period of the action. Up to that time, indeed, the damage done to the vitals of the battleships will not have been serious, but no doubt the internal economy of these vessels will have been impaired. The heavy quick-firers, judging from the Yalu, will not, at long range, inflict much injury on the water-line. It will be upon the upper works, superstructures, military masts, funnels, ventilators, chart-houses, bridges, and stacks of boats and top-hamper, that the hail of projectiles whether fired direct or ricochetting from the water, will descend. The battleship has to carry about with her all sorts of odds and ends which are essential to her in peace, but useless in war. It is difficult to know what should be done with the top-hamper.* When the ship clears for action the boats cannot be taken below, and must remain above to be shot to splinters and to cause fires. Equally dangerous and difficult to dispose of are wooden companion ladders, mess-tables, benches, and the various *impedimenta* usually found between decks; if of wood they will add to the risk of fire, which is very great. They can hardly be thrown overboard, though it is a point to be

* The presence of such top-hamper adds greatly to the difficulty of clearing for action. It is said by Lieutenant-Commander Wainwright (U.S.N.) that some of the ships of the English Mediterranean fleet took twenty-four hours to clear. The amount of time available may determine the tactics of a battle. The French instructions order boats to be filled with water and surrounded with splinter-proof material. The torpedo launches carried would most likely be lowered if time allowed.

noted that the Japanese crews did without them, and so kept their decks clear. It would seem best to leave all but the very barest minimum of boats on shore. It is stated that this course will be adopted by the French in war, and it is a proceeding which commends itself to common sense. No boats will be of much use for saving life after a battle. This is a duty which, as we have urged, should fall upon special vessels, protected by the white flag or the red cross.* There should, however, be some provision of life-preservers for the crew of the big ship, and as far as possible the preservers should be non-inflammable. With the ship's upper deck thoroughly cleared of wood there will be no wreckage to float and save the drowning. India-rubber distendable air-bags would seem, on the whole, the best suited, and there is no reason why one should not be supplied to each man, ready to be inflated. It would support him in the water till the special craft could pick him up.

Upon the upper works of the ships, then, will fall most of the damage inflicted during the preliminary cannonade. They will have been prepared for the strain in every conceivable way. Round the funnels sacks of coal will be placed, and near the quick-firers mantlets to catch splinters. The conning-tower and the positions from which the ship will be fought, will also, doubtless, receive attention. In this way the injury done may be reduced to a minimum, but it will still be extensive. The effect of even small shells charged with high explosives upon unarmoured structures is very deadly. Great holes will be torn in the outer plating; splinters and frag-ments of side and shell sent flying through the confined space within; and any wood that may be about, which has not been thoroughly drenched with water, will be set on fire. The funnels and ventilators may be riddled till they come down, and inside them, on the splinter-gratings, which commonly cross them at the level of the armour-deck, fragments of iron

---

* Page 107

and wood will collect and obstruct the draught. If the ventilators are blocked, and flow of air to the stokehold checked, the stokers and engine-room men will be exposed to terrible hardships—gasping in a hot and vitiated atmosphere for the air which cannot reach them. The boiler-fires will fail and the steam-pressure sink. It is true that nothing of this kind appears to have happened at the Yalu, but the fire maintained there was not so accurate as it would probably be with highly-skilled and cool Western gunners. The danger to the funnels and ventilators is, indeed, so great that it is strange that no attempt has been made to protect them by an armoured shaft rising at least as high as the upper deck. The American monitors had, one and all, armoured funnels. If the funnel is injured between decks there will be some risk of fire, and there is certain to be great inconvenience from smoke. The more improvised protection—provided the material used is non-inflammable—the better. Other points which will require protection, if any can be given them, will be the supports of the barbettes in ships such as the *Amiral Duperré* or the *Benbow*, where the bases of the heavy gun positions are left unarmoured.* If the iron-work under them is much damaged, there will be the risk of the barbette, with its ponderous weight, coming down and sinking the ship. In such ships, too, the auxiliary 6-inch battery must receive attention. We have seen the free use made of sand-bags on board the Northern ships in their attacks upon the New Orleans forts and Mobile. It is to be feared that with a full weight of coal, stores, and ammunition on board, modern ships could not load themselves with sand. Coal would be the only substance available. And here again it is impossible to overlook the fact that a fleet fresh out of port, with full bunkers, has a great advantage over one which has been cruising some time at sea, from the mere fact that in the former coal protection will be at its maximum, and plenty

---

* In the "Admirals," there are coal-bunkers under the barbettes.

of coal at hand for extemporised defences, whilst in the latter many of the bunkers will be empty. In theory, of course, a warship will always burn first the coal in those bunkers, which are least valuable for defensive purpose; she will avoid using the fuel from the water-line bunkers. But in practice, regard for stability will prevent the emptying of the lowest bunkers, whilst the upper ones are left full. Doubtless, if there is time, the ship can use water-ballast, and transfer as much of her fuel as is convenient to the positions where it will be most useful; but she may not always have the time required. And thus, in considering the relative value of coal and armour protection, it should not be forgotten that, however efficacious the former may be when in its place, it is as likely as not, not to be in that place, whilst armour cannot be burnt in the furnaces, and is always there.

During the preliminary attack fires are certain to be frequent, unless the ships engaging are of the very latest pattern. At the Yalu and at Lissa, as we have seen, they were numerous, and it is possible, though not certain, that they will be even more frequently produced by high explosive bursters. Such fires will greatly add to the difficulty of working the ship. The temperature in the *Lai Yuen's* engine-room is reported to have risen to 200° Fahrenheit, and the engineers to have been scared and blinded by the heat. Fires will certainly render yet harder the position of the stokers and engine-room complement, and may seriously interfere with the supply of ammunition at a critical moment. Though the hoists to the heavier guns are well protected on all ships, there are many cases when projectiles and charges have to go up to the smaller guns with very scanty armour to protect them against heat or the enemy's shells. The contingency of a shell on its way up dropping down the hoist through some damage to the apparatus or through accident, is guarded against in some ships by automatic brakes on the hoists, but there are other ships which are defective in this respect. To extinguish fires, good pumps placed out of reach of shells, and good discipline are necessary; but since prevention is better than cure, it will be better to use as little wood in the construction of the ship as is possible.

The number of hits which will be effected in this preliminary period, demands some attention. An English admiral has estimated it, throughout the battle, at two per cent: other writers place it as high as fifteen. The average of the Chinese and Japanese fleets at the Yalu, works out to twelve and a half, if we accept the estimate given by an eye-witness, but this is probably too high. At the same time, there seems little doubt that the shooting with quick-firers will be more accurate than it was in the past with the slow-fire muzzle or breech-loader. In practice, great feats have been performed. Thus, the *Royal Arthur*, at ranges varying between 1600 and 2200 yards, hit a target fourteen times out of sixteen shots, and this whilst steaming at eight knots. The French fleet has demolished targets at 4000 metres range. There is, however, a great difference between firing at a motionless target, and firing at a moving enemy who is returning your fire. In the old days, misses were frequent when the ships fought at very close quarters and it would be thought impossible for a shot to go astray. But generally speaking, the state of the sea will exercise most influence upon the quality of the marksmanship. If the water is calm, good shooting may be expected, and the percentage of hits will rise above ten or fifteen: if rough, it will fall very rapidly, perhaps below two per cent. The steadiness of the combatant vessels will in any sea-way become a factor of great importance, as two guns upon a stable and steady ship will beyond doubt effect more hits than as many weapons upon an unsteady and unstable one. In the days of top-heavy ships, it is well to bear this in mind.

We have already alluded to the huge military masts carried by most French battleships.* They would be of immense

---

* In the latest American design, the *Iowa*, the military mast has vanished, and there is not even the light mast of English pattern, but only signal-poles. So also in the *Brooklyn*. The French are removing the after military mast from value in close action, but it is doubtful whether they will survive the long-range cannonade. They tend to make the ship which carries them unsteady; they are very heavy, and their fall might do serious damage. The English military masts are much lighter and smaller. French officers are of opinion that their masts would stand a good deal of knocking about, and that a single 6-inch or 4.7-inch shell would not bring them down. At the Yalu the Japanese military masts on board the three great cruisers, *Matsushima, Itsukushima*, and *Hashidate*, apparently survived the conflict without receiving damage. On the other hand the *Akagi* lost one of her masts, which was not, however, of the military pattern, and the Chinese ships are stated to have been even more unfortunate.

The maintenance of communications within the ship during action is even more important than the maintenance of her communications with the admiral. How far is it possible to use voice-pipes in the turmoil of battle? We read how at the Yalu officers fought with their ears plugged and yet remained deaf for weeks. There will be, not only the tremendous din and concussion of the ship's own guns, but the not less disquieting crash of the enemy's shells, and the crunch and jar of the iron under the blows. In the ship's interior will be smoke from high explosive shells, smoke, perhaps, from the funnels, and smoke from fire. Voice-pipes may be severed by heavy projectiles impinging upon the tubes which carry them down, or perforating the armour. A trustworthy system of signalling from one part of the ship to another is most urgently required. It should be operated by electricity, have a reply-indicator, and should supplement the existing voice-pipes* Mischances with the present form

many of their ships. On the one hand, military masts are valuable for signalling and top-fire at close quarters: on the other hand it is doubtful whether they would survive the preliminary encounter.

* The telephone has been suggested, but in the uproar of battle might be difficult to use.

of engine-room telegraph, which is purely mechanical, are constantly occurring, and in the *Victoria* disaster the catastrophe may have been aggravated by its failure on the *Camperdown* to convey an order correctly at a very critical moment. There may be many critical moments in battle when such a failure would mean destruction. A string of men to pass orders is found not only on board the *Monitor* in 1862, but on board the *Huascar* many years later. Yet here again there is room for mistakes in the turmoil and excitement of the fray. These vast machines, with all their complex mechanism, where the want of simplicity is so painfully manifest, have, as in the past, to be handled by men who are human and liable to error, but a new danger has been created by our Frankensteins in the risk of the machinery's error.

It has been doubted whether it is advantageous to indicate to the enemy the precise position of the captain—the brain of the ship—by the conning-tower. Forward in most British battleships is a tower protected by 15-inch to 10-inch plating; aft, a second one, with 3-inch or 4-inch armour. The conning-tower is the centre of the ship's nervous system; all the communications are collected there; it is crammed with voice-pipes, steering gear, and firing keys: from it the outlook is very circumscribed in many cases, and too often there is a cumber of chart-houses and bridges above it, which do indeed screen it from view, but may yet be wrecked by a well-placed shell, and set on fire or brought down upon it, thus rendering it useless. Beyond all doubt a heavy fire will be concentrated upon it; experience shows us in the past a heavy roll of casualties in the conning-tower. In it Worden was blinded fighting the *Merrimac*; in it Rodgers was killed before Charleston; in it pilot after pilot was killed or wounded on the Mississippi; in it Grau was blown to pieces; in it another commander of the *Huascar* died; in it two of the *Tsi Yuen's* officers perished. The rain of splinters will make it very hard to see what is happening from the narrow opening, and any shutters which are left closed may be jammed. The concussion of a heavy shell upon the structure, though no doubt, it would not necessarily demolish it, might kill or injure those inside it, and would very probably destroy the communications. It would almost appear as if the best method of ensuring communication would be to have at least three or four stations,* protected by 4-inch or 5-inch Harveyed steel, with, as Admiral Colomb has suggested, a large voice-pipe leading straight down to a station below the armoured deck, the simplest telegraphic instruments, and the least possible number of gun and torpedo directors. The multiplication of positions whence the ship can be fought, gives a better guarantee against the destruction of its brain, than the provision of a single heavily plated shelter, as the enemy will be at a loss to know where to concentrate their fire. Moreover, if the captain delegates his authority, if the gunnery officer is given the control of the ship's guns, the torpedo officer of its torpedoes, and the conning-towers are four in number, dispersed lozenge-wise, three will always face the enemy; when captain, gunnery officer, and torpedo officer will each have a separate position, each commanding a good view of the enemy, and all three will not fall at one blow, as they would, if present together in one conning-tower. In case of injury to those in the captain's tower, the command can instantly devolve. For the ship to be straying masterless in a great fleet action—like the *King Yuen* or *Huascar*—describing erratic curves, would be fraught with the utmost danger not only to herself but to her fellows. She might, at such a time, ram or be rammed by her friends, and veering to and fro, though only for a moment, would throw any line into confusion.

* The more recent French ships have three such positions. Croneau, ii., 425. The Germans station the second in command on the lower deck, in a position of safety. But even so, if the captain is killed, his successor will take seconds or minutes to reach the tower, during which much may happen. With three towers, the command can pass at once, and only a message from one to the others is needed.

The paramount importance of preventing such an evil is evident. It may be that in the earlier stage it would not be likely to occur, though no one could say, for after all the conning-tower is a small target to hit. But should there be a close action hits must be more numerous, and in a close action a brainless ship will be most perilous. Many captains may, till the fleets close, decide to keep outside their shelters and choose a point of vantage on bridge or deck, where they can be seen, and whence they can encourage their men.* The hail from the quick-firers, however, killed or wounded all at the Yalu who showed themselves on the Chinese decks. A suggestion which has been made in "Le Yacht" is worth consideration if such a course is adopted. Cloth or canvas, of the same colour as the ship's upper works, should be hung round the captain's position, hiding all but his head. Where practicable the same protection should be given to the crews of the machine-guns and small exposed weapons. It is a protection, as it in some measure conceals from the enemy the exact position of men and guns; moreover, a screen between the gunners and the enemy makes them cooler. The only risk is that of its being fired by the enemy's projectiles. To prevent this, it can be drenched in alum or any other anti-combustion solution.

How far armour will be penetrated in a long range engagement is an open question. If we can judge from the Yalu, it will not be penetrated at all when it is of moderate thickness —twelve inches or thereabouts.† But guns are so rapidly increasing in power, that it is dangerous to dogmatise. A weapon such as the new Elswick 8-inch quick-firer, with a nominal perforation of twenty inches of steel at the muzzle,

* Or like some of the American monitor captains, stand to leeward of the conning-tower.

† The experience of the Yalu shows that thin armour is worse than useless. By thin armour, is meant plating less than 4 inches thick, which is the least thickness that, under service conditions, at long ranges, could be trusted to keep out the 4·7-inch and, possibly, the 6-inch shell.

should, under service conditions, in action, send its bolt through ten inches of compound, or eight inches of Harveyed, armour at 2000 yards, if it hits nearly at right angles to the target. The thickness of metal perforated by guns on the proving ground, is, of course, only useful to indicate their power under the most favourable circumstances, and to give some standard of comparison. And when the claims of moderate sized guns are urged, and it is said that the present English 29-ton gun, for instance, is quite heavy enough for work at sea, because it can pierce the thickest plate afloat, it should not be forgotten that whilst it can do it, it is never likely to do it, and that to decide a battle it may be necessary to be able to pierce the enemy's thick armour, and to possess guns which not only can, but are likely to do it.* The rapid progress of artillery is, however, giving us guns which will be able to do all that is required on a moderate weight. Such weapons must necessarily be long, but it is better to submit to some inconvenience than to sacrifice ballistics.†

* The endurance of a battleship must ultimately depend upon the endurance of her men, but there can be no doubt that if the ship does not capsize through injury on her water-line, all guns which are not protected by armour, may be put out of action in a close encounter. The thickly-armoured positions may hold out, and inflict much damage if they cannot be silenced, and the victor, without heavy guns, will be driven to torpedo or ram the ship. With heavy guns he can overpower the enemy. The recoil from the big guns seems to have reached its limit, and it does not appear that any nation will go much lower than 12 inches for the heavy armament. The United States are returning to the 13-inch gun for their new battleships. There are many advantages in the big gun: like the big ship, it can deal a crushing blow, and can fire a shell containing a large burster; its shell, too, is more likely to perforate. On the other hand, the larger the gun, the slower its fire, and the shorter its life. Fewer weapons can be carried, and the chance of a breakdown, where much machinery is carried, increases. The battle of the guns is no new one. In the past, there was the struggle between the 18-pounder and the 24-pounder, between the long 42-pounder and the short 32-pounder.

† The howitzer has received much support in France of late. Short, large-calibre guns might, perhaps, be combined with the longer weapons, for use, like the old carronades, at close quarters. But they would be of little value till the fleets closed. With long guns, the trajectory is flat, and there is less chance of missing the target through a trifling error in judging the distance. On the other hand short, large-calibre guns can fire a very heavy shell.

When the now battered hulls draw nearer, and hits, though from a diminished number of guns, become more frequent, the last phase of the battle begins. Probably the auxiliary armament of all the ships, where it is not thoroughly protected as in modern English and American types, will have been put out of action by the awful carnage wrought by the high-explosive shells in batteries which are not armoured, where the guns are massed together. This is the hour for the torpedo-boat, and it will dash to the attack, no longer to be pelted with light projectiles, for the ship it is assailing has reverted under fire to the conditions of 1870, when heavy guns behind thick armour were the only armament. The torpedo attack must be met by a corresponding defence of torpedo-boats, and a fresh struggle will begin, to be decided by a superiority upon the one side or the other, followed, perhaps, by a cruiser action, as these craft draw in to support their various boats. Mutual destruction or disablement may be expected to be the issue, when the battleships on each side will meet in the final collision, and reserves of uninjured ships will decide the day. The ram will now be used upon ships with engines disabled, if they will not surrender; the torpedo will also come into play. A duel between the heavy guns on either side will conclude the battle. We have some idea, from the effect of the Chinese 12-inch and 10-inch shells upon the *Matsushima*, what will be the effect of projectiles weighing 850 lbs. and 1200 lbs. crashing through thick steel, carrying inboard splinters and fragments of plating, and exploding with delayed-action fuses in the interior.* These deadly and stunning blows dealt on either side will rend and tear the ships perhaps past recognition. So in the uproar and confusion, the smoke and the fire, the long agony of the battle will draw to its close

* A fifty pound charge of melinite exploding against a steel armour-deck shatters it over a surface of one square yard, driving down fragments weighing 400 lbs. with a velocity of 210 to 200 foot seconds upon engines or boilers. A 12-inch shell might thus disable any ship. It is doubtful, however, whether the fuse has yet been devised which will take a high-explosive through even thin armour. Croneau, ii., 71.

THE END OF A BATTLESHIP.

Plate XXXIII.

on board the mastodons, which are now settling in the water, rolling terribly, and threatening to engulf their crews. The picture that rises before us is one of horror almost transcending imagination, a scene of bloodshed and destruction so fearful, that man's high purpose and devotion can alone redeem it from the ghastliness of the shambles. But it is neither profitable nor elevating to batten upon horrors.

But will the slaughter be great, it may be asked? Will the damage be so tremendous? And it is certainly the case that a ship can only take a certain number of hits without surrendering or sinking. The temper of the crews will largely determine the percentage of loss. Courageous, resolute, and devoted men will stand firm through slaughter from which weaker men will quail, and thus the braver the combatants the heavier the loss. The sailors on either side will be disciplined men, not as very often on both sides in the last French war, a conglomeration of merchant seamen, prisoners, landsmen, and genuine naval sailors. In the British Navy, they are now picked men—they may be said to be the flower of the nation. They are taken in youth, taught and trained to instinctive obedience, and high courage. They are not swept on board against their will by the arbitrary injustice of the press. They are regarded by the nation with the most absolute confidence. They are animated by the national spirit bred in men who know what England really is, and who day by day behold her power and the splendour of her Empire. They know that the race is for an object—

. . . οὐκ ἱερήιον οὐδὲ βοείην
ἀλλὰ περὶ ψυχῆς.

that the result of the battle will be life or death to that England. They have behind them a past of uniform success, and they will not be ready tamely to surrender it. We may expect from them a most obstinate resistance if the battle goes against them, and with an obstinate resistance the loss must be heavy. So, too, with our opponents there will be every probability of a determined resistance. The French sailor is a picked man; he does not want courage, discipline, or training, and he has a burning desire to revenge the defeats of the past. It is, of course, often asserted that war grows less bloody with time, and history does show it is so on land, but at sea there is small evidence to prove it. We have already reviewed the losses at the Yalu and Lissa, and have compared them with earlier battles. All the sea engagements of the period 1860-6, in which ironclads fought, show slight loss of life, because armour had then conquered the gun. It is now beaten by it,* and there is also the torpedo to be reckoned with. The slaughter on board the *Matsushima* drove her out of battle, but it amounted to more than one-third of her crew. So in the desperate actions of the war of 1812 British ships more than once held out till they had suffered similar loss.† The casualties may then reach as high a figure as thirty or forty per cent, including killed, wounded, and drowned. And as it is with loss of life, so it is with the loss of ships. Though the vitals are well protected in battleships, the deck may at close quarters be perforated by plunging shot, the hull below the water-line pierced as the ship rolls. Heavy losses in armoured ships may not be anticipated till the fleets draw near to each other,‡ but when

---

* The thickest armour is still impenetrable to the guns, under service conditions, or only barely penetrable. But only a small portion of the ship can be thus protected. The increasing power of the gun has compelled designers to leave many important parts of the ship unprotected.

† The following is the percentage of British loss in some of the hottest actions of this war. *Guerrière* and *Constitution*, 32; *Frolic* and *Wasp*, 67; *Macedonian* and *United States*, 37; *Java* and *Constitution*, 38; *Peacock* and *Hornet*, 33; *Reindeer* and *Wasp*, 67. In two of the bloodiest single-ship actions of the French war, the English losses reached 32 and 34 per cent.

‡ A few hits on the water-line may, however, lead to the loss of any ship. The catastrophe to the *Victoria*, as battleships go, a stable vessel, has shewn how slight injuries on the water-line may impair any ship's flotatory qualities. Professor Elgar, writing in Nature, xlix., 153, suggests that it would be well before action to fill unarmoured ends with water. There would then at least be no changes in the ship's trim. But if in the *Sanspareil* and probably in the "Admirals" the ends were thus filled, the top of the belt would be on or below

they do so approach, some are almost bound to go to the bottom. The torpedo and the ram will claim their victims; wrecked water-lines in a sea-way will lead others to capsize, perhaps with little notice. The losses in craft so ill protected, as most cruisers are, will probably be very severe. The old wooden ship of war, attacked by feeble smooth-bores, could stand a prodigious amount of battering, without very often being much the worse for it. Far otherwise is it with our delicate boxes of machinery, attacked by guns which can easily send a shell through them at two miles. Though the *Saikio* came off so cheaply at the Yalu, it was to chance she owed her escape, as the fate of better built Chinese cruisers showed. Again the wooden ship, when she did sink, sank slowly, giving her men plenty of time to escape. These iron hulls capsize, or founder in a minute.

It is difficult then to suppose that the loss, whether of men or *matériel* will be small. Nor will it, in all probability, be spread over a long period of time. On the contrary, the battle cannot last very long. The fleet actions of the past occupy a time which seldom exceeds five hours. Lissa was over in considerably less, and the Yalu in a trifle less. At the Yalu, however, there was no effort to come to close quarters, but merely a prolonged and distant cannonade. The increasing rapidity of gun-fire, the relatively small supply of ammunition carried, the potency of the implements of destruction, all point to a short and sharp struggle. Neither side has much to gain by prolonging the unendurable tension of the battle. There will be on either side an anxiety to bring the affair to an issue, and as soon as either sees a favourable chance, he will dash in.

Not that such a struggle need necessarily be decisive. If no mistakes are made, if men and ships on either side are equal, there cannot be great results to either. But so many

---

the water-line. Luckily the present form of projectile ricochets over the ship or goes to the bottom when it hits the water short of it, so that long range hits between wind and water will be rare.

ifs are not likely to be combined. One side may have more ships, better men, greater manœuvring skill, and abler commanders. War is more largely a personal matter than is often supposed. Not the best ships, but the best men will win. Only we must insist upon good ships and not in peace, at least, profess to regard ourselves as better than our neighbours. If at the close of the day one side has ships intact, and the other ships damaged, no power on earth can save the latter. There is no wind to suddenly blow him away from his foe, or to compel his enemy's retreat. Absolute, hopeless ruin for the beaten side is the prospect in a great battle where the result is not wholly indecisive. There does not seem to be any betwixt or between.

The losses of ships on the beaten side in the more important naval engagements of the last hundred years may be summarised as follows :—

Year.	Name of Battle.	Nationality of Defeated.	No. of ships engaged on beaten side.	Burnt or sunk in Battle.	Captured in Battle.	Destroyed or captured after Battle.	Total Loss.	Captured ships+sunk or destroyed after Battle.
1782	April 12th	French	30	—	5	—	5	1
1794	June 1st	French	26	1	6	—	7	—
1797	St. Vincent	Spaniards	25	—	4	—	4	—
1797	Camperdown	Dutch	16	—	9	—	9	—
1798	Nile	French	13	1	8	2	11	—
1805	Trafalgar	French & Span.	33	1	17†	—	18†	5
1827	Navarino	Turks & Egypt.	3	3	—	—	3	—
1866	Lissa	Italians	23	2	—	1*	3	—
1894	Yalu	Chinese	12	4	—	1	5	—

* Sank at Ancona after Battle.
† Of these three were recaptured and ten wrecked, scuttled or burnt. Five other ships were captured, October 24th, and November 4th.

The disappearance of capture may be due to chance and an insufficiency of modern instances, or may be a feature of naval warfare under the new conditions. It seems to have been replaced by the total destruction of the beaten ship.

All that has been said hitherto applies to day actions, in which neither fleet is surprised. If a fleet should be caught unawares, either by day or by night, it will be lost. But this is not at all likely if both sides have, as they probably will, plenty of scouts. A night action, in which both sides would be willing combatants, is hard to conceive. It might be preceded by a torpedo attack, in which one or more large ships having been damaged, the assailant's heavy ships come up and endeavour to capture or destroy them. The use of the search-light would be necessary upon either side, and very strict control over it would have to be maintained to prevent it from being flashed in the eyes of friends. But a night attack would leave so much to chance—if conducted by heavy ships—that admirals are not likely to run the hazard.

Boarding, as a feature of naval warfare, has vanished. It is only when the motive power of a ship is disabled that it becomes practicable, and a disabled ship is entirely at the mercy of an assailant who is free to move. Perhaps the action between the *Covadonga* and *Independencia* illustrates this most clearly. It is needless to waste life by boarding when the crew can be reduced to submission by the threat of a torpedo or the ram.

If the battle be as we have represented it, what type of ship will be best adapted for action in it? By comparing such an ideal vessel with the battleships now under construction, we may be able to verify the probability of our guesses at truth, since the acutest minds are everywhere brought to bear upon the problems of naval construction. Taking protection first, the parts most essential to the ship, and therefore requiring most attention, may be placed as follows: First, the lower-works, on the safety of which depends the safety of the ship, as, if they are shattered and torn open, in spite of compartments and water-tight doors, she must founder. Next come the engines, boilers, and motive power generally. As the lower portion of the hull cannot be hit directly, unless the ship rolls very much, it is left unarmoured on the exterior, but to protect the interior from harm a horizontal deck of armour will be necessary. The lower this is kept below the water-line the better, though to give space for engines and to assure flotation there are obvious limits.

On the side upwards from the water line, to as great a height as the weight at the disposal of the architect will allow, should be disposed plating, proof against the largest quick-firer under service conditions, to protect the ship's upper works from the ravages of all but the largest shells. The captain's position, as the brain of the ship, and the communications as its vital nerves, should be assured by duplication and moderate armour. The heavy guns should be mounted in separate and well-armoured positions. The quick-firers should, wherever possible, be in turrets, when awkward arrangements for housing them become unnecessary. It is well to allow for a possible growth in length in the near future, and it is manifest that a gun much over 20 feet long could not be stowed as on the *Royal Sovereign*. The thickness of armour will range from 6 inches or 8 inches of Harveyed steel on the quick-firer turrets, to double that amount on the heavy guns. On the side, 8 inches or 9 inches of steel will, at long ranges, exclude even the 8-inch projectile when it does not strike perpendicularly. Such a ship will have three or four positions whence she can be fought in action. As far as stability will permit the guns should be mounted high, when the command will be greater, and the difficulty of using the weapons to advantage in a sea-way less. The funnels and ventilators will be carried up a large, thinly armoured shaft, to some feet above the upper deck. The freeboard will be high, but the superstructure will be reduced to a minimum. Such a vessel will be very far from invulnerable—an impossible aim unless all fighting qualities are sacrificed.

Though water line hits may not be numerous, chance projectiles are nearly certain to strike the ship betwixt wind and water. It is to guard against such hits, which in so vulnerable a quarter might do immense damage, that a belt is carried by every battleship except the great Italian vessels. No amount of subdivision without armour can ensure the ship's flotation. A single heavy shell bursting in a mass of cells such as are found above the armour-deck of the Italian ships, would tear them open and shatter them severely, perhaps setting their cork packing on fire. Again, grave injuries can easily be inflicted upon the armour-deck, where it is placed low, and has no barrier of side-plating to explode shells outside the ship. It does not seem absolutely essential to carry the belt round the ends of the ship, as shell wounds there are not so serious as amidships, but it is undoubtedly better to have a complete belt. If the bows are much wounded, the rush of the ship through the water will tend to force the sea in, and impair the manœuvring qualities by depressing the forward portion. The screws will come nearer to the surface, and in a sea-way there will be risk to the engines and propellers from racing, and from the enemy's projectiles. Injuries astern are not so much to be feared.

The guns carried should be numerous, manageable, and powerful. Four heavy weapons is the number which experience accords to large battleships, though in some recent German examples there are six. Ability to pierce the thickest plating at short ranges must be demanded of such weapons, but at the same time the weight of the gun must not be extreme. The 9·45-inch gun* of fifty or fifty-five calibres can perforate two feet of wrought iron at 2000 yards, and weighs from thirty tons upwards. Such a gun would exhibit a perforation greater than that of the 68-ton weapons of the *Royal Sovereign*, with less than half their weight. Whilst it should be loaded and trained by electricity or hydraulics to ensure rapidity of fire, alternative hand-gear can be fitted. Each of the heavy guns, if weight allows, should have a separate armoured position, as in the French *Magenta* class, for why separate widely the secondary armament whilst concentrating the primary? The auxiliary armament should include as many 8-inch or 6-inch quick-firers as can be given,

---

* The Canet 9·45-inch gun, as long ago as 1890, could perforate 23·5 inches of wrought iron at 2000 mètres. This gun was fifty calibres long. See p. 250.

whilst to stop hostile torpedo-boats and riddle the enemy's unarmoured works at close quarters, the 12-pounder and 1-pounder are necessary. A certain differentiation of armament is as needful to-day as it was in Nelson's time. To give the battleship an armament composed of one single size of gun, would be sacrificing to an ideal simplicity the efficiency of the ship. The principle of dispersing the armament and concentrating its fire, should be carried as far as is possible. Torpedo attacks will be generally delivered from ahead, and demand a strong bow-fire.

The ship which we have sketched corresponds generally to the English *Majestic* class, though in certain features, such as the four gun-positions, it approximates to the earlier French type. The arguments against four gun-positions are strong,* but still stronger, it appears to us, is the argument that, with the heavy weapons mounted in pairs, a single hit might disable half the ship's primary armament.

In action at long ranges, the heavy guns would fire only an occasional shot, and the quick-firers would maintain a rapid and steady fire upon the enemy. When the time came for closing, the guns would attack, according to their size, the thick armour, the thin armour, or the unprotected portions of the opponent's side.† This, of course, involves familiarity on the part of the officers and gunners with the enemy's designs, but it is always easy now to obtain fairly accurate information on such matters. At this period the fire will be as rapid ‡ as possible, and efforts will be made to concentrate

---

* See page 269.

† At long ranges the diminutive size of the target prevents such discrimination.

‡ Rapidity of fire in practice will depend not only on the mount and breech action of the gun, and training of the gunners, but also upon the supply of ammunition. Power-hoists to the quick-firers are of great importance, as they reduce the number of men that will be required below, and they abolish the need for large emergency magazines on deck, which must prove a source of danger, especially where there are many guns close together. The accidents on board the *Palestro* and *Matsushima* warn us of the danger of such exposed

the fleet's weight of metal upon ship after ship in succession. But as each side will try the same game this need not lead to decisive results.

It is more than possible that the command within the ship may devolve with great rapidity. In her engagement with the Chilian ironclads the *Huascar* had in quick succession four commanders; on the *Akagi*, at the Yalu, commander after commander was injured, and any ship which comes to anything like close quarters may fare as badly. It seems, then, a matter of great importance that every commissioned officer on board should have practice in handling the ship at fleet manœuvres. So, too, the methods of fighting the ship, when she has lost heavily in men, must be studied. No modern battleship carries any spare men, yet to supply the place of those who have fallen at the guns, or in exposed positions, men will have to be drawn from somewhere at whatever sacrifice. If they come from the stokehold it will be at the expense of the ship's speed; if from the magazines, at the expense of her rapidity of fire. The importance of giving engine-room hands and stokers a training in gunnery, where practicable, is manifest. It is when the ship is *in extremis* that its value will be felt.

A small matter, but a very important one, is the adoption of some distinctive mark or colour for the ships of each side. Neither side is likely to gain much by a disguise. A distinctive colour, which is varied from week to week, with a broad stripe running right round the ship, will serve to show friends to friends. Individual ships can be marked on the Austrian plan, by belts of colour on the funnels.* In spite of all precaution, in a *mêlée* there might be great risk of accidental injury to friends. The ships would be much injured, perhaps veiled in a wreath of smoke, and if the

---

* magazines. On the *Tamandare*, in the Paraguayan war, there were three explosions arising from this cause. Still, as Captain Mahan has pointed out in the *Century* (August, 1895), it is better to risk an explosion than to concede to the enemy superior rapidity of fire.

morale of the gunners has been much shaken, they may not be too careful. The view from casemate or turret is not extensive, and the temptation to fire at an object crossing the field of vision must be strong. For to be active in battle is less trying to the nerves than to stand and look on. In such a matter as this, discipline and training will tell strongly.

This then is a forecast of the battle of to-morrow. Two great lines of monster ships steaming side by side, but far apart, whilst the uproar of the cannonade, the hail of shells, fills the air. As the minutes pass, funnels and superstructures fly in splinters, the draught sinks, the speed decreases, ships drop to the rear. The moment for close action has come, and the victor steams in on the vanquished. The ram and the torpedo, amidst an inferno of sinking ships and exploding shells, claim their victims. The torpedo-boats of the weaker side in vain essay to cover the beaten battleships. Beneath a pall of smoke, upon a sea of blood, the mastery of the waters is decided for a generation. Such an encounter will not lack sensation. To live through it will be a life's experience; to fall in it a glorious end. And that Heaven may send our fleet success, when the great day comes, is the ardent prayer of every Englishman. For though men can do much by the stoutness and constancy of their hearts, there are chances which lie evermore on the knees of the gods.

CHAPTER XXIV.

IRONCLAD CATASTROPHES.

THE first of the three great disasters which have made Englishmen look with some apprehension upon the ironclad, did not occur till 1871. For ten years our experience of the new type of warship was untried by any serious misadventure. And strangely enough this first great catastrophe had been all but foretold at the Admiralty,* and was the fault, not of Whitehall, but of the British public and press, which had persistently urged the construction of a certain demonstrably unsound type of vessel.

The ill-fated *Captain* was an iron armoured, turret-ship, of 6,900 tons, designed by Captain Cowper Coles, the English inventor of the turret. He had converted the three-decker *Royal Sovereign* into the first English turret-ship, but she was not a sea-going vessel, and was fit for little more than harbour defence. For the latter she was excellent: her turrets were ingenious and gave complete satisfaction. Captain Coles, however, dreamed of yet greater triumphs. He had set his heart upon a sea-going, masted turret-ship, with low freeboard, and, much against the will of the technical advisers of the First Lord,† chiefly through the influence of the press, he was at last permitted to build such a ship. Messrs. Laird were the contractors, and needless to say their work was well done.

---

* Parliamentary Papers, 1871, xlii.; 677, "utterly unsafe;" 678, "cannot possibly prove a satisfactory sea-going ship;" 892, "the danger to be apprehended from these [fully rigged] monitors is very great."

† Parliamentary Papers, 1871, xlii., 668, 673.

When completed in 1870, she was considered by all except a few experts, who had misgivings, the finest fighting ship in the fleet, and was to be the type of future battleships.

Her length was 320 feet, her beam 53 feet, and her draught 25 feet 9½ inches. As designed by Captain Coles she was to have had a freeboard of a little more than 8 feet 6 inches.* A curious error of her designer had reduced this to 6 feet 8 inches, so that, if unsafe in embryo, she was still more unsafe in her completed state. Indeed, Messrs. Laird would seem to have been by no means easy about her, as, when they handed her over, they requested the Admiralty authorities to test her stability by inclining her. This was done with fairly satisfactory results.† The ship carried four 25-ton guns in two turrets, placed fore and aft in the keel-line. She had a high forecastle and poop, which were connected by a hurricane-deck running above the turrets. The armour on the turrets was 13 to 8 inches thick, and on the water-line 6 to 8 inches. There were three tripod masts, with full sail-power, a sail-power which was greater than that given to her safer competitor, the *Monarch*, in proportion to her size. There was one funnel. The complement consisted of 500 officers and men, and the supply of fuel was 500 tons, though Captain Coles had undertaken to give her 1000 tons.

As the finest ship in the fleet she was commanded by a most able and promising officer, Captain Burgoyne, whilst on board her, in various capacities, were sons of Mr. Childers, Lord Northbrook, and Sir Baldwin Walker. Everyone had absolute faith in her, and she was in due course sent to sea with the Channel squadron. In May she faced a heavy gale in the Bay of Biscay. On this cruise her fighting capacities were well tested. In a heavy sea she fired her great guns without any difficulty, making good practice. Under sail she stayed and wore beautifully, beating the *Monarch* with great ease.

* The *Monarch's* freeboard was 14 feet. The *Captain*, in spite of Coles' criticisms of the *Monarch*, was only an indifferent replica of that ship.

† Her metacentric height was 2·6 feet.

Admiral Symonds, after noting her behaviour and inspecting her, reported: "She is a most formidable vessel, and could, I believe, by her superior armament, destroy all the broadside ships of the squadron in detail." A second successful cruise across the Bay of Biscay confirmed this good opinion of her, and even her detractors were forced to confess themselves mistaken.

A third time she went to sea with the Channel squadron under Admiral Milne; and Captain Coles, her designer, sailed in her, to observe her behaviour. The vessels cruising with her were the *Lord Warden*, the flagship, the *Minotaur, Agincourt, Northumberland, Monarch, Hercules, Bellerophon*, and the unarmoured ships *Inconstant* and *Bristol*. To test the turret-ship thoroughly they crossed the Bay of Biscay, and on September 6th, 1871, were near Cape Finisterre. That day a heavy sea was running, and, on Admiral Milne visiting her, he pointed out to Captain Coles that the lee side of the deck was under water when she rolled, and said that it looked ugly. Captain Coles assured him that it made no difference, and mattered nothing. Both Coles and Burgoyne were anxious that the admiral should spend the night on board, but, fortunately for him, he declined. The *Captain* was under sail, but with steam up, ready to be used if required. She was rolling heavily, the angles averaging twelve-and-a-half degrees and sometimes reaching fourteen degrees.

At 8 p.m. that evening the sea was high, and it was cloudy to the west, but there was as yet no indication of a gale. The ships were in station, the *Captain* astern of the *Lord Warden*. At 11 o'clock there was a fresh breeze and some rain. At midnight the barometer dropped, the wind rose, and, as it became evident, that dirty weather was at hand, sails were reefed. A little before 1 p.m. a furious gale set in from the south-west and sails were furled. The rest may be told in Admiral Milne's words: "At this moment the *Captain* was astern of this ship, apparently closing under steam. The signal, "Open order," was made, and at once answered; and at 1.15 a.m. she was on the *Lord Warden's* lee quarter, about six points abaft the ship; her topsails were either close reefed or on the lap; her foresail was close up, the mainsail having been furled at 5.30 p.m.; but I could not see any fore-and-aft set. She was heeling over a good deal to starboard,* with the wind on her port side. Her red bow light was at this time clearly seen. Some minutes after I again looked for her, but it was thick with rain, and the light was no longer visible. The squalls of wind and rain were very heavy, and the *Lord Warden* was kept by the aid of the screw and after-trysails with her bow to a heavy cross-sea, and at times it was thought that the sea would have broken over her gangways. At 2.15 a.m (of the 7th) the gale had somewhat subsided, and the wind went round to the northwest, but without any squall; in fact, the wind moderated, the heavy bank of cloud had passed off to the eastward, and the stars came out clear and bright; the moon, which had given considerable light, was setting: no large ship was seen near us where the *Captain* had been last observed, although the lights of some were visible at a distance. When the day broke the squadron was somewhat scattered, and only ten ships, instead of eleven, could be discerned, the *Captain* being the missing one."

The dreadful truth dawned upon the admiral. The splendid, the trusted ship, was gone, and how no man knew as yet. The vessels of the squadron scattered and searched in every direction, but it was not till the afternoon that the foreboding became certainty. Some portions of her hurricane-deck, a spar with a handkerchief tied to it, the body of a seaman, told the tale. The *Captain* had foundered in one of the heavy squalls soon after one o'clock, when a heavy cross sea was running, and had taken to the depths her crew. The *Inconstant*, the fastest ship of the squadron, was ordered to steam at her fullest speed for Plymouth and carry home the terrible news.

* Eyewitnesses placed her heel at fifteen degrees.

And what had been her end? The survivors, who in evil plight struggled ashore to the sullen coast of Finisterre, could alone tell the story. About midnight the wind was very strong, and the ship was then under her three topsails, in each of which double reefs had been taken. Steam was up, but, apparently, the screw was not working, and she was making little way, rolling very heavily. The starboard watch had been called a few minutes after midnight, and had just mustered on deck. As they were called there was a very heavy lurch, but the ship righted herself again. When the men came on deck* they heard Captain Burgoyne give the orders, "Let go topsail halyards," and then "Let go fore and main topsail sheets." Before they got to the sheets a second and more terrible lurch began. In quick succession the angle of the heel was called, in answer to a question of Captain Burgoyne, "Eighteen degrees! Twenty-three degrees! Twenty-eight degrees!" At the sheets the heel to starboard was so great that some of the men were washed off the deck. The ship was now on her beam-ends, lying down on her side, slowly capsizing, and "trembling with every blow which the short, jumping seas, white with foam, struck her." It was a dreadful moment. The steam, escaping with a tremendous roar from the funnel, did not drown the cries of the stokers, which came up from the bowels of the ship. The boilers were fired athwartship, and when the *Captain* was on her beam-ends the furnace doors in the port row of boilers would no longer be able to resist the pressure of the glowing coal inside, but would be forced open, and thus would discharge their contents upon the hapless men, flung in a heap upon the fronts of the starboard boilers. And, as the draught failed, and the water descended by the funnel to the furnaces, upon the torment of fire would come a rush of flame and steam, till death by drowning ended the tortures of that inferno. It is sometimes said

---

* Court Martial, H.M.S. *Captain*, p. 124. This fixes the time and shows that Admiral Milne must have been mistaken when he thought he saw the *Captain* at 1.15.

that the men in engine-room and stokehold are safe in war. But, when they die unseen and unsung, at moments such as this, without a chance of escape, we must own that they confront a danger more unnerving than is faced by the sailor fighting on deck, and that they deserve special honour.

As the ship heeled, and it became evident that she was capsizing, many of the men ran forward to the weather-forecastle netting and leapt overboard, still hearing the roar of the steam, even when the funnel was below the water. Others climbed up the tilting deck with the help of ropes, and got out on the port side, and then as the *Captain* slowly turned over, walked up her bottom. One man caught his foot in a Kingston valve, and finally reached the place where the keel would have been, had there been one, when the ship fell suddenly away from under him. The gunner had a very narrow escape. He had been asleep in his cabin, when some marines awakened him by the noise they made. Noticing that the ship was rolling heavily, he got up and went to the turrets to see that the guns were properly secured. He visited the fore-turret, and was in the after-turret when the fatal lurch began. As the heel grew greater, he climbed out of one of the sighting holes, and was just clear when the ship went down. The last seen of her was the prow. The few survivors, all, except the captain and the gunner, of the starboard watch, climbed on to the launch and the pinnace, which were floating about. The second launch was cleared, and the men set to work to row her to the help of the pinnace, upon which, as it swam bottom upwards, were Captain Burgoyne and several men. Many of the men jumped to the launch, but Burgoyne would not, and as the heavy sea prevented close approach, and all but swamped the launch, he had to be abandoned. He refused an oar, telling the men they would want all those they had. Eighteen survivors reached dry land, after in vain hailing the *Inconstant*, which passed close to them, without hearing or seeing them in the uproar and darkness of the squall.

The news was at first received with incredulity in England, and then with grief and indignation. The loss of so fine a ship, with so many promising officers and nearly 500 men, was a national disaster. At the court martial which sat to try the survivors the verdict was a vindication of the Admiralty. "The *Captain*," it ran, "was built in deference to public opinion, and in opposition to the views and opinions of the Controller of the Navy and his department." Her heavy masting, her sails, her low freeboard, far lower than her designer had intended, and her great top-weights, in the shape of her hurricane deck and turrets, were the causes of the disaster. She might have been a safe ship without her masts, or with them she might have been a satisfactory coast-service vessel. But there is no doubt that she was too unstable for work at sea. Her loss has not been without effect. In the first place, it has led our constructors to pay great attention to stability, a virtue the value of which would be felt in war, though it does not necessarily make a great show in peace. In the second, it has for ever warned off amateur designers. The art of designing a ship is so intricate, and needs such a deep technical knowledge, if the product is to be satisfactory, that in this there is only cause for satisfaction.

The second disaster was happily unattended with loss of life. The Channel Squadron, consisting of the five ironclads, *Warrior, Achilles, Hector, Iron Duke,* and *Vanguard,* left Kingstown for Queenstown on September 1st, 1875, at 10.30 a.m. On reaching the Kish lightship the *Achilles* left the squadron to steer for Liverpool, whilst the other four proceeded on their course, formed in line ahead. At about 12.30 a very thick fog came on, and it was not possible to see more than fifty yards ahead. The positions of the ships were now as follows: First came the *Warrior* and *Hector*; then, a mile or two astern of them, the *Vanguard* and *Iron Duke,* the former ship leading.* The speed, which had been ten or

* The *Vanguard* and *Iron Duke* were three cables instead of two cables, the right distance, apart at the time of the collision.

twelve knots before the fog came down, was reduced to seven or eight.*

Soon after half-past twelve a large sailing vessel crossed the *Vanguard's* bows, and compelled her to sheer from her place in the line. Her helm had been put hard-a-port, and her way thus checked, when suddenly the *Iron Duke*, which had first sheered, for no adequate reason, and then had come back to her course, loomed up through the fog, not one cable distant, with her ram pointed at the *Vanguard's* broadside. Simultaneously the *Vanguard* was seen by the *Iron Duke*, and Captain Hickley, the commander of the latter, who was on deck, at once ordered his engines to go astern, but too late to avert a collision. Steaming at a rate of something less than seven knots, the *Iron Duke* struck the *Vanguard*, which was steaming about six knots, four feet below her armour, just abaft the mainmast on the port quarter, abreast of the engine-room. A very large rent, twenty-five feet square, was made in her, and the water came through in a torrent. Unfortunately, the ram had damaged the ship at her most vulnerable point, tearing a hole in the athwartship bulkhead, which parted the engine and boiler compartments—the two largest compartments in the ship. The shock was very violent. The armour-belt of the *Vanguard*, here 8 inches thick, was driven in more than a foot, but the inner skin was not actually pierced by the ram.† Other bulkheads in the ship were so much damaged that they leaked badly, and on deck spars and blocks fell from her masts. Immediately the collision occurred the water-tight doors were closed. There was no panic, and the discipline maintained was excellent. The engine-room, stokehold, and alleys were quickly filled, the

* The " Fog Signal Instructions," however, laid down the rule that in fogs the speed should not exceed three or four knots.

† The inner skin of the double bottom only went as high as the lower edge of the armour belt. Thus for some feet below the water-line there was nothing behind the side-plating and its supports. Had the *Vanguard* been built with a wing-bulkhead as are all modern ironclads she would certainly have floated, notwithstanding her severe injuries.

boiler drowned out, and the steam-pumps left without steam. An artificer, with great presence of mind, opened the boiler valves, and allowed the steam to escape, thereby averting an explosion. The men were mustered on deck in order, and no attempt was made to save the ship, all the energy of Captain Dawkins, her commander, being devoted to the saving of life. A certain want of promptitude, resolution, and resource was perhaps visible amongst the officers, but looking at the case of the *Vanguard* in the light of later catastrophes, it is doubtful whether much could have been done. The *Iron Duke*, which had disappeared in the fog, came up as close as she could with safety, and as quickly as possible, but, with perfect order, the men were transferred to her. Within twenty minutes they had all been taken off, the captain, as usual, being the last to leave the sinking ironclad. One hour after the collision, at 2.15 p.m., the *Vanguard*, which was heavily down by the stern, whirled round two or three times and went to the bottom in nineteen fathoms.

The *Vanguard* was a second-class battleship of 5899 tons and 3500 horse-power.* She was one of a class of six ships designed to meet the French *Alma* class, and was primarily meant for service on distant stations. She carried ten 12¼-ton muzzle-loaders and two 64-pounders. She had a complete armour-belt, 6 to 8 inches thick on the water-line, and a central battery protected by 6-inch plating. She had a speed of 14·9 knots at her trial, and carried a crew of 450 men. On her steam trials she was found to be defective in stability, and her double-bottom had been filled with bricks and cement. There were seven athwartship bulkheads dividing the hold into eight compartments, and it was calculated that any one of these might be breached without disaster to the ship. Unfortunately, the possibility of a blow being struck over one of the bulkheads, and thus laying open two compartments to the sea, had been overlooked.

* See Plate xxxvii. p. 220 for elevation of her sister-ship *Audacious*.

The court, which tried the officers of the *Vanguard*, came to the conclusion that the disaster was due, firstly to the high speed which the squadron was maintaining in spite of the fog; secondly to the fact that Captain Dawkins, of the *Vanguard*, though leader of his division, and though the weather was foggy, had left the deck; thirdly to a reduction of speed by the *Vanguard* without any signal to the *Iron Duke* astern; fourthly to an increase of speed by the *Iron Duke* in spite of the fog, and in spite of the fact that the speed was already high; fifthly to the *Iron Duke's* improperly sheering out of line; and sixthly to the absence of any fog-signals on her part. Captain Dawkins was censured and dismissed his ship for neglecting to get the pumps to work, and instead hoisting out the boats. The court held that he should have tried to cover the rent in the side with sails. The navigating lieutenant was censured for not endeavouring to run the ship into shoal water, when she might have been recovered, even if she had sunk. It was held that the *Iron Duke* should have made some effort to tow her into shallow water. The commander of the *Vanguard* was reprimanded with the chief-engineer and carpenter. The *Iron Duke's* watch-officer was dismissed his ship.

The *Iron Duke*, which was precisely similar to the *Vanguard*, suffered no injury of any moment. Her ram projected but very slightly below her armour-belt, and could have repeated the blow without danger. The accident produced in England a tendency to favour the ram, whilst, seeing how easily the largest ship could be destroyed by it, various impracticable suggestions for an unsinkable ship were put forward.

Three years later a similar disaster occurred to the German fleet, but this time there was grievous loss of life. On May 6th, 1878, a squadron of three ironclads—the *König Wilhelm*, carrying the flag of Admiral von Batsch, the *Grosser Kurfürst*, and the *Preussen*—went into commission at Wilhelmshaven. After completing their crews and fitting out, on May 29th they left the port on their way to Plymouth. On May 31st, they were in the Channel off Folkestone. The formation adopted was a triangular one, the *König Wilhelm* leading, and the *Preussen* following astern, in line with her. To starboard of the flagship, slightly abaft her beam, was the *Kurfürst*. Her distance from the *König Wilhelm*, had originally been 440 yards, but an hour before the collision, she had been ordered to draw closer, till only 110 yards parted the two vessels, and from the shore it was noticed that they were in dangerous proximity. Whilst steaming thus, two sailing vessels, hauled to the wind on the port tack, crossed the bows of the squadron. In obedience to the rule of the road, the *Kurfürst* ported her helm, and turned to starboard to clear them. Having done this, she turned sharply to port to recover her original direction. The *König Wilhelm* at first tried to pass ahead of the sailing vessels, but finding this impossible, turned to starboard, and found the *Kurfürst* lying across her bows, at right angles to her course.* To avoid the collision now imminent, the *Kurfürst's* captain went full steam ahead, and tried to cross the bows of the oncoming ironclad in time to clear her. Seeing that this was impossible, he essayed to turn to starboard, hoping to come round on a course parallel to the *König Wilhelm*, or at least to receive only a glancing blow. On board the *König Wilhelm*, both admiral and captain were below, and in these few critical instants there was not time to summon them on deck, or for them to do anything if they had been summoned. The helm was in charge of a petty officer, a one-year volunteer of no experience, and six raw recruits. When the watch-officer gave orders for the helm to be starboarded, to bring the ship round to port, the men got confused, and instead of obeying the order, did the exact opposite, and ported the helm, thus swinging the ram more round to starboard, whilst the *Kurfürst's* stern swung round to port to meet it. As the

* *See* Plan, p. 194.

collision was now inevitable, the order on board the *König Wilhelm* was given to reverse the engines, and they were actually going astern at the moment of impact.

With a speed which, in spite of the reversal of the engines, reached six or seven knots, the *König Wilhelm* crashed into the *Grosser Kurfürst*, which was steaming at nine or ten knots, between the main and mizzenmasts. The ram ploughed up the armour as if it had been orange-peel, whilst a sound of crunching and rending filled the air. The angle of impact was more than forty-five degrees and less than ninety. On board the *König Wilhelm* there was no shock, but a gentle trembling. Glasses of water on her tables were not upset, nor was the water spilled. On the *Kurfürst* there was a violent shock. The ship lurched to starboard, away from the *König Wilhelm*, but kept her way, and twisting and breaking the ram cleared it, and grated alongside. The bowsprit of the *König Wilhelm* caught the *Kurfürst's* rigging and brought down her mizzen topgallantmast, before it was broken off. The boats on the rammed ship's quarterdeck were shattered or swept away. The water poured through the great breach in the side and down the stokehold, flooding the furnaces, and driving the stokers up the hatchways and steps inside the ventilators, whilst the steam escaped violently. A heavy list to port laid the doomed vessel on her beam end, and prevented the crew from getting out the boats, which were smashed on the port side, and lying on the side to starboard. There was little time to do anything, but an effort was made by the captain, Count von Montz, to run her into shallow water. Before she had moved any distance, five minutes from the time of the collision, she sank in fifteen fathoms of water, sucking down many of the crew. Her hammocks had been stowed away in an unusual position—between the boom-boats—so that they could not float away and act as buoys. Of the men on board, most jumped into the water when the end was at hand. Thirty sailors met a dreadful fate. In spite of the entreaties of the boatswain, they leaped over the bows, and

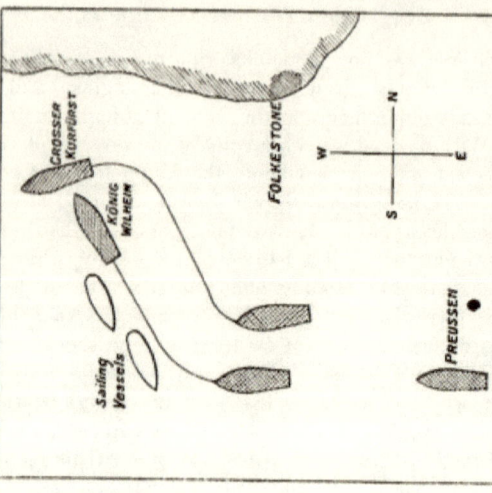

Sinking of Grosser Kurfürst.

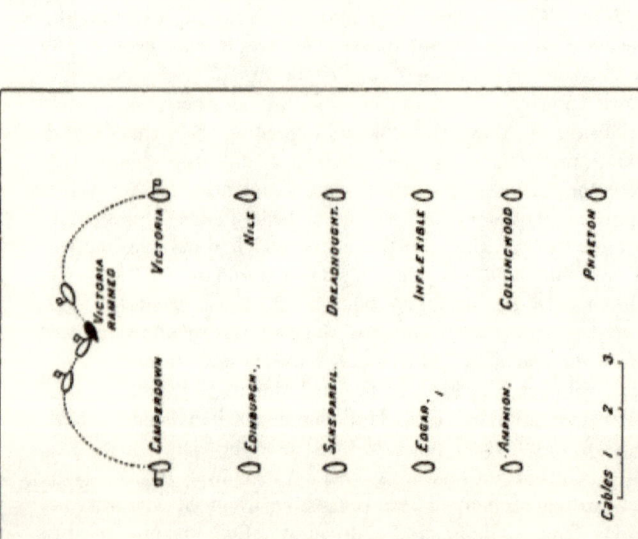

"Sinking of the Victoria."

Map XXX.

were caught in the netting under the jib-boom, entangled, and carried down. The first lieutenant felt himself sucked in when the *Kurfürst* foundered. There was a sensation of a tremendous pressure upon his ribs, as if the water was forcing him down. Then a minute later the pressure was reversed, and drove him to the surface, where he caught a spar and saved his life. The captain was similarly carried down, but came again to the top, and was saved. Fishing vessels, and boats from the *König Wilhelm*, were quickly on the spot, though the *Preussen* was very slow in getting her boats out. Of a crew, which numbered 497, 216 were picked up, of whom three afterwards died from exhaustion. Twenty-three officers were saved, and six drowned, amongst whom were an engineer and a paymaster.

The ram of the *König Wilhelm* was greatly damaged. The stem was broken in two places, and twisted over to port at an angle of forty-five degrees.* All the rivets near it were sheared, or broken away. The sea rushed into her fore compartment and filled it, heavily depressing the bows. There was great excitement on board, as it was at first thought that she too was going to founder, and her captain prepared to beach her, but, finding that the pumps could keep the water down, abandoned the idea, and returned to succour the *Kurfürst*. A sail was placed over the bows, whilst the four side-boats, the cutters, gigs, and one steam-launch, were lowered to save the drowning men. When the *Kurfürst* sank, a cloud of steam, caused probably by the bursting of her boilers, was seen to rise from the water. The *König Wilhelm* and her consort, after cruising about in the neighbourhood of the sunken ship till the afternoon, went to Portsmouth, where the damage was repaired.

This accident, whilst it showed the dreadful efficacy of the ram, showed also that its use was attended with much danger to the assailant. In a heavy sea, the injuries to the *König Wilhelm* might have caused her loss. Her bows, however,

* The stem was a solid forging 4 inches thick.

were not particularly strong, as she was a comparatively old ship, and built before the value of a ram was fully understood. Though armour-belted, her plating did not descend to the extremity of the ram, and there was no support against a transverse strain. The *Grosser Kurfürst* was a turret-ship of 6600 tons, resembling the English *Monarch*. In her two turrets she carried four 24-centimètre Krupps, and on her upper deck two 17-centimètre guns. Her armour was from 7 to 10 inches thick.

The last and most tragic of all these misadventures, was the loss of the *Victoria*. She was a single-turret battleship of the first class, and the most recent construction.* Completed in 1890 at a cost of £724,800, exclusive of guns and gun-mountings, she had a speed of 16·7 knots on the measured mile, and a tremendous armament, included in which were two 110-ton guns, one of 29 tons, and twelve of 6 inches. She carried a belt of armour from 16 to 18 inches thick, extending for about half her length on the water-line; and forward, was her single turret, with its two huge guns. Her original name had been the *Renown*, but on the stocks it had been changed to *Victoria*, in honour of the Queen. The total strength of the crew including officers was 659. On board her, as the finest ship of the Mediterranean fleet, Vice-Admiral Sir George Tryon, the commander-in-chief on the Mediterranean station, had hoisted his flag.

At 10 a.m., on the morning of June 22nd, 1893, the fleet left Beyrout for Tripoli. The vessels present were the armoured battleships *Victoria*, *Camperdown*, carrying the flag of Rear-Admiral Markham, the second in command, *Collingwood*, *Sanspareil*, *Nile*, *Edinburgh*, *Inflexible*, and *Dreadnought*, with the cruisers *Edgar*, *Amphion*, *Phaeton*, *Barham*, and *Fearless*. The order was line abreast, and the speed eight knots. Five miles off the intended destination, at 2.20 p.m., the order was changed to columns of divisions line ahead, disposed abeam to port, the two columns being

* *See* page 232

H.M. Battleship Victoria.

six cables* apart. This brought the fleet into two parallel columns, the starboard one headed by the *Victoria*, containing six ships, and the port column headed by the *Camperdown*, containing five. The *Barham* and *Fearless* were not formed up with the heavier ships. A few minutes earlier, Admiral Tryon had sent for the *Victoria's* commander, Captain Bourke, and her navigating officer. To them he explained the manœuvre, by which he proposed that the fleet should take up its position preparatory to anchoring. The two columns, only six cables apart, were to turn inwards sixteen points,† towards each other. This half-turn would exactly reverse their direction and leave the ships still in a double column, but extremely close together.

The danger of the proposed manœuvre was at once realised by both Captain Bourke and the staff-commander. The space between the two columns was wholly insufficient, for in manœuvring with other ships the tactical diameter‡ of the least handy ship must govern the movements of the handiest. With twenty-eight degrees of helm, which was for manœuvring purposes the limit of the *Victoria*, and without "jockeying" with the screws, driving one ahead and the other astern, a practice which Admiral Tryon discountenanced, the diameter of the *Victoria's* circle was 800 yards, or four cables. The *Camperdown* with about the same turning circle was therefore demonstrably bound, if both ships started turning inwards at once with a distance of only six cables between them, to ram or to be rammed by the *Victoria*. They must collide unless quite exceptional measures were taken. The staff-commander suggested eight cables as a better distance, and the admiral accepted the suggestion, remarking, "Yes, it should be eight cables." The surprise of the commander was therefore great, when at 2.20 the admiral sent orders to signal the

---

* A cable is 200 yards.

† There are thirty-two points in the whole circle of the compass. One point is eleven degrees fifteen minutes.

‡ Tactical diameter is the diameter of the circle which a ship describes in making a complete turn.

distance between the columns as six cables, and as he apprehended that there must be some mistake, the admiral's flag-lieutenant, who had carried the order forward, went below, and asked once more whether the distance was to be only six cables. Once more he was told by Admiral Tryon to "leave it at six cables." Captain Bourke, who was with the admiral, reminded him that the turning circle of the *Victoria* was 800 yards in diameter, but to no purpose.

Admiral Tryon was a masterful, as well as an able officer. He was, says Captain Bourke, "always ready and glad to discuss any manœuvre after it had been performed, but I never knew him to consult anyone before. He loved argument, but he was a strict disciplinarian. He always used to say that he hated people who agreed with him, but that again was different from arguing against a direct order." Captain Bourke left the admiral with an uneasy confidence in him; he was uneasy because the manœuvre was manifestly dangerous, confident because he was serving under a commander of vast knowledge, immense experience, and great caution. The discipline of the service, which to obtain great results must necessarily be strict and exacting, forbade further action on his part. He had done his best to point out the extreme peril of the evolution, and Sir George Tryon would not understand. Therefore, he probably thought, the admiral must have some other intention than that to which the signal appeared to point.

An interval of an hour passed, during which a remonstrance might have averted the terrible disaster which was impending. But no further remonstrance was possible on Captain Bourke's part without something verging upon insubordination. At 3.28 the fatal signal was made in the following terms:—

Second division alter course in succession sixteen points to starboard, preserving the order of the fleet.

First division alter course in succession sixteen points to port, preserving the order of the fleet.

The signal was received on board the *Camperdown* and other ships. Admiral Markham was at once seized with the same misgivings as Captain Bourke. "It is impossible as it is an impracticable manœuvre," was his remark to his flag-lieutenant. He ordered him to keep the signal, which he was repeating, at the dip, to show that it was not understood. On this the *Victoria* signalled to him to know why he was waiting. He answered: "Because I do not quite understand the signal." Unhappily his reply, which might even now have saved the *Victoria*, was not received on board her, and, as there was no answer, Admiral Markham came to the conclusion that the commander-in-chief must intend the second division to turn first, whilst he with the first division circled round outside it. It was a most unfortunate error, but there is only one law at sea, for the junior officer to obey. There was hardly one of the captains of the other ships who did not think the manœuvre fraught with the utmost danger, yet all complied with the signal.*

The signal was passed down the two lines, and the fatal turn began. The *Victoria* and the *Camperdown* at the head of the two columns led the way. On board the *Victoria* the helm used was thirty-five degrees, the extreme limit possible, and when the ship had swung but a very little distance to port, it became evident that a collision was at hand. Captain Bourke, the staff-commander, and Midshipman Lanyon were close to the admiral on the *Victoria's* flying-bridge above the chart-house. The first remark of the captain was: "We shall be very close to that ship (the *Camperdown*)," and, turning to Lanyon, he ordered him to take the distance. This occupied some seconds, during which the two great ships were swinging rapidly towards each other. Meanwhile, Captain Bourke asked the admiral to permit him to go astern with his port screw, and so help the turn. Three times he asked in quick succession before the admiral, after a glance at the *Nile*, the next astern, consented. A very short time afterwards both screws were put astern full speed, but it was

* Dangerous manœuvres, it must be remembered, may be necessary as a training.

now too late. No power could avert the collision, and the two ships drew closer and closer. A minute before the actual crash came, the orders "close water-tight doors" and "out collision-mat" were given. At the former order the crew would go at once to their collision-stations and fasten every door and hatchway, thus isolating every compartment and flat. The order for collision-stations is given by a "G" on the bugle or by the ship's foghorn.

The *Victoria's* crew was a new one, and therefore had not had time to become fully acquainted with the ship. The time in previous drills, occupied in closing water-tight doors, was three minutes, therefore at the moment of the collision, they could not have been secured. The discipline was admirable ; there was everywhere steadiness and obedience, no hurry and no confusion, but the time given was not sufficient. At "out collision-mat" a large mat is brought to the neighbourhood of the leak by a party on deck, ready to be placed over it, if possible. Thus the last seconds passed on board the *Victoria* in preparing for the now imminent disaster.

Four minutes at the most after the signal, the *Camperdown* struck the *Victoria* very nearly at right angles, just before the armoured breastwork which encompasses the base of the turret. The ram ploughed its way in about nine feet, shattering a coal bunker and breaking a man's leg. A petty officer, standing in his mess, looked up and saw the nose of the huge ship come right in amidst a cloud of coal dust. The water could be heard pouring into the ship below. The deck and ironwork buckled up before the ram, and there was a dreadful crunching sound.* The shock was tremendous, if indeed it could be called a shock, for the *Victoria* was forced bodily, sideways, a distance of 70 feet. No one was thrown down, but the wrench was violently felt throughout the ship. For an appreciable time the two vessels remained in contact, and the way on them gradually swung their sterns together whilst

---

* The blow was struck just over a water-tight bulkhead, which was probably destroyed.

the *Camperdown's* ram, still in the breach, worked round and perhaps enlarged the hole. Then as Admiral Tryon hailed the *Camperdown* and ordered her to go astern, that ship cleared the *Victoria*. The *Camperdown's* engines had for some seconds been moving astern, and since her speed, which was only five knots at the moment of the collision, had been further checked by the collision, it would not be long before they began to drive her backwards. The water at once began to pour into the *Victoria* by the breach, which measured about 125 square feet.

On board the *Camperdown* the turn had been executed with twenty-eight degrees helm, instead of thirty-five degrees, the extreme angle possible. Though both Admiral Markham and Captain Johnstone had expressed, as we have seen, their opinion that the manœuvre was a dangerous one, the fullest helm was not used, nor were the screws "jockeyed" to diminish the turning circle. Both watched the *Victoria* attentively till, when it was seen that she was end-on to the *Camperdown*, approaching her and not circling outside her as had been expected, the orders were at last given to close water-tight doors and go astern with the starboard screw. An instant later both screws were ordered to go astern at full speed, but, through some defect of the engine-room telegraph, the order when it reached the engine-room was only for three-quarter speed astern. This could not have made much difference, as the time was too short for the reversal of the engines to have much effect upon the speed. At 3.34 the *Camperdown* delivered the blow, and about two minutes afterwards cleared the *Victoria*, when the flagship steamed ahead. The collision-mat was placed over the *Camperdown's* bows, whilst she too filled forward, and was heavily down by the bows. The crew had been prevented by the inrush of water from closing all the water-tight doors forward.

Meantime on board the *Victoria* the men, closing the doors forward, were driven up by the water and gathered on the upper deck, above the auxiliary battery. The party with the collision-mat could do nothing ; the water rose steadily forward, and when the mat was taken to the forecastle the upper deck was so far below the surface that the men at the foremost stations were wet to the waist. As the bow was depressed the stern rose out of water, and the port screw could be seen from the other ships racing in the air. The effort had to be abandoned, and, as the men of the forecastle party reached the upper deck, the water was up to the turret-ports and was beginning to wash in at the door in the screen, at the forward end of the auxiliary battery.* There was a steadily growing heel to starboard, the side on which were the injuries. The engines, which had been stopped at the moment of the collision, were going ahead, in the vain endeavour to run the ship into shoal water, and this, in no small degree, tended to force the water into the breach, and also to depress the ship's bows by the leverage of the water acting upon the inclined plane of the deck. The steering engine would not work, as the hydraulic machinery had been disabled, and the same was the case with the hydraulic boat-hoists, when they were tried.

Admiral Tryon, the staff-commander, and Midshipman Lanyon were on the top of the chart-house ; whilst Captain Bourke had gone below, at the time of the collision, to see that the water-tight doors were closed. The first recorded remark of the admiral was "It is all my fault." He then asked whether the ship would float, and the commander assured him that in his opinion it would. No one appears to have expected the sudden disaster which followed. The *Dreadnought* having prepared to send boats, a signal was made directing that they should not be despatched, but kept ready. Below, as Captain Bourke passed through the passages and flats of the huge ship, where the electric light was now burning faint and dim, the men were coming on deck without haste or hurry. Looking down into the starboard engine-room,

* This is the condition represented in the diagram.

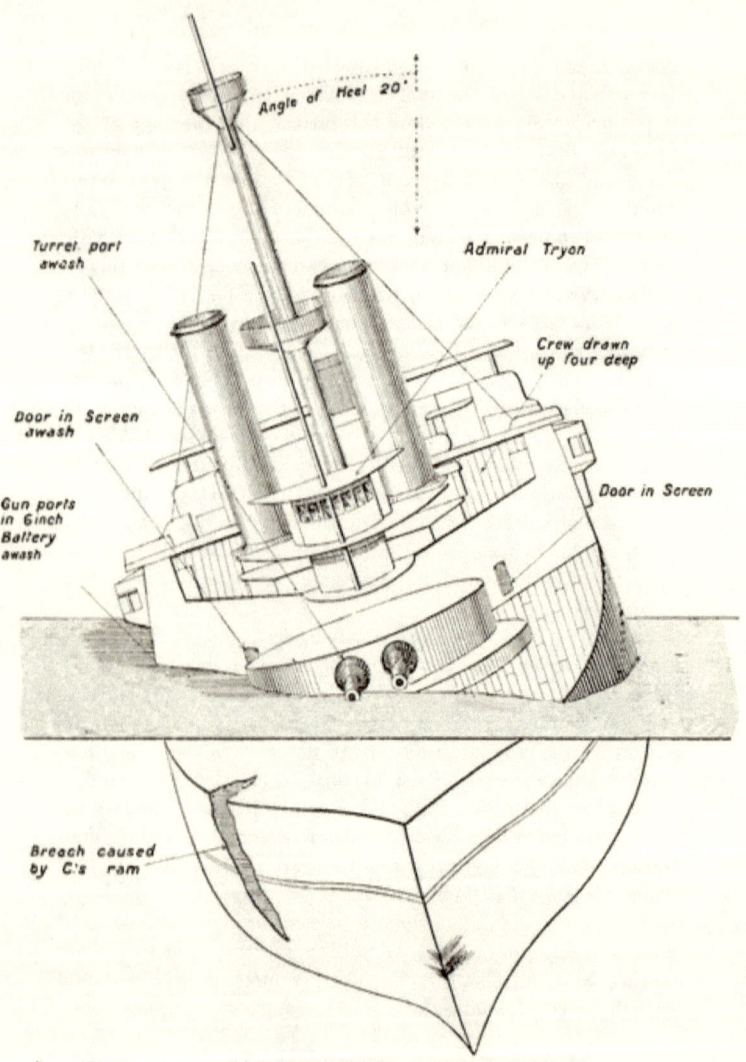

PLATE XXXV. DIAGRAM OF THE VICTORIA ON THE POINT OF CAPSIZING.

he spoke to the engineer in charge, and heard that there was no water in the engine-room. He heard, also, the gongs of the telegraphs, and could see that the engine-room men were steady at their posts. Meeting, in the main passage, the fleet-engineer, he was told by him that the boiler-rooms were water-tight. The trend to starboard was great when he reached the upper deck. There, on the port side, with their faces away from the ship's side, were drawn up, four deep in line, the whole of the *Victoria's* crew, except the doomed men below in the engine-room and stokehold. It was a memorable sight. Steadiness the most perfect, obedience the most unwavering, discipline the most admirable, held the company. Though it must have been evident to all that the ship's end was at hand, there was still no panic. Not a man moved to the side, not a sailor stirred. All in the presence of this overmastering catastrophe demeaned themselves like men, as if to prove to this nation what their bearing would be before the enemy. For many there these were the last few minutes of life. And deep and tender though our sorrow must be for noble lives thus lost in peace, we can yet feel something akin to exultation at the thought that at this supreme moment another imperishable tradition was being added to the glories of our sea service, and that the children of Nelson met death in a manner not unworthy of their past.

The tilt of the great ship grew. The deck heeled towards the perpendicular, and the order "Jump," was given. The line of men broke and made for the side, but with difficulty, owing to the growing angle of the list. Some threw themselves over, others were thrown through the air by the ship. Those astern had to leap past the whirling screw, which is said to have killed or injured no few. There was no want of gallant deeds in the awful instants which succeeded. On the chart-house Midshipman Lanyon remained by Admiral Tryon, and the staff-commander, to the last, though ordered by the admiral to the boats. The final lurch came at 3.44½, just ten minutes after the collision. There was a tremendous roll to starboard, the stem of the *Victoria* dived, a crash of boats and top-hamper falling into the water filled the air, and then the flagship went to the bottom amidst a cloud of smoke and steam. The last seen of her was the stern, with the screws still revolving. A great uprush of air, a violent upheaval of the surface followed, and spars and fragments of wreckage carried up by it injured many of the men in the water, whilst others were sucked down in the whirling vortex. The men were so close together that it was hard to strike out, and as there were many who could not swim, or who had been injured, in the water, these clutched hold of the others and carried them down with them. The officers and crews of the other ships were horror-struck witnesses of the scene, and, as quickly as they could, sent their boats to the help of the drowning men. In from five to ten minutes they were near the place where the *Victoria* had foundered, and rescued between them 338 lives. Admiral Tryon was never seen again after the final lurch.

Nothing was wanting to complete the tragedy. There was amongst the spectators of that scene the same foreboding of evil which was present in the mind of the Greek in his theatre. The play must inevitably end in tragedy. They saw evil coming and could not avert it; they saw their chief seized with "God-sent madness" and stood powerless. The commander-in-chief was guilty of a grave error, and by that error he doomed 321 officers and men. He refused to listen to a suggestion or remonstrance, and his order was obeyed. He expiated his fault by death, and with a noble magnanimity acknowledged his fault. The second in command was guilty of an error of judgment in obeying, without a question as to the real meaning of the signal. There is little doubt that on board the *Victoria* the officers had a presentiment of the collision which would follow. They may be blamed for doing nothing which might have averted it, for not making their remonstrance felt. But in justice we must own that it was discipline which was to blame, not the officers. Without discipline, without prompt and unquestioning obedience, a

PLATE XXXVI.

THE LAST OF THE VICTORIA.

navy would be worthless. And so, as it has been well
said, by one of those antinomies which occur more often
in fiction or the drama than in real life, the collision was
precipitated. The finest ironclad, or all but the finest, in the
squadron, before the eyes of all, on a clear summer day, went
to the bottom as the result of a touch of the ram. All the
skill and ingenuity which had been lavished upon her
proved of no avail to keep her afloat; perhaps they were even
a snare, as they led to excessive confidence in her stability.

The *Camperdown* was very severely damaged in her bows.
A great hole, ten feet by six feet, was torn by the sharp edge
of the *Victoria's* armour-deck, and the stem was broken above
the ram and twisted to port. The water shipped brought up
her draught forward from 27 feet 9 inches to 32 feet, an
increase of over four feet. She could not, therefore, have
repeated the blow, and must have been in great danger if
a storm had arisen.

The great loss of life on board the *Victoria* was due in the
first instance to the number of men who could not swim, and
would go down at once; in the second, to the violent swirl and
subsequent agitation of the water; and in the third, to the
number of men below in the engine-room and stokehold. No
order was given to them to come on deck, and they died like
Englishmen, with a poetic splendour, doing their duty, however useless and hopeless a one. Upon their fate we scarcely
can dare to speculate. The rush of water down the funnels
would be followed by a burst of steam from the boilers.
Probably, as the ship touched the bottom, the pressure, in
eighty fathoms of water, drove in the sides, and so killed any
who had survived the scalding steam. The water above the
ship was greatly agitated by escaping air for some considerable time after she had gone down. There was no explosion
as she sank, but only steam, probably from the funnels.

The court martial which met at Malta found that Sir
George Tryon was to blame for the collision, acquitted
Captain Bourke and the *Victoria's* officers, and praised the

order and discipline maintained on board. It regretted that
Admiral Markham had not signalled to the commander-in-
chief his doubts as to the evolution. An Admiralty minute
blamed Captain Johnstone for not making preparations in
view of the collision which he expected. The same minute
pointed out that the foundering of the *Victoria* was due, not
to defective construction, or instability, but to the fact that
many of the water-tight doors and scuttles forward had not
been closed, and thus the water, instead of being confined to
two compartments at the most, filled the forward part of the
ship. This weight of water depressed the bow, and brought
the ventilators on the upper deck, some of which could not
be secured, below the water-line, and admitted the sea to the
mess deck. Next the turret-ports, the door in the screen, and
the forward portholes of the 6-inch battery, all of which were
open, became awash. Flowing through them into the angle
formed by the side, and the now sharply sloping deck, the
water lodged there, and capsized the vessel, as all her stability
was gone. Had all watertight doors been closed, had the
ports and ventilators been secured, the *Victoria* would beyond
a doubt have floated, though with a heavy list. But we have
seen that the men at collision-stations had only one minute
instead of three to do the work.* It may be said again that
Captain Bourke should have caused the doors to be closed
earlier. To have done so, however, would have been almost
insubordination, a direct reflection upon the admiral. It may
be said, again, that they should not have been open at all.
But if the doors are there they are meant to be used, and it
is extremely difficult, indeed impossible, to work a ship
minutely subdivided, with every door closed. Even in battle
it would not in practice be found feasible thus to isolate every
compartment. Human foresight and ingenuity can do much,

* There were a great many doors in the forward bulkheads, and there were
numerous openings in the decks fitted with water-tight hatches, all of which had
to be closed. Some of the doors could only be secured by entering the flooded
compartments, a manifest impossibility.

but cannot do everything. The ship which no injury of the ram shall be able to sink has yet to be designed.

One point of some importance is dwelt upon by the minute. The presence of an armour-belt would not have saved the ship, as it could not have resisted the terrific force of the *Camperdown's* blow, and must have been crushed in. This has been questioned both in England and in France, and does, indeed, seem doubtful. The force of the blow struck by the *Camperdown's* ram was about that of the 45-ton gun projectile at the muzzle. If 16-inch armour will keep out such a projectile where all the energy is concentrated upon an extremely small space, it should have been able to keep out the ram, or at all events to limit the damage done. The loss of the *Victoria* has not increased the confidence in the English type of battle-ship with its large unarmoured ends.*

"The order and discipline maintained upon the *Victoria* . . . was in the highest degree honourable to all concerned, and will ever remain a noble example to the service," concludes the minute. It is a noble epitaph, and that order and discipline converted a great disaster into a greater triumph.

Last of all, and still unexplained, is the loss of the Spanish cruiser *Reina Regente*, with all hands, during March, 1895. She was conveying the members of the Moorish mission from Spain to Tangiers. She had thus no great distance to cover, and she was not, like the *Captain*, exposed to the fury of the Bay of Biscay. A violent storm, however, arose, and in it she must have foundered. No trace of her, whether wreckage or bodies of her crew, has as yet been discovered. She was a very heavily armed vessel for her size, and it is therefore conjectured that her stability was deficient, and that she capsized. She carried a crew of over 400 officers and men, besides the members of the mission. In general outline and size she was similar to the English belted cruisers of the

---

* If the compartments above the armour-deck were riddled the *Victoria* and similar ships would sink 5 feet by the head. They should, says Professor Elgar, be classed as protected ships, not as armour-clads. *See also* pages 233, 174.

Australia type. She carried four 7·8-inch guns, two forward and two aft, with six 4·7-inch guns in the waist of the ship. She was a comparatively new ship, having been launched in 1887 at Messrs. Thomson's yard at Clydebank. Her displacement was 5000 tons.

Catastrophes of less importance were the loss of the Japanese cruiser *Unebi* in some unexplained way at sea, the foundering of the French floating battery *Arrogante* off Hyéres, the loss of the British gunboats *Wasp* and *Serpent*, the boiler and gun explosions on board the *Thunderer*, in all of which there was heavy loss of life.

CHAPTER XXV.

THE DEVELOPMENT OF THE ENGLISH BATTLESHIP.

THERE is, perhaps, no period in the history of the human race which has seen changes so numerous, so startling, so far-reaching, as the present century, or we might almost say the half-century in which we live. Fifty years ago ships, guns, and the art of war were much as they were left at the close of the Dutch wars of the seventeenth century. All was clear and all was certain, for the implements of battle had been constantly tested, not only in peace but in war. What was found by practical experience to be useless was quickly eliminated, because there was practical experience. Structural modifications of ships were numerous between 1690 and 1840, but at the same time they involved no radical changes. There was a slow and steady progress, not an advance of lightning speed. Warships were divided into distinct classes with distinct objects: line-of-battle ships, frigates, corvettes, sloops. For the one class to engage the other was, if the assailant was of superior class, almost a breach of the conventions of war, if of the inferior, an act of foolish temerity. The line-of-battle ship of 1840 was an implement in which practical efficiency and beauty of form were combined in the highest degree. Yet, propelled by the force of the winds, in a day when men had not mastered and applied the forces of nature to their own use, the *Victory* or the *Agamemnon* was helpless and motionless in a calm. Her movements had, to a certain extent, to be governed by the direction of the wind, which cannot be foretold or foreknown. Whilst there was certainty as to the value of ships and the methods of tactics, there was yet uncertainty in regard to strategical combinations. It was often impossible for fleets to effect a junction, starting from different points, on a pre-determined day. The first requirement of the sailor was ability to sail a ship. He had to be a good seaman, to be a smart topman or yardsman, alert, agile, and courageous. The merchant-sailor and war-sailor were of one and the same trade; and each could learn his vocation in either service. The only weapon carried by ships was the gun, and the gun was of a rudeness and simplicity which rendered naval gunnery a science easily acquired by any man who had sea-legs. There was no breech action with its elaborate mechanism, to be understood:* there were no sights to be mastered; there were no complicated gun-carriages and mountings; no hydraulic or electric motors. The torpedo existed only in the imagination of sanguine inventors; the ram was an impossibility with sails. Battles were fought and won by concentrating a superior force upon a detail of the enemy—a task rendered moderately easy where the only motive agent was sail-power. The ships then drew as close as possible and battered each other, till the critical eye of the captain told him that the time had come for boarding, when a rush of seamen and marines carried the enemy's deck.

The steam engine has changed all this, and changed it in two ways. First, by rendering the ship, when it is placed on board her, capable of defying the winds and following her own course; and, secondly, by enabling men to employ mechanical implements of extraordinary power. Till machinery was introduced it was impossible to forge or handle large masses of iron. It was impossible to obtain accuracy and exactness in matters like the rifling of a cannon or the fit of a breech-plug. The machine is unaffected by external causation to a degree unattainable in a human being. That is to say, a

---

* The return of simplicity is a gratifying feature in our new quick-firers.

lathe can be set with accuracy to plane a shaft to a certain dimension, and it will continue to do this till it wears out or the power is exhausted. A man with a file varies from day to day. He will approximate to the required standard, but will never attain to it, and his personal equation, profoundly affected by a vast train of causes ceaselessly in operation, must be taken into account. This is the real explanation of the failure of early ideas on subjects such as the use of turrets, or the employment of the breechloader. They were sound enough in themselves but the means necessary for their manufacture were wanting.* So iron ships could not be cheaply constructed before the days of rolling-mills and steam-hammers, if, indeed, they could be constructed at all. And when we plume ourselves upon our enormous advance, let us remember that our ancestors were, after all, step by step building up the means which should take us forward.

Steam was at first applied to the propulsion of ships by means of paddles, which were clumsy and exposed. It was impossible to mount a heavy battery on the broadside with such interruptions amidships, and thus the new motor could not well be employed on board anything larger than a frigate. Accordingly, sailing line-of-battle ships were to be towed into action by tugs when the wind was against them. But the advent of the screw changed everything, as, from the date of its application to the steamer *Stockton*, in 1837, by Ericsson, it steadily made its way. It lay below the water-line and was not exposed to hostile shot, whilst it could be readily fitted to ships of the accepted pattern. By the date of the Crimean war the fighting ships of the world were screw-propelled.

The adoption of steam had numerous effects on the construction of the ship. Whereas with the warships of the past speed had been obtained by diminishing size, thus making the frigate faster than the line-of-battle ship, it could now only be

* Lord Armstrong, even as late as 1854, could not obtain satisfactory steel forgings for a gun-barrel only 1·88 inches in diameter.

secured by giving increased weight and space to machinery, engines, and boilers on board the fast ship. Moreover, the discovery that the larger the ship the more economically is the power exerted, also tended to increased size. The principle would not apply to sailing ships whose motor was the wind, where not economy of power, but the direction to the required end of what power existed, was sought. Extreme fineness of lines again was dangerous where a vessel had the leverage of masts upon the hull, tending, if the latter was not broad and steady, to capsize her. The incompatibility of steam and sails, which lies in the different form of hull required with each to give the best results, has led to the gradual supersession of the latter by the former, though at first both were applied conjointly, and though there are great advantages in the possession of sails as an auxiliary. Thus steam, at first a mere auxiliary, becomes next the main, and finally, the sole motive force. The growth of displacements begins forthwith. From the days of the *Henri Grace à Dieu*, of 1488, to the *Queen*, of 1839, the increase had been only threefold, from 1500 tons to 4500 tons. From the *Queen* to the *Italia* or *Majestic* is a two-fold increase in the half-century, a growth not of 3000 tons, but of 6000 tons.* Steam and machinery have made the advance possible and necessary. A 15,000 ton sailing ship would require enormous anchors, huge cables, a vast sail area, with gigantic masts and steering apparatus to match. Without steam, and machinery to apply the force of steam, the management of such a vessel would be a business of extraordinary difficulty.

Steam having thus vastly enlarged displacements, began to increase the speed of ships. The *Warrior* achieved a great feat when she covered a little over fourteen knots in an hour. The United States' Government, in the Civil War, projected cruisers of seventeen knots, the famous *Wampanoag* class. In our own day most of the advance has been made, since we

---

* Allowance being made for the different systems of measuring tonnage.

are now building for our service vessels of thirty knots, a gain of sixteen knots upon the *Warrior*, and an enormous advance upon the mean rate of progression between two points in the case of the sailing ship. The steamer goes in a straight line where it lists; the sailing ship must humour the winds, and zigzag for days, or for days lie motionless. Hence, though at times, with a favourable breeze, the vessels of oak and hemp could obtain fourteen or even fifteen knots, they cannot be matched with even the "tramp" for continuous speed. Nelson, in his three months' chase of Villeneuve only averaged ninety-three miles a day.

Another effect of the introduction of steam has been the growing tendency to fill our warships with machinery of kinds other than those necessary for their propulsion. The characteristic of the nineteenth century is to employ mechanical agencies to do the work of man, and it has run riot on our ships. The motive force was present in the boilers which supplied the main engines with steam. It was certain then that secondary or auxiliary engines would be introduced, and all the more, because the increasing size of the weights to be handled, the increasing amount of work to be done, as the ship grew larger, rendered it inconvenient to use hand-power.

As a consequence of this the crews carried are smaller on the modern ship than they were, ton for ton, in the old days. In 1793 a 120-gun ship of 2508 tons (old measurement) carried 841 officers, men and boys; a thirty-eight-gun frigate of the largest class, 1063 tons in displacement, 277 men; and a twenty-gun ship of 432 tons, 138 men, whilst the French ships of like size carried, as a rule, one fourth more. In 1895 the *Royal Sovereign* of 14,150 tons carries 720 officers and men; the *Devastation*, of 9300 tons, 400; the *Cambrian*, a cruiser of 4360 tons, 265; and the *Dryad*, a torpedo-gunboat of 1070 tons, about ninety men. It is possible that these crews are not large enough for battle purposes, but the economy in human labour is enormous. To gain some idea of the work done on board ship by mechanical power, we may observe that the *Royal Sovereign* class carry eighty-six engines, the French *Magenta* over 100, whilst there are at least as many on our largest cruisers. Taking the *Terrible*, for instance, we find her equipped as follows:

SETS OF ENGINES.

```
 2 working the screws (main engines).
 2 reversing the main engines.
 2 turning the main engines.
 4 driving main circulating pumps.
 2    ,,    auxiliary circulating pumps and air pumps.
 6    ,,    main feed pumps.
 8    ,,    auxiliary    ,,
 4    ,,    fire and bilge ,,
 2    ,,    distilling    ,,
 1    ,,    drain-tank    ,,
 8    ,,    air compressors for air jets in Belleville boilers.
12    ,,    fan compressors for stokeholds.
 2    ,,       ,,        ,,       ,, engine rooms.
 4    ,,       ,,        ,,       ,, ventilation.
 3    ,,    electric light dynamos.
 4    ,,    air compressors for torpedoes.
 2    ,,    steering apparatus.
 2 for boat-hoisting.
 2 ,, coal      ,,
12 ,, ash      ,,
 1 workshop engine.
 2 capstan engines.
```

The engines are in most cases duplicated to prevent disablement by accident, but even so, dividing by two, forty-three odd sets of engines are necessary to the economy of a large cruiser. The warship, then, approximates to a floating factory, her decks and hull crammed with machines and guns. The *Terrible* further carries forty-eight boilers, occupying, with the engine rooms, 252 feet of her total length, and that amidships, where the breadth of the vessel is greatest.

Growth in size, advance in speed, development of machinery, are all directly traceable to the influence of steam. Due in part to this cause, in part to the influence of the past, is a growing tendency to specialisation in the various types of ship. Our ancestors, after long experience, found that a certain differentiation was necessary. They had the line-of-battle ship, the frigate, the corvette, the sloop, the bomb-ketch, the fire-ship. But all, except the last, were constructed to use guns, though the armament varied in power according to the ship's sphere of action. The appearance of armour for a time destroyed this specialisation. For the first few years after the introduction of the ironclad, we built little but large ironclads and small ironclads. The latter were intended for coast-defence, or for service upon distant stations. They were to the typical battleship of the day much as the frigate was to to the line-of-battle ship. The difference between the two was in degree, not kind. Our ancestors, however, did not build vessels for the express purpose of coast-defence, since their strategical wisdom led them to aim at the command of the sea and the blockade of the enemy's ports, for which coast-defence ships, or ships, to call them by their true name, of indifferent sea-keeping quality, were quite unsuited. When new weapons, in the shape of the ram and torpedo, came into favour, and when it grew increasingly evident that slow, small ironclads would be eaten up by fast, large ones, a new era began. Though all three weapons, the gun, ram and torpedo could be combined on board one ship, there were yet advantages to be gained from the use of a particular type of ship for each. As a high speed demanded heavy engines, and without it the small ship would be left a prey to the large, the required weight was provided by denuding her of armour. The cruiser began to differ in kind from the battleship: she had no longer thinner armour, but no armour at all.* On the other hand, she could decline battle and run. The torpedo could be employed by a small boat, and it was fondly imagined by enthusiasts that with its advent the day of great ships had

---

* The frigate, at any but the very closest range, was impervious to even the line-of-battle ship's guns. Now the cruiser is vulnerable, at the extremest ranges, to the guns of even her own class of ship.

passed away. Special craft were constructed to use it, in which diminutive size and high speed were combined. These, again, led up to special craft constructed to combat them. The cruiser, with the lapse of time, has differentiated within itself to a degree unknown in the frigate. The gap between the *Powerful*, of 14,250 tons, and the *Forbin*, of 1850, is immeasurably greater than that which parted the forty-four-gun and twenty-gun frigate. We have this cruiser for ocean work; that for work on the narrow seas; a third for scouting; a fourth, though this type does not appear in our fleet, for commerce-destruction. Our specialisation may be excessive, as we can never be certain that our special ship will meet the enemy's special ship, and the observance of the uniform mean, whilst adopting a certain number of types, might be best. A Nelson might look aghast at the motley array of battleships, rams, large cruisers, small cruisers, dynamite ships, torpedo gunboats, destroyers, torpedo-boats, coast-defence craft, and torpedo-boat carriers which this decade has produced.

A point also to be noticed is the difficulty with modern ships of estimating a vessel's fighting force, when she is seen at a distance. With the old sailing ship it could be told at once. Now a change in the armament in the case of a vessel well known, the substitution of long for short guns, of quick for slow-firers, may have doubled her power, and may enable her to crush an unwary adversary. War may bring many surprises of this kind.

The last point to be considered before we pass to a general view of the changes in our period, is the increased attention which is everywhere being given to naval matters. Fifty years ago there were only two fleets in the world, the English and French. Russia was quite a minor power, though making great exertions, and whilst the United States built very fine ships, they had very few of them. England and France are still first to-day, but in the intervening years, strong navies, some of which aspire to the front rank, have sprung up in other quarters. Germany, Italy, Japan are well to the fore in the second class, and the United States shows signs of embarking upon a great naval programme. The minor states of the world are just as active. This phase is probably due in part to the increasing fierceness of competition between nations as between individuals, in part to the clearer apprehension of the value of sea power.

The application of armour to ships was an idea of considerable antiquity. In the history of human progress, frequent instances will be found where suggestions are made, to be regarded at the time as wild chimeras, and adopted with enthusiasm by posterity. All the old line-of-battle ships were in a certain sense armoured, since their sides and timbers were made very heavy and solid on the water-line with the express object of keeping out shot. Thus, a ninety-gun ship had 16-inch oak timbers spaced at intervals of five inches, with planking inside seven inches thick and outside eight inches. Even at close quarters, the projectiles from the old pattern smooth-bores would not always go through, especially when fired with reduced charges, as was the custom on the guns growing hot. The Spanish floating batteries at the siege of Gibraltar were in like manner protected by an enormous thickness of timber and hide, with bars of iron interspersed. Still, earlier lead p ating had been employed by the knights of St. John, though of all metals this would seem most unsuitable. But it was not till General Paixhans had produced a gun which could fire shells, that governments began to seriously dream of iron-plated ships. General Paixhans himself suggested the efficacy of armour as a protection against these new and terrible projectiles. Admiral Page assures us that the French government knew its value, and that Admiral Mackau, Louis Philippe's minister of marine, had actually tested its resistance, but kept his knowledge a dark secret, wishing to employ it as a trump card in the event of war with England, and so to get the upper hand. In 1842, the British Admiralty fired at a shield of iron plates, joined together to form a thickness of six inches, and supported upon the scantling of an eighty-gun ship, with the result that no shots perforated it. At the same date, the inventor, R. L. Stevens, was busy in the United States constructing a shot-proof frigate plated with 4½-inch iron. This vessel, however, was never completed, and it appears that, if it had been, it could not have floated. Ericsson, in 1841, pointed out that the shot from a 12-inch gun which he had designed would pierce its plating, and some years later, during the Crimean War, submitted to the Emperor Napoleon a rough sketch of an impregnable armoured monitor. Between 1849 and 1851, a most important series of experiments were conducted in England, on the iron ship *Simoom*. Two iron plates ⅜-inch thick, placed some distance apart, were fired at with the 32-pounder, when it was found that the shot in passing through produced a most deadly hail of splinters. With wood backing to the iron it was still the same, but when the wood alone was used, there were far fewer splinters. One noteworthy fact, however, escaped the consideration of the experts. It seems to have been conclusively proved that these thin plates shattered shells before they could burst. The importance of this result does not appear to have been realised, nor were the experiments pushed further to ascertain whether by increasing the thickness of plate the shells might not be altogether excluded. The results of 1842 were apparently overlooked. Iron was condemned as a material for warships, and a general prejudice was created against armour.

The Crimean War over, in which the value of the armoured floating battery had been proved, France began to construct sea-going ironclads. The *Gloire*, the first of these, was a two-decker, from which the upper deck was removed. The weight thus gained was applied in 4½-inch armour-plating to her sides. She had no ram, but the end-to-end belt of iron made her bow strong. She was masted and rigged, but was also fitted with steam power, which gave her a trial speed of thirteen knots. In general outline she was like any other frigate of her day, but slightly longer and heavier in appearance. Her gun-ports were only 6½ feet above water. Her armament was composed of thirty-four 50-pounder guns on her main deck, whilst on her upper deck were two heavy shell guns. The English Admiralty, alarmed at the prospect of having possibly to meet a French ironclad fleet with an English unarmoured one, ordered the construction of the armoured frigate *Warrior*, in 1859. But whereas the *Gloire* was a wooden ship converted to her new shape, the *Warrior* was of iron, and designed expressly to carry armour. M. Dupuy de Lôme was the architect of the French. Messrs. Scott Russell and Isaac Watts, of the English vessel. The *Warrior* is remarkable for the fineness of her lines. Her great length, in proportion to her breadth, gave her a speed extraordinary for her day, since on her trial she achieved fourteen knots. Unlike the *Gloire* she was not armoured from end to end, but had a large patch of plating, 218 feet long and 4½ inches thick, over her battery and waterline amidships. Athwartship she had armoured bulkheads, enclosing a portion of her battery. As her total length was 420 feet over all, only about half her side was protected. In exchange for this loss, she was a better sea boat than the *Gloire*, but her rudder head was completely exposed to the smallest shot, and she never steered really well. She was fully rigged, having three masts; and her two funnels could be telescoped when under sail. Forward, she had a projecting figure-head, which hid a slight spur designed for ramming. Her battery at first consisted of twenty-six 68-pounders on her main deck, behind armour, and twelve which had no protection, outside the armour; whilst ten more guns of the same size were disposed on the upper deck.* Forward and aft of her armour she had water-tight compartments, which, with those amidships reached a total of ninety-two.† In

---

* Later altered to two 110-pounder rifled Armstrong pivots; four 40-pounder rifles, all on upper deck; thirty-four 68-pounders, smooth-bore, on lower deck.

† Viz.; In hold space, 35; in double bottom, &c., 57. The *Royal Sovereign* has 145 compartments.

the water her appearance was most imposing; indeed, as a conception, she was excellent in every respect. Her designers rightly saw that, for England to retain her command of the sea, her ships must be sea-going and sea-keeping. She was no radical departure from the established form of ship, but descended lineally from the frigate, yet in her there was an increase in displacement, a concentration of armament as yet hardly dreamt of, although her fire right ahead and right astern was nil. The weight of armour was 975 tons, as compared with 350 tons in the floating batteries of Louis Napoleon. A sister ship, the *Black Prince*, of the same size and armament, was constructed at the same time. Both ships long outlived the *Gloire*.*

In 1860, Captain Coles, who had, during the Crimean War, prepared a raft with a shot-proof iron shield four inches thick, to carry one 68-pounder gun, brought forward a design for a turret-ship—the first such design that was made public, antedating by over a year Ericsson's *Monitor*.† Coles' ship was to carry nine conical turrets, each containing a pair of guns. The guns were to be "self-acting," running out after their recoil down a slope. As early as 1855, the great engineer Brunel had told Coles, with true insight, "You only need a breechloader to make your shield perfect." A remoter ancestor of the turret may be found in Captain Waymouth's proposal in the sixteenth century, to mount "murtherers" in turrets on the upper decks of ships, the turrets to revolve upon swivels. In 1862, the first English turret-ship was commenced, but to her and her progeny we must recur later.

The design of the *Warrior* was reproduced in an improved form in the *Achilles*, which was larger, belted from end to end with 4½-inch armour, and slightly faster. She carried no less than 1200 tons of plating. Simultaneously a large number of wooden ships were cut down, converted after the pattern

---

* The *Defence* and *Resistance*, iron ships very similar to the *Warrior*, but a little smaller, followed her immediately.

† *See* i. 33.

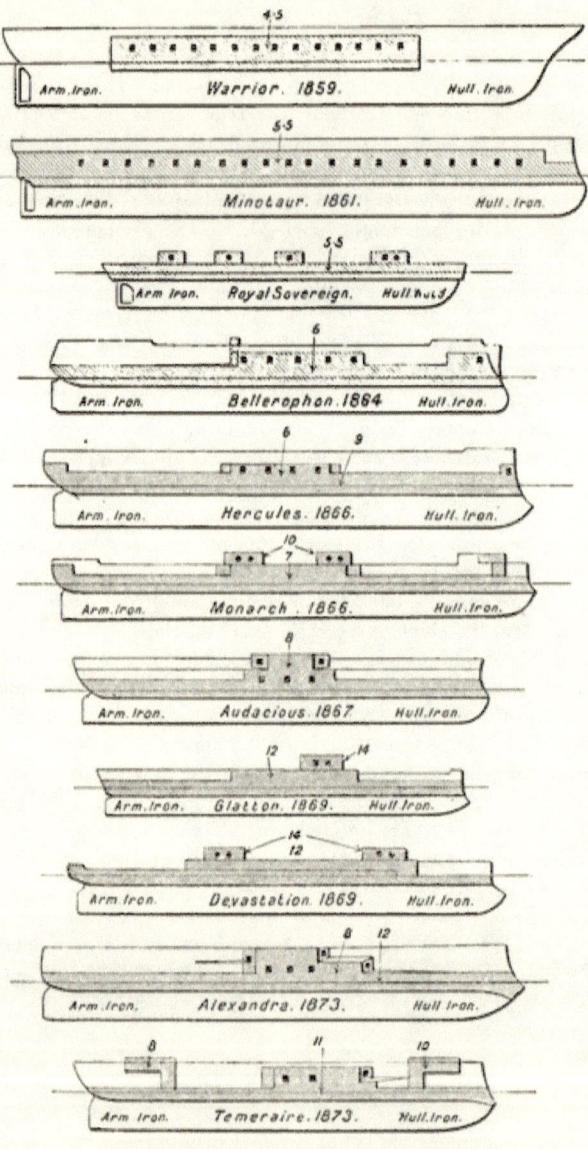

British Ironclads. 1859 — 1873
Figures give the thickness of armour in inches.

Plate XXXVII.

of the *Gloire*, and given an armour belt. Amongst these were the *Zealous, Repulse, Ocean, Research, Royal Alfred, Royal Oak, Prince Consort,* and *Caledonia*.* Their upper decks having been removed, they were lengthened like their French prototype and plated, but were not altogether a success. They were cheaper in the first instance, but far less durable than iron ships, and vanished from the fleet at an early date. The series of broadside ships was further developed in the *Minotaur, Agincourt,* and *Northumberland*, with an end-to-end belt, an increase of displacement, and a more powerful armament than the *Warrior*. They were all ships of enormous length, and were somewhat deficient in handiness. With the advance in artillery, their armour was made $5\frac{1}{2}$ inches instead of $4\frac{1}{2}$ inches thick, but the backing was thinner.

In 1862, Sir. E. J. Reed was appointed to the post of Chief Constructor, and signalised his appointment by designing a series of ships, of moderate size, well-armoured, handy, and fairly fast, which if re-armed, would be capable of rendering great service to us to-day. The first of these was the *Bellerophon*, with plating 8 inches, thick on the water-line. Her guns were concentrated amidships, in a small central battery protected as heretofore, not only on the two sides, but by stout bulkheads athwartship, which prevented her from being raked. In her, for the first time, the bracket-frame system of construction was employed. She mounted the 12-ton and $6\frac{1}{2}$-ton rifled muzzle-loader,† but had no bow or stern fire from guns behind armour. The *Hercules* was of the same class, but a vast improvement. In her, the expedient of recessed ports, which had been tried in the *Pallas* of 1866,‡ received

---

* Wooden ironclads of a more powerful type were the *Lord Clyde* and *Lord Warden*, with 7-inch armour on the water-line, 1300 tons of plating, and a strong bow fire. They carried each sixteen 8-inch muzzle-loaders, and four 100-pounders.

† Ten 12-ton and two $6\frac{1}{2}$-ton.

‡ A small armoured ship for service in distant seas. Similar to her were the *Enterprise, Research,* and *Favourite*.

its final sanction, and henceforward appears, whether for the heavy or the auxiliary armament, in most of our ships, though the sponson* competes with it as a means of giving increased arcs of fire. Her armour was at its thickest nine inches, whilst two 10-inch muzzle-loaders could fire within a few degrees of the keel-line. She was at her launch a splendid ship, superior to anything afloat, and she is still a favourite in our service. At the same time, six smaller central-battery ships of the *Audacious* class were designed for work on distant stations. Two thousand five hundred tons smaller than the *Hercules*, they had 8-inch armour and a central box battery.† But above this was a second battery, very slightly projecting from the side, at each corner of which was a recessed port, giving right ahead or right astern fire. They thus fulfilled, for the first time in an English ship of high freeboard, the ideal of all-round fire. They were the first English ironclads to carry an upper deck battery.

Following the *Hercules* came the first-class battleship *Sultan*, still of the central-battery type, but with considerable improvements on the *Hercules*. The latter ship was extremely deficient in axial fire; this defect was to some extent remedied in the *Sultan* by an upper-deck battery, in which two 12½-ton guns were carried. Forward, quite unprotected, are two more 12½-ton guns. In the *Hercules* a 12½-ton gun had been placed forward, and a second aft, with very slight armour protection. These exposed weapons at the ships's extremities henceforward disappear. The speed was a little higher than in the *Sultan's* predecessor. With the growing tendency amongst naval men at this period to favour attack

---

* A sponson is a curved projection from the side; for example, the framework which on paddle-steamers carries the paddle-wheels. Other examples can be seen in the illustrations of the *Chih Yuen* (p. 114) amidships, just below and a little to the left of her funnel, and in the illustrations of the *Royal Sovereign* (i. 56), *Blenheim* (i. 174), and *Dupuy de Lôme* (i. 310).

† They were armed with ten 12-ton guns.

with the ram, it was now decided to give a better bow-fire to battleships. The last but one of our central-battery ironclads, the *Alexandra* exemplifies that decision. Her guns are mounted amidships in a central battery, which rises perpendicularly from her greatest breadth at the water-line, whilst fore and aft her sides "tumble home," or fall in towards the upper deck, thus allowing four heavy guns to fire very nearly right ahead,* whilst two heavy weapons can be trained astern. No high-freeboard ship had as yet mounted such heavy guns or carried such thick armour. She had two 25-ton and ten 18-ton muzzle-loaders, and her plating reached a thickness of 12 inches. Her belt was carried down over her ram to strengthen it for shock tactics. Her speed was fifteen knots.

Last of this great group, and standing midway between the central-battery and the barbette ship, came the *Téméraire*. She was commenced, like the *Alexandra*, in 1873, but her design differed considerably from that ship. Amidships, as in the *Alexandra*, was a box battery with axial fire, mounting two 25-ton* and four 18-ton guns. But the feature which differentiated her from the *Alexandra* was the introduction of two barbette towers placed fore and aft, each containing one 25-ton gun,† mounted upon a disappearing carriage. After each shot, the recoil upsets the gun and brings it under the shelter of the armour. The gun, when loaded, is carried back to the firing position by hydraulic power. These guns had an all-round fire, and were placed at a great height above the water, but the protection given to them was not very satisfactory, as the barbettes were open underneath. There was an armoured ammunition trunk leading down from them to the magazines, but the smallest shell exploding under the guns would put them out of

* The two 25-ton 11-inch guns fire right ahead; two 18-ton weapons within three degrees of the keel-line.

† Eleven-inch in calibre. There is another pattern of 25-ton gun, 12-inch in calibre, carried on board the *Monarch* and other ships.

action. As in the *Alexandra*, the armour-belt was carried down over the ram.*

We must now return some years to the *Royal Sovereign*. She was a three-decker, cut down on Captain Coles' plan, to the level of the lower deck, ten feet from the water. The hull thus left was plated on the water-line and above it with 4¼-inch iron, whilst upon the deck four turrets were placed. The foremost carried two 12½-ton guns, the others one each. The turrets had armour 10 inches thick near the gun ports, and elsewhere 5 inches. They were rotated by hand-power applied by rack and pinion or winches. The bases of the turrets rested upon the lower deck, and the weight, instead of being supported upon spindles, as in the American monitors, was carried on a roller-way. The ship had only three light signal-masts and no rig. Her crew consisted of 300 officers and men. She had a lightly-armoured conning-tower placed just forward of her one funnel. On her trial, in July, 1864, she proved herself to be a most satisfactory vessel.

The *Prince Albert* was contemporary with the *Royal Sovereign*, and like her carried four turrets placed in the centre-line, with five guns, but differed in being built of iron, expressly to suit Captain Coles' designs, instead of having been merely adapted to them. She was, then, the first English ship built to carry the turret. Like the *Royal Sovereign* she was designed for coast defence, and was not a sea-going ship. Coles, emboldened by his success, was anxious to see the turret system applied to sea-going battleships, and as the result of his energetic insistence the *Monarch* was laid down. She was a moderate-freeboard turret-ship of 8300 tons, carrying two turrets placed on the centre-line, and containing

* In 1878 when war with Russia was thought imminent, four ironclads were purchased by England. They had been designed and built in England for foreign powers, and thus are not necessarily of English type. The following were the ships: The *Superb*, generally resembling the *Hercules*, but more heavily armed; the *Neptune*, a rigged turret-ship, similar to the *Monarch*, but more recent, and armed with the 35-ton Whitworth breechloader, which was exchanged for the 38-ton muzzle-loader; the *Belleisle* and *Orion*, small central battery vessels of limited sea-going power.

each two 25-ton guns. Forward was a forecastle which cut off the ahead fire of the fore-turret, but contained two 6½-ton guns*; these were protected by a screen from ahead fire. Astern she had a poop which mounted one gun of similar pattern. Thus one great advantage of the turret, the all-round fire which it gives, was lost. The vessel was fully rigged and had three masts. Under steam she made very nearly fifteen knots. Her armour on her turrets was 10 inches at its thickest point, on her side 7 inches. Her forecastle made her a tolerable sea boat, and in her day she was considered the finest ironclad afloat.

Captain Coles, however, had never considered her as an ideal ship. He criticised her continually whilst she was on the stocks, and complained of her high freeboard and her loss of all-round fire.† Supported vehemently by the press, he was permitted in 1867 to design the ill-fated *Captain*.‡ He gave her less than half the freeboard of the *Monarch*, and quite as much rigging. Her displacement was 7900 tons, and her armour was a little thicker than the *Monarch's*. The speed was half a knot or more slower than the *Monarch's*. The vessel capsized, in 1870, in the Bay of Biscay, and no more low-freeboard monitors, fully masted and rigged were designed for our service. The *Monarch*, and the purchased *Neptune* are the only survivors of the type.

Instead, the class of mastless turret-ships for sea-going purposes, was developed. The first exemplar of this type is the admirable *Devastation*, designed by Sir Edward Reed in 1869. She was of 9,300 tons, the largest battleship built for the Navy since the broadside ironclad went out of fashion. She had neither forecastle nor poop, but was frankly and entirely a sea-going monitor. In two turrets, placed fore and aft, she carried four 35-ton guns. The thickness of her plating

---

* Afterwards changed to two 12-ton guns.

† It is somewhat strange to find that Coles' own ideal ship, the *Captain*, had a forecastle and poop in spite of these criticisms.

‡ *See* Chapter XXIV., pages 183-4.

ranged from 10 to 14 inches. Between her turrets was a light superstructure with a flying deck. On the loss of the *Captain* deep misgivings were aroused as to the value of low-freeboard turret-ships, and a searching inquiry was made into her stability and seaworthiness. She emerged triumphant, and though she is not altogether a comfortable ship, as her decks are awash in any sea, she has earned golden opinions as an almost invulnerable, powerful and handy ship. She has recently been reconstructed and re-armed with the 10-inch breechloader, and though now an old, is none the less a fine, and valuable, vessel. She is impervious to the deadly hail of the quick-firer, and might fare better in an action at sea than many larger and more modern battleships.

Following her, with considerable improvements, the *Thunderer* and *Dreadnought* were laid down for sea service, whilst contemporary with her, or earlier, were the coast-service ships, *Hotspur*, *Glatton*, the four *Cyclops*,* and the *Rupert*. These ships constitute a very formidable group, apart from the rest of the fleet, and the *Rupert* and *Hotspur* are something better than mere coast-defence ships. They were the ultimate development of the low-freeboard turret-ship which Ericsson and Coles had devised. The *Dreadnought* differed little from the earlier *Devastation*. Her armour was thicker, her guns heavier,† and her displacement larger by a thousand tons. The freeboard was also increased, and hydraulic gear was fitted to her turrets and guns.

The steady advance of artillery had now developed the gun till none but the thickest armour would exclude the newest projectiles. Hitherto it had been possible to design a ship, which should be plated from stem to stern with mail impenetrable, under the conditions of the battle, to any projectile, though it might be easily perforated on the proving ground. Henceforward, the great extent of moderate armour disappears,

---

* Their names were: the *Cyclops, Gorgon, Hecate, Hydra*.

† 38-ton instead of 35-ton.

to re-appear in 1893, and is replaced by a small surface protected by very thick plating. The vitals only of the ship are thus sheltered: the engines, the boilers, the heaviest guns, the captain's station, the lines of communication. All else is left open, on the principle that no resistance is better than a weak one. The ships of the period which we are now approaching have never given entire satisfaction. In their own day they offered very nearly the whole area of their side to the smallest gun's attack. Before the development of the quick-firer and the appearance of high explosives this was not, perhaps, a matter of great importance. But, when great attention was paid to auxiliary armaments, and the number of moderate-sized shells that could be projected in a given time increased, it became evident that these ships could be disabled without a hit upon their thick armour.

The removal of armour from the ship's side is contemporary with the introduction of horizontal plating in the shape of a protective deck.* In the earlier turret-ships of the *Devastation* and *Rupert* period, low-freeboard vessels, which would be much exposed to a plunging fire, we find 2 or 3 inches of armour placed above the belt, horizontally, protecting the deck. In the *Alexandra* and *Téméraire* an armour-deck is also present, but again above the belt and above the water-line. Henceforward we shall see it employed, often without the belt and below the water-line, for a considerable extent of the ship's side, serving to divide the ship into two portions: the one below the water-line containing the vitals, protected from gun-fire by the water on either side and the stout deck of iron above, off which projectiles would glance; the other above the water-line open to every shot and shell over the greater part of its surface, but containing nothing, which, if injured, might compromise the ship's safety. In theory,

* The Mississippi gunboats, designed by Mr. Eads in 1861, had curved decks, plated with thin iron. Ericsson's *Monitor* was similarly protected, but these were low freeboard vessels. The step forward lay in applying such a deck to vessels of high freeboard. The *Comus*, designed in 1875, was the first English unarmoured ship in which it appeared.

this part might be riddled and shot away without damage to the flotatory or fighting qualities of the ship.*

The first armourclad of this description was the *Shannon*, a vessel intended for cruising purposes. The first battleship was the Italian *Duilio*, which was copied in the English *Inflexible*. Her displacement was 11,800 tons, or more than 2000 tons in excess of the *Dreadnought*, which was her immediate predecessor in the line of turret-ships. Instead of a water-line belt of plating running right round the ship, the armour was stripped off the extremities, and applied to increase the thickness on a square citadel placed amidships. This carried two thicknesses of iron armour, together reaching from 16 to 24 inches of metal,† and occupied less than one-third the ship's length. Strong bulkheads ran athwartship, whilst an armour-deck forward and aft, below the water-line, protected the vitals. The two turrets, each containing two 81-ton guns, and plated with compound armour, were placed upon this citadel, *en échelon* amidships. In this way, great concentration of fire was obtained. Four guns could bear ahead and four on either broadside, though the arc of two of the four guns was limited ; but astern only two could be fired, inasmuch as of the two superstructures fore and aft, which gave comfortable quarters for the crew, the after one was broader, and obstructed the inmost gun in each turret. The breadth was extreme, and the speed low. Mr. Barnaby, who had designed this vessel, was denounced in no measured terms by numerous critics, and the ship was christened a coffin, owing to her want of armour on the water-line.

A wholly anomalous vessel, neither battleship nor cruiser, was the *Polyphemus*, designed in 1873 and launched in 1881. She has a steel cigar-shaped hull, plated on its upper portion with 3-inch armour. To accommodate her crew and add to her seaworthiness, a light superstructure has been added.

---

* *See* the transverse sections, Plate XXXVIII., which will make the text clearer.

† An additional defect in the *Inflexible*, *Ajax*, and *Agamemnon* is that their armour is not rolled in one thickness, but is made up of two plates.

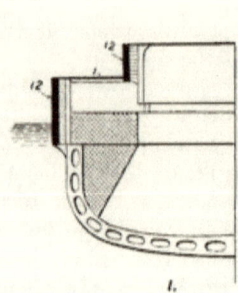
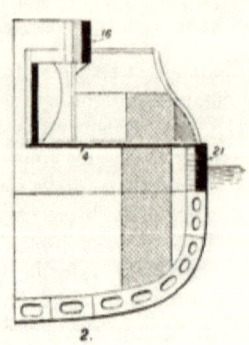

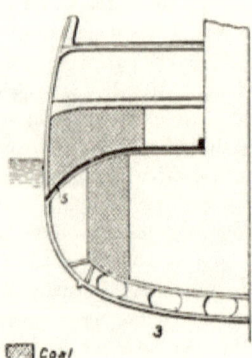
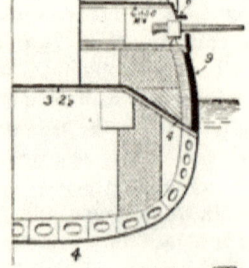

Coal
Water excluding Material
Backing
Armour

Half-Sections showing System of Protection

1. Low freeboard turret-ship with armour deck
   Devastation type
2. High freeboard barbette-ship with armour deck
   Admiral Baudin type
3. Deck protected cruiser no armour belt.
   Edgar type.
4. Deck below belt armour.
   Majestic type

Figures give thickness of plate in inches.

PLATE XXXVIII.

Her primary object is the use of the ram and the torpedo. She carries no heavy guns, and only a few light quick-firers. She cost £200,000 or more, though only a small ship, and her design has not been repeated in England, but in America her chief features have been embodied in the *Katahdin*.

Two vessels of smaller size but similar features, the *Ajax* and *Agamemnon*, followed the *Inflexible*. They carried the 38-ton gun and were a decided failure; slow, ill-armoured, ill-armed, unsteerable. They are the last vessels built for our fleet which were equipped with the heavy muzzle-loader. Henceforward, all guns mounted are breech-loaders. An auxiliary armament—guns of moderate power and penetration, midway between weapons whose work it is to perforate armour and the smaller anti-torpedo-boat artillery makes its appearance upon the *Inflexible* type, though it is at first very weak. On board the *Inflexible* it was composed of eight 4-inch guns; on the *Ajax* of two 6-inch. On this latter ship, steel-faced or compound armour was adopted throughout.

The next battleships of the "écheloned" turret type, the *Edinburgh* and *Colossus*, launched in 1886, are an enormous advance. Not only are the guns twice as powerful, new-pattern breech-loaders, but the speed is high, 16.5 knots instead of the wretched 13.25 of the *Ajax* and 13.8 of the *Inflexible*. Instead of four 38-ton guns, there are four of 45-tons in the two turrets. Instead of the two 6-inch guns, there are five in the auxiliary battery. The armour is still eighteen inches thick, and still concentrated amidships, leaving the ends exposed. But the proportions are not so unfavourable to speed as in the *Ajax* or *Inflexible*; the length is greater and the beam less.

Contemporary with these two ships was the coast-service turret-ram *Conqueror*, which, launched in 1882, was followed six years later by the precisely similar *Hero*. They are useful little ships of 6200 tons, with a low freeboard forward. A little before the ship's centre is a single turret containing two 45-ton guns, with a good arc of fire ahead and on the broadside.

Astern of this is a high superstructure to house the crew, in which are mounted four 6-inch guns. The turret armour is twelve inches at its thickest, and very nearly all the water-line is protected; the belt is carried down to the ram. Both vessels are bad sea boats, and they are very weak in stern fire, but they are perhaps stronger ships than the *Ajax* or *Agamemnon*. In a sea-way their speed falls from its nominal fifteen-and-a-half knots to nearer eleven. They are not suited to work on open waters, from their low freeboard.

Whilst the *Colossus* was in hand, another and a very different type of battleship was designed and commenced. This was the first of the "Admiral" class, the *Collingwood*. In proportions and displacement she was closely similar to the *Colossus*, but the disposition of her armour and armament was widely contrasted. The central citadel, with its vicious concentration of armament, was abandoned, and the four heavy, 45-ton breechloaders were placed fore and aft in two lofty barbette towers, protected by thick armour. The guns were left very much exposed, but perfect shelter was given to the gunners, except from shells bursting underneath the barbettes. To load, the breeches of the gun are depressed till they come below the roof of the barbette, when the hydraulic apparatus, which is placed behind armour, drives the shot and powder into the gun. Forward and aft, the *Collingwood* has a moderate freeboard. Amidships is a superstructure carrying six 6-inch breechloaders behind one inch of steel plating. Against raking fire there are 6-inch bulkheads fore and aft. The auxiliary armament is thus about the same as that of the *Colossus*, but has a certain measure of protection given it, which, as we shall find, tends steadily to increase. With great prescience a good speed was assured to the ship, and on the measured mile, under forced draught, she accomplished 16·8 knots. Her armour extends for 140 feet in a narrow belt on the water line; the barbette towers are thickly plated, and from each of them a well-protected shaft runs down to the armour deck, which is placed below the water-line.* Strong bulkheads at each extremity of the water-line belt run athwartship, giving security against raking fire. There is a well-plated conning-tower and one mast with two fighting-tops. The thickness of the armour varies between eighteen and twelve inches, and it is compound.

This, then, was a ship which could steam fast and hit hard, which was, moreover, well adapted for fighting in line ahead. The end-on fire is not that of the *Colossus*, but the broadside fire is better, as the guns can be trained through a wider angle. Yet the unprotected ends are long, and the barbettes could be put out of action by lodging shells beneath them. Still the *Collingwood* was a great advance upon her predecessors, and in general outline differs little from our newest ships. She was completed in 1886, and has proved very satisfactory.

Five similar ships, the *Anson, Benbow, Camperdown, Howe,* and *Rodney* followed her. They were a little larger and a little faster, exhibiting that continuous progress which marks our battleships. In the disposition of their armour and armament there was no change. The 45-ton gun on four of them was replaced by the far more powerful 67-ton gun, which for ten years was to be the standard heavy gun of the fleet. The auxiliary and anti-torpedo-boat armament suffered no change. The weight of armour carried exceeded 2500 tons, whereas in the *Colossus* it had been 2360 tons, and in the *Ajax*, 2220. The *Benbow* differed from the other four considerably. Instead of the four 67-ton guns she was given two of 111-tons, one fore and the other aft, and her auxiliary armament grew from six 6-inch to ten 6-inch guns. The speed of these ships was seventeen knots, and they were faster than any French battleship of their day. Their defect was still the insufficient protection given to the water-line, and the absence of strong armour over the auxiliary battery.

* Except amidships where it is carried across between the upper edges of the belt. As in Fig. 2, Plate XXXVIII.

In these six vessels there was a visible tendency to abandon the development of bow fire, and to increase broadside fire. The next pair of battleships exhibit no such tendency. Indeed, in them bow fire attained its greatest proportions, though at the expense of stern fire. They may be described as greatly enlarged *Conquerors*. They retain the essential features of that ship: the single turret, the weak stern fire, the low freeboard forward, the high superstructure aft; but they carry a shorter and thicker armour-belt. The displacement rises from 6200 tons to 10,500 tons. Forward, in a single great turret, are two huge 110-ton guns, firing right ahead or on either broadside. The turret, unlike the barbettes of the "Admiral" class, has its base well protected by a large armoured redoubt. Astern is a single 29-ton gun, sheltered only by a shield. The superstructure contains twelve 6-inch guns, placed behind steel plating 3 inches thick, with a bulkhead of the same thickness across the stern of the battery, and one of 6 inches, protecting the guns from raking fire, forward. The thick armour-belt on the water-line is 162 feet long, and there are the usual English athwartship bulkheads. The redoubt is pear-shaped and contains the loading apparatus and the base of the turret. The conning-tower is heavily armoured. Forward and aft of the belt there is an armour-deck below the water-line 3 inches thick. The speed is seventeen knots or more. In these vessels for the first time triple-expansion engines appear on board a British ironclad. The machine-gun armament is unusually powerful, including twenty-four 6-pounder and 3-pounder quick-firers, and many smaller guns. These vessels were christened *Victoria* and *Sanspareil*, and were the last ships armed with the 110-ton gun. Completed in 1889 the *Victoria* same to an early and tragic end in 1893.

The appearance of high explosives and the growing potency of the quick-firer, which, though as yet it only projected shells of 6lbs., was still capable, in the opinion of many experts,

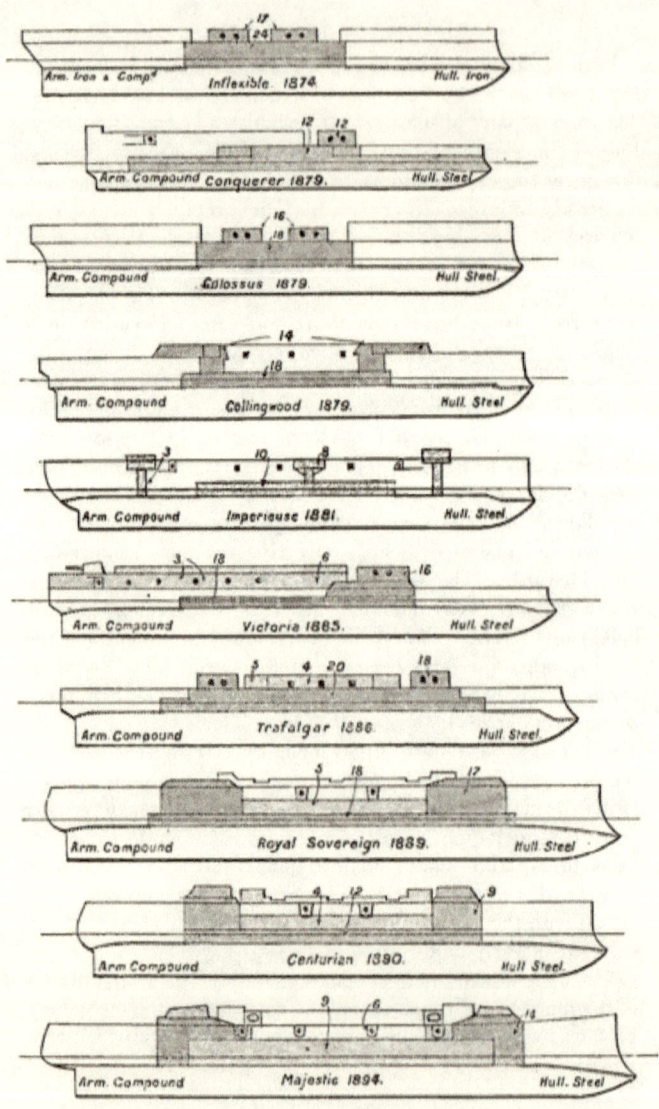

British Ironclads. 1874.-1894.
Figures give thickness of Armour in inches.

PLATE XXXIX.

of riddling the unarmoured sides and ends* of the "Admirals" and *Victorias*, and impairing their flotatory qualities, led to a distinct change in the *Nile* and *Trafalgar*, which followed. The general idea of their design was a *Devastation* improved and brought up to date, and there can be no doubt that the idea was realised, and that two remarkably formidable ships were the product. They show a great increase in size: length, beam, and displacement are all enlarged. The armour-belt on the water-line is 230 feet instead of 162 feet long, and rises to a great height above the water. It is no longer a narrow streak 3 feet above the surface, but ascends 10 feet to the level of the upper deck, thus forming, with the athwartship bulkheads—16 inches and 14 inches thick—at each end, a great citadel, encased with plating from 14 inches to 20 inches thick. At each end stands a turret, similarly protected, and containing two 67-ton guns. There are two steel armoured screens crossing the citadel, to prevent splinters from raking it. The unarmoured ends are comparatively insignificant, and their loss would not damage the ship's flotation. In the superstructure between the turrets are six sponsons, three on each side, in which are mounted six 4·7-inch quick-firers, behind 4-inch steel, a thickness sufficient to withstand machine-gun bullets, and projectiles from the small quick-firers. There are 5-inch bulkheads to protect the men from a raking fire. The weight of armour reaches the high figure of 3400 tons, exclusive of the deck, which would add another 1000 tons; and thus these two ships are, for their size, the best protected in the service. The speed is not so high as in their predecessors, but it is by no means inadequate, as sixteen-and-three-quarter knots were accomplished on the measured mile. In general appearance the two *Niles* recall the *Devastation*.

* The elevations of the *Collingwood* and *Victoria* (Plate XXXIX.) show the small extent of surface above the water-line protected by armour. The " Admirals," with their unarmoured ends riddled, sink 15 inches. *See* also page 174-5.

It was asserted in England that these two ships were the last ironclads that would be laid down, as the torpedo was, at the date of their commencement, growing in favour, and the torpedo-boat was threatening the large ship. But a year before their completion this prediction was falsified in the most singular manner, by the great Naval Defence Act of 1889. This provided for the construction of eight large battleships, the *Royal Sovereigns,* and two smaller ones, the *Centurions.* The *Royal Sovereigns* show a yet further growth of displacement, which becomes in them 14,150 tons.*
They are longer and broader than the two *Niles.* Their primary armament is still the 67-ton gun, carried in two pairs, forward and aft, but the secondary artillery attains an extraordinary importance. The six 4·7-inch quick-firers, whose collective discharge weighs 270lbs., give way to ten 6-inch quick-firers, discharging in one round from each gun 100olbs. weight of metal. The machine-gun armament has grown, whilst the coal supply is enormous. Seven† of the group are high-freeboard barbette-ships, and the eighth, the *Hood,* a moderate-freeboard turret-ship. But whilst the barbette reappears, it is no longer the barbette of the "Admirals," a shallow steel cylinder' standing upon the upper deck, and undefended below. The thick plating is in the *Royal Sovereigns* carried down to the armour-deck, forming two great redoubts at each end of the ship. The bulkheads below the turrets were somewhat thin, at their lower edge, in the *Nile*; here they are thick enough to give thorough protection. The water-line belt is 250 feet long, against 230 feet, but its maximum thickness is reduced from 20 inches to 18 inches, and its breadth amidships from 16½ feet to 8½ feet. As the guns of all ironclads are differentiated, some mounted to attack the enemy's armoured positions, some to wreck his

---

* The progress in size of the English battleship, is tabulated in Table XXII.

† *Royal Sovereign, Resolution, Revenge, Repulse, Royal Oak, Ramillies,* and *Empress of India.*

superstructures and defenceless sides, some to pour a hail upon port-holes and gunners, so in the *Royal Sovereign* the armour is differentiated. For $6\frac{1}{2}$ feet above the thick belt, between the heavy-gun barbettes fore and aft, is a thinner plating of 5 inches of steel, thus giving $9\frac{1}{2}$ feet of protected side above the water amidships. The thin plating of steel is in two thicknesses, an outer of 4 inches, and an inner of 1 inch. Behind this side are 10 feet of coal. At each end of the light citadel, which has a length of 145 feet, is a 3-inch bulkhead, rising to the upper deck.

The auxiliary armament is thus carried. Four 6-inch guns are on the main deck in armoured casemates, two on each side. On the front of these casemates the plating is 6 inches thick, and on the rear 2 inches. The other six guns are on the upper deck, protected only by shields. Five 6-inch guns fire on each broadside, two ahead, and two astern. The speed is high, ranging from eighteen knots in the *Royal Sovereign* to seventeen knots in the *Hood*. At sea, with natural draught, the *Royal Sovereign* steamed to Gibraltar, 1081 knots, in seventy hours, burning 487 tons of coal. This gives her a sea speed of 15·4 knots. She carries enough coal in her bunkers for three such trips.

Her barbette guns are, of course, much exposed. On the other hand, the men working them are most admirably protected. There is one very serious defect, however, in their mounting, since they require to be brought back to the fore and aft position before they can be loaded.* The conning-towers are two in number; on the forward one are 14 inches of armour, on the after one 3 inches. Numerous search-lights are provided, and there are two military masts. In sea-worthiness, comfort, armament, armour, speed, coal endurance,

---

* If the enemy is, for instance, to starboard, the heavy guns, after being fired, cannot be kept trained in his direction, when the only target is their muzzles, but they must be revolved back till they are parallel with the keel line, when they practically present their whole length as a target. Modern French ships, in most instances, have a central load, which allows the guns to be kept trained on the enemy.

and homogeneity of structure, there is nothing like this splendid group of ships outside our Navy. Other powers may have ships better in one or two points: the *Royal Sovereigns* are generally excellent, and combine power with grace of form. The gulf between them and the *Inflexible*, though it does not represent more than fifteen years' progress, is profound.

Little less excellent are the two *Centurions*.* With 4000 tons less displacement than the *Royal Sovereigns*, so much power cannot be expected. But they are admirable compromises, and contain all the elements of a good fighting ship. Their speed, as they were designed for service on foreign stations, is slightly higher than that of the *Royal Sovereigns*, reaching 18·5 knots. Their coal supply is the same. The heavy armament consists of four 29-ton guns mounted in pairs in barbettes fore and aft. The guns have the advantage of a very strong nickel-steel shield 6 inches thick, which revolves with the gun. Somewhat unwisely this shield has been left open at the rear, with the result that, if the ships found themselves engaged, with an enemy upon either beam, the heavy guns would be quickly out of action as the gunners would be exposed to every projectile.† The armour on the barbette is 9 inches as against the *Royal Sovereign's* 18 inches; on the belt 12 inches, where the *Royal Sovereign* has 18 inches. The belt is only 200 feet long, being thus 50 feet shorter than that of the *Royal Sovereign*; on the other hand, the heavy guns are carried two feet higher above the water-line. As in the larger battleships, there is a lightly armoured citadel above the thicker belt. The quick-firers, ten 4·7-inch guns, are carried, four on the main deck in casemates, and six on the upper deck. In general appearance, the *Centurions* closely resemble the *Royal Sovereigns*. They have two military masts and two funnels. Their freeboard is high, and they are good sea boats.

* The *Centurion* and *Barfleur*.

† The open rear of the barbettes must in this position always be turned towards one enemy when the guns are trained on the other.

The *Renown*, which was laid down in 1893, is a greatly developed *Centurion*. Her armour is rather more evenly distributed, and the differentiation of plating is beginning to disappear. She has a water-line belt of 8-inch Harveyed steel, with, above it, 6 inches of the same material, instead of the *Centurion's* 4 inches. Her two barbettes placed fore and aft, and each containing a pair of 29-ton guns, have 10 inches of plating upon them. There is a novel arrangement of the deck which greatly adds to the strength of the vessel amidships. In the *Centurion* and all other ironclads, the armour-deck runs across the ship from the upper extremities of the thick belt, only dipping below the water-line forward and aft. In the *Renown* it is curved, and arches down to the lower edge of the belt. Thus any projectile which perforates the side, has to encounter this further obstacle before it can do vital damage. The auxiliary armament is not only a very powerful, but also a very well-protected one. Instead of the *Centurion's* ten 4·7-inch guns, the *Renown* carries ten 6-inch guns, of course quick-firers, and all the ten are in casemates, on the faces of which are 6 inches of Harveyed steel. There are no large weapons exposed upon the upper deck. Four of the casemates are located upon the upper, and six upon the main deck. The light armament is also extremely powerful, including eight 12-pounder quick-firers. The speed is to be eighteen knots, and the draught is such as to enable her, like the *Centurions*, to pass through the Suez Canal. She is a far more formidable ship than the *Centurion*, and might even venture to face the *Royal Sovereign*, but her displacement is 12,350 tons instead of 10,500, as in the *Centurion*.

The nine huge battleships of the *Majestic* class* are rather developed *Renowns* than *Royal Sovereigns*. The water-line protection consists of a belt of Harveyed steel 9 inches thick and 16 feet deep, extending for 220 feet amidships. Bulkheads, with an outward curve at either end, enclose a great

---

* Their names are: the *Majestic, Magnificent, Cæsar, Hannibal, Illustrious, Jupiter, Mars, Prince George,* and *Victorious.*

citadel a little over 300 feet long. With their bases protected by this citadel, stand at each end of it two barbettes, containing each two 46-ton guns. The barbettes carry 14-inch armour. On the main deck are eight casemates, each containing one 6-inch quick-firer, and on the upper deck are four, similarly armed. Each casemate is protected on its outer face by 6-inch Harveyed steel. As in the *Renown*, the deck springs from the lower edge of the side armour. The speed is to be that of the *Royal Sovereigns*, but we may hope that, on trial, the anticipated rate may be exceeded.* Certainly the only weakness of our new battleships seems to lie in their comparatively low boiler and engine power. The length is 390 feet, and thus the *Majesties* are the largest battleships afloat in our service. The displacement is 15,000 tons, and no less than 1850 tons of coal will be carried. The weight of armour, excluding the protective deck, is 2800 tons, or a little less than is carried by the *Royal Sovereigns*, but the improved quality of the mail gives far greater security for less weight. The minor armament includes sixteen 12-pounder quick-firers. In the design the effect of the quick-firer is plain, since the main effort of the architect has been to render the side impervious to the 6-inch and 4·7-inch shell, whilst abandoning all effort to make it proof to the heaviest projectiles.

We may next consider in general outline our thirty-five years of progress in battleship construction. And first, whilst English ironclads, as a rule, show a steady advance, each being better than its predecessor, and closely related to it, there are yet certain types which appear from time to time and die out, because they prove to be unfitted for the conditions of war, or because the advance of naval opinion discards them. The descendants of the *Royal Sovereign*, the first English turret-ship, display a great mortality. Four varying types appear from time to time, and three of these may be said to have died an early death: the low-freeboard masted turret-

---

* The *Magnificent* has done 17·6 knots, and the *Majestic* 17·8 knots.

ship, which expired with the *Captain*; the single-turret-ship, which disappears with the *Victoria*, and the "écheloned" turret-ship, of which the *Colossus* is the last example. The mastless turret-ship of the *Devastation* type, on the other hand, has a singular vitality. After the *Dreadnought*, launched in 1875, it does not appear for twelve years, when just as it might appear defunct, it turns up again in the *Nile* and *Trafalgar*, and from them hands on some of its features to the later *Majestic* class. The battleships of the present day may thus be said to be sprung from two distinct ancestors, the *Warrior* and the *Royal Sovereign*.* The extensive side-plating, the high freeboard, the great length of the latter-day *Royal Sovereign* recall the *Warrior*, whilst the barbette system of mounting guns is a variation upon Captain Coles' turret.

The tendency towards high freeboard and good speed in our recent battleships is well marked. Both these features characterised our older ships, but the rage for impregnability drove them out. We had to learn that, admirable though our mastless monitors of the *Devastation* type were, as fighting-machines, a ship has to do other things besides fighting. She must be fairly comfortable, if her crew are to retain their health, and without health the sailor must necessarily lose a great deal of his nerve. High-freeboard ships, in which the crews need not be battened down in a moderate sea, become essential, when it is desired to maintain in good physical condition the men who have to fight and work the ships. Nor is this the only gain which a high freeboard gives. When the guns are placed very low, in rough weather the waves may cut off the enemy's hull from sight, and seas breaking over the forward part of the ship may bury the forward turret or barbette in spray and foam. On the other hand, the low-freeboard ship is a small target to hit, a fact which was remarked by the *Shah's* gunners when they faced the

---

* The older turret-ship of 1864.

Huascar. Still the advantages of a high freeboard are greater than the disadvantages, and the high freeboard appears to have come to stay. The last low-freeboard vessels in our fleet were the two *Niles* and the single turret-ship of the eight *Royal Sovereigns*, the *Hood.*

The early ironclads had to face guns of comparatively feeble power. The round shot of the 68-pounder would not perforate the *Warrior's* plating, even upon the proving ground. The subsequent advance in the thickness and resisting power of armour was entirely due to the advance of artillery. Successively 4½ inches of wrought iron gave place to 6 inches and 9 inches. But the gun kept pace with this progress, and before a ship had left the stocks had usually rendered it, in a sense, obsolete. The first armour was uniform in thickness. Then, as the attack grew more formidable, and the weight of iron required to give protection heavier, a greater thickness was given to the vital parts of the ship. None of our older vessels were really "ironclad," and it goes without saying that none of our modern battleships are such. The low-freeboard turret-ships, of the period 1870-5, are the only vessels to which the term can in strict correctness be applied. It was found necessary with high freeboard ships to denude a great portion of the side of armour, in order to increase the thickness over the battery and vitals. This denudation reached its extreme limits in the *Inflexible* and the "Admirals," where there are absurdly small patches of plating, and where by far the greater portion of the side is open to the smallest shell. It was the aim of naval architects to keep out the heaviest projectiles from certain portions of the ship, but in their effort to ensure this they went too far. The secondary armament of the French ships was formidable even in the period 1875—1885, and would have wrought terrible havoc on the unprotected upper works and water-line, forward and aft. Quite possibly the *Inflexible* or *Colossus* would have been put out of action without their thick armour being struck. Sounder counsels appear in the *Royal Sovereigns*, where there is a combination of thick and thin armour. The purpose of the thin armour is to exclude high explosives which have a devastating effect upon the ship's interior. But even in the *Royal Sovereign* there is very much ill defended, the loss of which might cause the vessel harm. In the *Majestics* thin side-armour, except over the auxiliary guns, disappears, and there is one moderate thickness.

But why, it may be asked, is not the ship made invulnerable? Of course, it could be done, but not probably with a displacement smaller than that of the *Great Eastern*. The danger of destruction by the ram or torpedo forbids such a monster, and as the ship has after all to fight other ships, she will find them, if they are of her date and if she is well planned, as vulnerable as herself. If she cannot resist every projectile, neither can they; if her upper works can be riddled, so can theirs; if she is open to the deadly assault of ram and torpedo, so also are they. The naval architect's business is, given a certain displacement, to effect the best compromise between the warring factors, speed, stability, power to wound, invulnerability, coal-endurance; or given the required degree in which these factors must be present, to produce them on the lowest displacement. A battleship B, of 15,000 tons, may look no better than one C, of 10,000 tons, to the casual eye, but as constructors are not born idiots there is something somewhere which will give the larger ship the advantage. It may be a heavier and stronger hull, which will wear better and stand the tremendous concussion of the guns longer; it may be a surplus of ammunition or coal. We may quarrel with the constructor for giving insufficient attention to one factor or another in his compromise, but we can rarely say that with a higher displacement he has produced an inferior ship.

The growth in displacement is one of the most striking features of our more recent battleships. Seeking perfection, striving to improve each type before it has entered upon service, we have been driven to greater and greater size. It is inevitable, but it is perhaps regrettable, as numbers are shown by the history of the past to be a more decisive factor than the size of individual ships. There must be a limit to this increase of size, and we may have reached it. Overgrown ships are not less objectionable than overgrown guns and overthick armour.

The system of mounting guns has changed widely since the days of the *Warrior*. We began with the broadside battery, in which the guns were ranged side by side bearing on the beam, with no effort to obtain axial fire. We advanced from this to the central battery, in which the guns were larger and fewer, concentrated in a small box amidships with a varying amount of axial fire. The turret was the next improvement, and this in turn begot the barbette, the difference between the two lying in the fact that in the turret the armoured wall revolves with the gun, whilst in the barbette the armoured wall is fixed and the gun revolves inside it upon a turntable. Our latest ships exhibit a combination of broadside and barbette or turret mounting, as the heavy guns are placed fore and aft in barbettes or turrets, whilst the medium weapons are disposed on the broadside.

If we compare the latest English type of battleship, the *Majestic*, with similar French, German, and American ships as the *Charlemagne*, *Brandenburg*, and *Iowa*, we shall find that foreign architects have apparently produced ships as good as ours on a smaller displacement. But, as we have said above, this can be easily explained. The English, French, and American ships all agree in the disposition of the heavy guns, which are mounted in pairs fore and aft. The Germans have preferred three pairs of heavy guns, all placed on the keel-line. The *Majestic's* guns are 27 feet above the water-line; the *Charlemagne's* forward pair nearly 29 feet, and her stern pair 21 feet, the *Iowa's* are nearly 18 feet. The forward pair of guns in the *Brandenburg* have a good command, but the other two pairs are mounted low. The French and Americans have preferred the turret, whilst the

THE DEVELOPMENT OF THE ENGLISH BATTLESHIP. 243

English and Germans adhere to the hooded barbette.* In the German ship the barbettes are unprotected underneath; in the other three, the armour runs down to the deck. The auxiliary armament differs widely in the four vessels. The *Majestic* has twelve 6-inch quick-firers in as many armoured casemates, each protected by six inches of steel. The French ship has ten 66-pounder 5·5-inch quick-firers, eight of which are mounted behind 3-inch armour.† The *Iowa* has eight 8-inch guns carried in pairs, in four turrets armoured with 7 and 8 inches of steel. She also carries six 4-inch quick-firers behind thin armour. The *Brandenburg* carries a very feeble battery of quick-firers, as she has only six 30-pounder and eight 20-pounder Krupps of this pattern. The four, however, agree curiously in the weight of the heavy gun adopted as the primary armament. The English ship carries the 46-ton gun; the French the 48-ton gun; the German the 42-ton gun; and the American the 45-ton gun. The weight of the English broadside, from guns above the 20-pounder, is 4000lbs; of the French, 3293lbs.; of the German, 4730lbs.; and of the American, 4532lbs; but the English ship is superior to any in the number of large quick-firers carried, and would in a given time discharge as great a weight of metal as any of the other three. In gun-power the English, German, and American ships are about equal, and the French a little inferior.

For protection, the *Majestic* has a broad but incomplete belt of 9 inches uniform thickness; the *Iowa*, a narrow incomplete belt of 14 inches maximum thickness with a strake of 5-inch armour above it; the *Charlemagne*, a narrow end-to-end belt, which tapers from 16 inches to 9 inches, and above it again 3¼-inch armour; the *Brandenburg*, a narrow end-to-end belt 16 inches to 12 inches thick. The armour upon the heavy gun positions is 12-inch in the *Brandenburg*, 14-inch in the

---

* The English hooded barbette, however, differs very little from the French and American turret, except that the armour is thinner.

† She has also six 3·9-inch quick-firers, which are mounted on her superstructure.

Majestic, 15-inch in the *Iowa*, and 16-inch in the *Charlemagne*. The latter ship, the *Iowa*, and the *Brandenburg* expose a considerable extent of side below their quick-firers, which might, on being riddled, render the quick-firers unworkable. The American and German ship have each one armoured position from which the ship can be fought; the *Majestic* two, and the *Charlemagne* three. The *Iowa* has the lower freeboard, and would be at a great disadvantage in a sea-way; the other three ships rise well out of the water, though the German is inferior in freeboard aft to the English and French.

As at present designed, the *Charlemagne* has two heavy military masts with two tops and an officer's position on each; the *Majestic*, two of much lighter type, each with two tops; the *Brandenburg*, two light masts with one top for guns, and one position for officers on each; on the *Iowa*, the cumbrous military mast vanishes altogether.*

In speed, the *Charlemagne* is expected to cover eighteen knots on the measured mile; the *Majestic*, seventeen-and-a-half; the *Iowa*, seventeen; and the *Brandenburg*, sixteen-and-a-half. In coal endurance, the *Iowa* and *Majestic* are about equal, with 2000 tons and 1850 tons respectively; the *Charlemagne* comes third with 1100 tons, a very big drop from the *Majestic*; and the *Brandenburg* last. The English ship could keep the sea for a month, steaming continuously at ten knots; the *Iowa*, five weeks; the *Charlemagne*, eighteen days; and the *Brandenburg*, twelve or fourteen. In practice, however, as a large reserve must be maintained, this time should be reduced by at least a quarter.

The deck of the *Charlemagne* is double, $3\frac{1}{2}$-inch above and $1\frac{1}{2}$-inch below, over the machinery. The *Majestic's* deck is single, 4-inch on the slopes and a little less on the flat, but, as has been said, it springs from the lower edge of the armour-belt, and thus has the full advantage of the protection which the latter gives. In the *Iowa*, the greatest thickness is 3 inches, and in the *Brandenburg*, $2\frac{1}{2}$ inches.

* There is a short steel tower, with a top a few feet above it.

In displacement, the English ship is largest, as her tonnage is 15,000; the *Iowa* follows with 11,500 tons; then the *Charlemagne*, with 11,200; and last comes the *Brandenburg*, with 9850. Of the four ships, the *Brandenburg* is an older design than the other three, and therefore lacks the extensive side-protection which is given in them. In offensive qualities the English and American ships would seem to excel, but as far as armament goes there is not much to choose. If the *Iowa* mounts 8-inch quick-firers, she will be able, at close quarters, to bring eight guns, capable of piercing thick armour, to bear on either broadside, and six ahead or astern.

The development of the ironclad has proceeded side by side with the development of artillery. At the date of the introduction of armour, during the Crimean War, the guns in existence were of very moderate power. Artillery had made no great advance, and for all practical purposes there was no difference between the guns of 1856 and those of 1806, except that the former fired shell and were slightly larger. The carronade of 68lbs., which was the heaviest weapon of Nelson's day, had given way to the 68-pounder long gun, a smooth bore of 8 inches calibre, and 4 tons 15 cwt. weight. This gave a muzzle velocity of 1579 feet per second to its round shell, whilst the total energy of the projectile was only 452 foot tons.* The gun was of cast iron, mounted upon a clumsy wooden carriage, which rendered accurate shooting very difficult.

The rifled gun had already been tried in the Crimean War, in the embryo form of the elliptical Lancaster gun, which gave its shot a twist, but it was in the Armstrong form that it entered the British service. The first Armstrong gun, completed in 1858, had a forged steel barrel, and was built up of a number of wrought-iron cylinders fitted closely over this by shrinkage. Thus, at one bound, the distinctive features of our newest guns, the use of steel, and the building up of the gun from a

* At 1000 yards.

number of parts, instead of casting it whole, were anticipated. The projectiles for this weapon were of 1·8 inches diameter, at first entirely of lead, and then of iron, lead-coated to take the grooves of the rifling. The gun was a breech-loader, and the breech action consisted of a plug which, working in a slot in the breech, was held tight by a hollow screw, which fitted inside the bore.* To prevent the escape of gas the breech-block had a copper bush, and the gun a copper face, which, by slightly yielding when the breech screw was tightened, hermetically sealed the gun.

These early breech-loaders were before their day. Accidents occurred with them, and men, who were not accustomed to complicated weapons, preferred the apparent simplicity of the muzzle-loader. The breech-loading system had been successful with guns of 110lbs. and 40lbs. shot, and was being widely adopted abroad, but still, at that time, the arguments for it were not as overwhelming as they have since become. The first breech-loader had only a short life, from 1860 to 1865. From the latter year to 1880, for a period of fifteen years, the muzzle-loader reigned in our fleet.

The resistance of the first ironclads to artillery was surprising. At Lissa, where two large fleets met, on neither side was the armour perforated. In America the *Monitor* and *Merrimac* cannonaded one another for the whole of a morning without anyone being much the worse for it. The monitors before Charlestown in 1863-4 received a great number of hits, yet their efficiency was little impaired. The *Montauk* was struck 214 times; the *Ironsides*, a broadside vessel, 193; the *Weehawken*, 187; the *Patapsco*, 144; the *Passaic* 134; the *Catskill*, 106; the *Nahant*, 105; the *Nantucket*, 104; the *Lehigh*, 36. Yet the gunners were not long before they re-established the ascendancy of their weapon. In quick succession rifled guns followed the feeble smooth-bores; the rifles grew heavier and heavier, whilst armour, to resist the

---

* *See* Fig. 1, Plate XI.

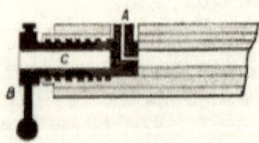 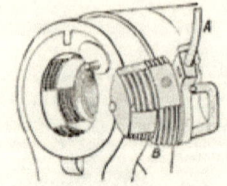

Armstrong Breechloader of 1859.
A. Breech plug. B Screw for tightening
or loosening A. C Hollow in B to
introduce charge when A is removed.
Movable parts black.

Armstrong Breechloader of 1889.
A Lever to release B by a fifth turn
B Breech plug with interrupted screw
Movable part shaded

PLATE XL. ENGLISH BREECHLOADERS, 1859 AND 1889.

increasing vehemence of the attack became thicker and thicker. In England the 110-pounder screw breech-loader was followed by muzzle-loading 97-pounders, 112-pounders, 174-pounders, 250-pounders, 400-pounders, 550-pounders, and so forth. The perforating power of the gun, which in the first muzzle-loaders was very low, rose till the 35-ton gun could pierce 16 inches of wrought iron at the muzzle. But with this increase in power came a great increase in weight, till instead of three or four tons the gun reached 35 tons. Such guns could not be mounted on the ordinary wooden carriage, but required excellent mechanical contrivances, if they were to be handled at sea, and such contrivances were quickly forthcoming. In 1864 iron had supplanted wood in gun-carriages; slides and traversing platforms appeared, and the recoil was controlled not by ropes or breechings, but by compressors which exerted great friction. Minor improvements were introduced as time went on, but the 35-ton gun was, in muzzle-loaders, the largest hand-worked weapon.

As the development of armour and artillery proceeded side by side, a new phase of evolution appeared, the phase of the monster machine-worked gun. The 12-inch 25-ton muzzle-loader gave place to the 35-ton, which could only be man-handled with difficulty. In the *Devastation* the gun was worked by hand, but the turret was rotated by steam; in the *Thunderer* the after-turret had man-handled 35-ton guns, the fore-turret hydraulic-handled 38-ton weapons. The hydraulic system of working guns proved, after extensive trial, to be altogether satisfactory, and was henceforward adopted for all British heavy guns, till the time when it gave way in its turn to electricity. With the appearance of hydraulics, the compressor, as a means of arresting recoil, yielded before the hydraulic brake, when the strain upon the mounting was lessened, and the power of the recoil used to run out the gun after the discharge. Hydraulic mountings were adopted in the French navy on board the *Dévastation*, and as in England, appear in every subsequent battleship, till the advent of electricity.

But the progress of the gun did not stop when the 12-inch 35-ton and the 12¼-inch 38-ton muzzle-loaders appeared. They were succeeded by the last of the muzzle-loaders, the 81-ton of 1875, and the Armstrong 100-ton of 1876. These were huge and unwieldy weapons, far longer in proportion than muzzle-loaders had been hitherto, but the craze between 1875 and 1885 was for heavy guns which should deal crushing blows, and abroad ordnance was increasing in length, with the substitution of slow-burning powders for the older quick-burning powder. The difference between these two kinds of powder may be roughly expressed as this: the quick-burning powder started the projectile with a violent jerk, and all its force was exerted in the neighbourhood of the breech, thus necessitating a very great thickness of metal round the breech, and rendering a long gun unnecessary; the slow-burning powder, on the other hand, started the projectile more gently, straining the breech less, and gradually increased its push, rendering a long gun, fairly strong in the chase as well as the breech, essential. A muzzle-loader of any length must necessarily be difficult to load, as it has to be run in till the muzzle is inboard. Thus, here was one cause which would compel England to adopt the breech-loader, unless she was ready to be distanced in ballistics by foreign competitors.

We clung, however, to our muzzle-loaders, which, though good enough in the sixties and early seventies, were now being beaten abroad, and not till a lamentable accident on board the *Thunderer*, when on January 2nd, 1879, a 38-ton muzzle-loader gun burst, through double loading, killing two officers and eight men, did we begin to waver. Such an accident could not occur with breech-loaders. In 1878 the Armstrong Company had turned out a 6-inch and an 8-inch breech-loader, using slow-burning powder. In practice these guns behaved admirably. They gave muzzle velocities of 2000 feet per second, and were vastly better for their size than any muzzle-loader. In 1880, the English Government decided to return to the breech-loader, and the muzzle-loader was for ever abandoned.

The breech action of the new guns differed considerably from the earlier and cruder type. The interrupted screw gearing into the breech of the gun was adopted.* Originally of American invention, it was the system employed in France, and is at once simple and safe. To enable the plug to be quickly withdrawn, the thread is not continuous either in the breech-plug or the gun, but is in each cut away at intervals. Thus, one quarter or one-fifth turn, according to the number of such intervals, will disengage the plug, when it is withdrawn and swung clear. To give a gas-tight joint, at first a steel cup with its concave side towards the charge, so that the explosion would press it against the breech-plug and rear of the gun, was employed, and later, a pad of asbestos encircling a "mushroom stalk" of steel. The "mushroom head" is driven back by the explosion towards the breech, compressing the pad and driving it outwards against the bore of the gun. A thoroughly tight joint is thus secured.

With breech-loaders, there was at first the same craze for monster guns. The 8-inch gun of 14 tons was followed by a 12-inch of 45-tons, a $13\frac{1}{2}$-inch of 68 tons, and a $16\frac{1}{4}$-inch of 110 tons. But with the latter was reached the limit in size, and efforts were now made to increase the power of guns in other directions, without sensibly adding to their weight. The great object of the heavy gun is to fire as many projectiles as possible in a given time, and to fire them through as great a thickness of plating as possible. By constant improvements in the mountings, the rapidity of fire of heavy guns has been raised till it stands as follows † :

Gun	Handling	Rate		Weight of metal fired per minute
110-ton gun	Machine-handled	3 shots in 6 minutes starting loaded	900 lbs.	
67 "	" "	7 shots in 12 minutes, 6 hits on target	700 lbs.	
45 "	" "	4 shots in 6 minutes	470 lbs.	
46 "	" "	1 shot in $1\frac{1}{2}$ minutes (estimated)	560 lbs.	
29 "	Man-handled	1 shot in 2 minutes 10 seconds	230 lbs.	
22 "	" "	5 shots in 6 minutes	340 lbs.	
14 "	" "	1 shot in 1 minute	210 lbs.	

The gain in rapidity, which results from employing machinery, is sufficiently obvious. In perforating power,

---

* See Fig. 2, Plate XL. The illustration of the 8-inch quick-firer also shows the present system of breech-action well. Page 250.

† With one exception these are all actual performances at sea.

great progress has been made by increasing the length of the gun. Our older muzzle-loaders were of 12 or 13 calibres; our early breech-loaders of 25 calibres*, whilst in our newest heavy gun, the 12-inch 46-ton, we have gone to 38 calibres. The French have advanced even further, and are actually mounting great guns of 50 calibres. The gain through increasing the length, and thus giving time for a slow-burning powder to exert its full power, is evident from these figures of Canet, which show the perforation with guns of the same calibre, but of varying length:

375-pounder, 9·45-inch gun of—

	25 cals.	30 cals.	36 cals.	43 cals.	50 ca's. in length.
Weight of gun	14 tons	19 tons	22 t ns	30 tons	34½ tons
,, charge	127lbs.	161lbs.	198lbs.	231lbs.	266lbs.
Inches of wrought iron perforated at 1000 yards	14·6	18·1	21·5	23·9	27·0
Muzzle velocity, ft. secs.	1772	2001	2231	2428	2624

By this expedient, with a gun of quite moderate size and proportions, enormous power can be obtained. Instead of guns growing larger in calibre, they are now tending to grow smaller with increased length, increased strength, increased charges, and increased muzzle velocities.

Whilst heavy guns have thus progressed, there has been not less striking advance in small and moderate-sized weapons. The appearance of the torpedo-boat, which could not be readily followed in its rapid motion by a heavy gun, necessitated a light quick-firing weapon which should get off so many shots during the period of the boat's approach, as to make certain of hitting her. The Hotchkiss and Nordenfelt, 3 and 6-pounders, were the result of this demand. They were placed upon mountings which gave no recoil, and were able to fire ten to twenty aimed shots a minute. It was not long before heavy guns followed upon the same road. In 1886 appeared the Armstrong 30-pounder quick-firer, which, after winning universal approval, became the 4·7-inch 45-pounder

---

* A gun of 25 calibres, is a gun the length of which is twenty-five times the diameter of the bore. Thus a 12-inch gun of 25 calibres is 25 feet long in the bore, i.e., 25 times 12 inches.

8-INCH QUICK-FIRER AND MOUNTING.

quick-firer. This gun can in a given time fire five times as many rounds as a breech-loader of similar size, but older pattern. It is marked by a simplicity and strength which are essential in war material. Next, a year or two later, came the 6-inch quick-firer, and last, and latest of all, the 8-inch gun of the same type, in which the breech opens automatically by the force of the gun's recoil.* It is probable that the gun of the future, whether large or small, will be wholly and entirely automatic.

In the so-called quick-firers, rapidity of fire is obtained, firstly, by the simplicity and rapidity of the breech action; secondly, by the use of recoilless sights, which do not necessitate re-laying after each shot, and enable the gunner to aim his gun whilst the weapon is being loaded; thirdly, by the use of a cartridge case,† which abolishes the necessity of sponging out the gun and is easy to manipulate; fourthly, by the introduction of a mount, in which friction is all but abolished, and recoil controlled and used to bring the gun back to firing position.

It is difficult to see how the rate of fire can be further increased, unless automatic guns of the Maxim type are introduced; otherwise we have neared finality. Without a mass of machinery it is impossible to fire more than five or six shots in the minute from a 6-inch gun, but it is satisfactory to note that England, whether in moderate artillery or in heavy guns, is leading the world in 1895 as she led it in 1865. The English 12-inch gun of 46 tons is superior to any weapon of equal size whether in perforation or rapidity of fire. Unfortunately we have still a large number of old pattern muzzle-loaders afloat, which reduce the average of our artillery.

* *Vide* Plate XLI., which shows the gun with breech open. The breech can, if necessary, be opened by hand. The large shield gives good protection to the gunners; and the ammunition comes up an armoured hoist to a door in the gun-pivot. One shot in fifteen seconds is the greatest rapidity of aimed fire. Even if this be halved, the gun fires 560lb. weight of metal a minute, or 750lb. starting loaded.

† The Whitworth breech-loader of 1860 appears to have been the first gun to use fixed ammunition. The powder was contained in a tin cartridge.

In projectiles and powder there has been a great progress, as in guns and armour. The old cast and wrought-iron round shot have given way, first to elongated Palliser projectiles, of iron with chilled heads, then to forged steel shot, and finally to nickel steel for armour-piercing purposes. The old lead covering which took the rifling was replaced, first by studs fitting into grooves in the bore, then in projectiles for the breech-loader, by copper rings which are cut into by ridges in the bore of the gun. To enable shot to perforate the new and very hard armour, produced by the Harvey process, the point of the shot, which is liable to fracture upon impact, is capped with soft iron, which enables the projectile to bite. Gunpowder is no longer the only explosive agent with which shells are charged. High explosives have been discovered which are infinitely more powerful. Melinite, a preparation of picric acid, cordite, gun-cotton, nitro-benzols, even dynamite, have been tried, with many others. Experiments with these terrible substances have been conducted in secret by most European states, but certain details have leaked out. The effect of their explosion is terrific, and some at least of them produce dense choking fumes which will suffocate those who are not blown to pieces.

Gunpowder which is a stable and trustworthy substance, but has the disadvantage of producing thick smoke when fired in the gun, is giving way to smokeless powders. In the English service the quick-firers of various size use cordite ammunition, which makes next to no smoke, and the new 46-ton gun is constructed to burn cordite. The time is at hand when no gun will fire gunpowder. A ship wreathed in smoke could be attacked with comparative impunity by the torpedo-boat, but the case is very different when smokeless ammunition is used. Thus the tactical effect of the introduction of the new powders has been to handicap the torpedo-boat, and to make the issue of naval battles less a matter of chance. Is progress the elimination of chance?

The quality of armour has also improved, whilst the gun and its accessories have been improving. The Kinburn batteries had plating of a very inferior quality, as the metallurgical science of the day was not sufficient to avoid burned metal and layers of scoria. The grade of the iron used was poor, whilst steel could not be produced in a trustworthy form at a moderate price. Gradually the iron improved; the mechanical agencies for preparing and rolling it were perfected; and its resistance steadily rose, till, with advances in the art of preparing steel, it gave way to steel. Yet, at first, solid steel armour did not win approval, and it was thought better to face soft iron with hard steel. This gave the compound plate which was first adopted by England for the *Inflexible's* turret. Italy and France preferred solid steel, which, though more brittle, had greater power of resistance. In the United States, nickel-steel was adopted for the new battleships of the 1891 programme, and was employed in England for the decks of the *Royal Sovereigns,* and the thin plating on the side of the *Centurions.* The nickel fills up the pores of the steel and gives great homogeneity and toughness to the mass. Tungsten is said to give even more satisfactory results. Last came the Harvey process of hardening steel or nickel-steel plates. By this, the steel, after being rolled to the required thickness, is heated, face downwards on a bed of charcoal for a fortnight. This done, it is bent to shape, heated again, and hardened again by the application of water. In the finished state, the surface is so hard that drills will not bite upon it, and special arrangements, whilst hardening, are necessary to leave soft places for rivet holes. The first English battleships in which it appears are the *Majesties* and the *Renown.* A plate thus hardened, 18-inches thick, has in the United States defied the attack of the 13-inch 66-ton gun. The projectile, weighing 1100lbs., and striking with a force sufficient to lift a weight of 1000 tons twenty-five feet, crushed in the backing of oak, but only dented the plate.

Various improvements in the method of applying the armour and building up the ship's side, have also to be chronicled. The *Warrior* carried her 4½-inch plates upon a cushion of oak from 10 to 18 inches thick. She had not a strong iron inner skin. But as Mr. Chalmers demonstrated that the application of iron stringers, placed horizontally between the timbers of the backing, and a thick skin inboard, behind the backing, gave better power of resistance, this system was adopted. In the *Bellerophon* were 6 inches of iron upon 10 inches of oak, with three thicknesses of ⅝-inch plate as the inner skin. The framing of this ship was far stronger than that of her predecessors. In later ships, steel has replaced iron in the ship's structure, teak has replaced oak in the backing, and many minor improvements have been introduced.

As to engineering progress, the *Penelope* was the first British ironclad to be fitted with twin screws; the *Alexandra* the first to carry compound engines; and the *Victoria* the first to carry triple-expansion engines. Forced draught, which consists in pumping air into the furnaces from below, appears in the "Admirals"; induced draught, which consists in sucking air through the furnaces from above, in the *Magnificent*. The water-tube boiler was introduced on board the French battleship *Brennus*. No English ironclad has as yet been fitted with it, but the great cruisers *Terrible* and *Powerful* have it.

Turning now to the cruiser, the progress has been immense. The cruiser as a distinct conception is the descendant of the frigate, and does not appear in the earliest days of ironclads. During the Civil War, the North, to protect its commerce, laid down a class of wooden vessels, of great length, the *Wampanoags*, which were to have a speed of seventeen knots, and a coal endurance of 5600 knots. As a matter of fact, they never did more than fifteen knots, and could hardly be considered a success. In 1866, before they were completed, the English Government replied to them with the *Inconstant*, and some years later with the *Shah* and *Raleigh*. These were noble ships, fully rigged, comfortable, and fast for their day, as the first two did sixteen-and-a-half knots on the measured mile, but they were entirely destitute of protection other than that which was afforded by the arrangement of their coal bunkers. They were followed by the *Bacchante* class, launched in 1875-7, and the *Active* class launched in 1869, which were slower and smaller. The "C class" launched in 1878-1881, were still smaller, but embodied one new and interesting feature which henceforward appears in all large cruisers—the armour-deck. The *Canada* and her sisters carried a $1\frac{1}{2}$-inch steel deck over engines, boilers, and magazines at a level of three feet below the water-line. These ships were, however, too slow for cruising purposes, as their speed was not sufficient to enable them to escape the battle-ship, and they were far too weak to encounter her when they were overtaken.

Between 1877-1880, were completed three large cruisers, the *Shannon*, *Nelson*, and *Northampton* which had partial belts on the water-line and athwartship bulkheads. They may be described as dismal failures—large, costly, slow, and vulnerable. 1877-8, however, saw the launch of two fast and lightly armed cruisers, which, built expressly for the purpose of scouting, have done good service with the fleet, the *Iris* and *Mercury*. In 1883 were launched four fine vessels of the *Amphion* class, with a speed of seventeen knots.

In 1884, an immense step forward was taken. That year, the Armstrong company launched the famous *Esmeralda*, a vessel which on a very small displacement, carried a tremendous armament. Her speed on the mile reached the figure of 18·28 knots, phenomenal at that time. She had an end-to-end steel deck 1 inch thick, curving up amidships. In her, forced draught was employed, air being pumped into the stokehold and driven into the furnaces, thus greatly increasing the rate of combustion. The influence of her design is visible in at least three classes of English ships, or indeed, in every cruiser we have built since her day. The four second-class cruisers of the *Mersey* type, launched in 1885-6, are an improvement upon her. They are larger, have far stronger end-to-end decks, and an armament in which, whilst the heavy guns are not so large, the auxiliary guns are much more numerous. In 1885, too, were commenced seven "belted cruisers," which might again be called greatly enlarged and improved *Esmeraldas*. They had been preceded a year or two, by the two *Imperieuses* which are French in type and design, and approach the battleship more closely than the cruiser. The new belted cruisers of the *Aurora* type were of 5600 tons, and were more heavily armed than the *Merseys*. They carried amidships, a belt of 10-inch armour, and were protected by bulkheads against raking fire. A third class of cruisers, designed more especially to combat the torpedo-boat, appeared in the *Archer* and her sisters—small ships heavily armed.

Cruisers having diverged and developed as a class apart from the battleship, now begin to subdivide, and to develop classes amongst themselves. In first-class cruisers designed for ocean work, with a powerful armament to fight all comers, the *Blake* succeeds the "belted cruiser." She is far larger, far faster, and is heavily armed, whilst some of her guns are behind thin armour. The only protection on the water-line is a stout armour deck. She and her sister *Blenheim* are followed by the nine *Edgars*, a little smaller and slower, but none the less splendid ships. Then in 1894, follow the two largest unarmoured cruisers which have ever been laid down, the *Powerful* and *Terrible*, each of 14,200 tons, with water-tube boilers, phenomenal speed, and armament almost wholly behind armour. Following these again in 1895, are four rather smaller vessels of 11,000 tons, with an armament wholly quick-firing.*

In other navies, the "belted cruiser" has persisted and developed, and there are signs that we shall recur to it.

---

* For tabulated details of the leading English cruisers *see* Table XXIII.

Russia has in hand, or completed, three huge vessels of this type, whilst France has launched six very remarkable ships almost wholly covered with thin armour. These are the *Dupuy de Lôme* and her daughters; and their thin mail might render them awkward antagonists to our unbelted ships.

In second-class cruisers we dropped a little in size, with a reduction in armament, from the *Merseys* to the *Medeas*, which are now rated third-class. But from the date of the *Medeas* begins a steady rise in displacement, armament, freeboard, and coal endurance. The *Apollo* class of 1889, eleven in number, are the parents of the larger *Aeolus* class and of the *Astreas*. The *Astreas*, again, lead up to the *Minerva* class, which are of the displacement of our first-class "belted cruisers," the *Auroras*, and they are, in their turn, followed by larger vessels.

Of third-class cruisers, descended from the *Medea* or *Archer*, the chief types are the *Blanche* and *Pearl* class, with the newer *Pelorus*. Here, also, the tendency to increased displacement is visible, though not so plainly.

In smaller craft the torpedo-boat has grown and progressed till it has attained extraordinary speeds, and is capable of keeping the sea in moderate weather. Between the French *Chevalier* of 1893, steaming twenty-seven knots, and the small launch of 1877, which could not exceed seventeen knots, there is an immense difference. But in its progress the torpedo-boat has produced new types of ships expressly designed to combat it, and harry it. The first type is the torpedo-gunboat, of which the French *Bombe* of 1885 was the predecessor. In England it appeared with the *Rattlesnake*, and as usual, grew rapidly in size, till our later vessels of the class approach in size the third-class cruisers. Experience, however, showed that these vessels could not, on the open sea, run down a hostile torpedo-boat, as their speed was not sufficiently great. Some kind of craft which should be able to deal with the torpedo-boat was urgently required; and the torpedo-boat destroyer appeared to serve the purpose. These little vessels are very large and very fast torpedo-boats, and are not only well adapted for the task of harrying the torpedo-boat pure and simple, but are also capable of acting as torpedo-boats themselves. The speed which they have attained is extraordinary; on the mile, the *Daring* achieved twenty-seven knots, and thirty knots are promised in those now under construction. Here at last progress seems to be reaching its limit, as the screw is not adapted to give a higher speed than this. Other methods of propulsion may, however be perfected in the future.*

The torpedo has not remained stationary amidst the whirl of change. We have seen it in its crude form scoring successes in the American Civil War, when it may be said to have been brought to birth. Stationary "infernal machines" were succeeded by "infernal machines" which were towed by ships; these were by no means satisfactory, and were abandoned in the seventies. The spar-torpedo, carried on a boom, which could be run out from a boat or ship, lived longer, and it is not certain whether at the present day, in the hands of cool and determined men, it might not claim as many victims as the Whitehead. The latter weapon differed from its precursors in being automobile. It was first tried in 1868, in a crude and imperfect form. Its distinctive feature lay in this, that it was a small ship propelled by engines, driven by compressed air, and carrying a heavy charge of gun-cotton forward. It was of 16 inches diameter, the speed was nine knots, and the charge carried 117lbs of explosive. England purchased the right to manufacture it, and was followed by most European powers. The early English type of torpedo was 14 inches in diameter with a speed of eighteen knots, and a

---

* Hydraulic propulsion by means of a jet of water directed astern, was tried on the *Waterwitch* in 1866, but proved a complete failure, owing to the excessive power required to obtain a very low speed. It was again tried in 1878 by Sweden, and in 1881 by England. As compared with the screw there is a loss of 60 per cent. of the power. It may, however, come back, but the difficulties to be faced with it are great. A wheeled ship has been projected by M. Bazin in France, but has not been tried except in a model, and could be of little use in war.

charge of 32lbs. Gradually the speed and charge rose till the newer varieties of 14-inch torpedo run thirty knots and carry 65lbs of gun-cotton. As it was doubtful whether even this amount of explosive would fatally injure a large and modern ship, it has been succeeded by the 18-inch torpedo with 190lbs. to 200lbs. of gun-cotton.

The range of a Whitehead of the latest pattern does not much exceed 350 yards when the ship which fires it is in motion. A vessel at rest may make hits at a range of 800 yards, but at such a distance practice is very erratic, and 500 yards is perhaps the extreme limit for ordinary purposes.

There are numerous varieties of automobile torpedo other than the Whitehead, but none are so perfect, or have been so widely adopted. Last of all has appeared the steerable or controllable torpedo, which is guided in its course from the ship or the shore. Of this type are the Brennan, Halpine, and Nordenfelt torpedoes, and though they have not as yet won favour, they may do so with gradual improvements.

Whilst dealing with the torpedo we must notice the various attempts at submarine navigation. The Confederate "Davids" were the first attempt at warfare below the surface. The French *Goubet* was a small, egg-shaped craft, propelled by electricity at a speed of only five knots, with a torpedo attached outside by a bayonet catch. This vessel was launched in 1888 and is of little value. Italy claims in the *Pullino* to have a thoroughly satisfactory submarine boat, and the United States are building a vessel which can be driven awash, or below the surface, by steam or electricity. But the problem of steering a boat below the surface has yet to be overcome, and there is some possibility that a completely immersed vessel might be very seriously affected, even at some distance, by the explosion of her own torpedo. Moreover, most craft of this kind display a dangerous tendency to dive, when their sides would be crushed in by the tremendous pressure. None the less submarine attacks upon ships in harbour are a possibility of the future.

APPENDIX I.

THE DEVELOPMENT OF THE FRENCH NAVY, 1855—1895.

We have seen that it was France who led the way in the adoption of armour, whether for such harbour service craft as the Kinburn batteries, or for the sea-going battleship in the shape of the *Gloire*. The lead that she obtained in 1858 she has on the whole maintained since, and there is no country where more ingenuity and audacity have been displayed in the designing of warships. A short summary of French naval progress will best enable Englishmen to check their own advance.

Contemporary with the *Gloire*, and precisely similar to her, were the armoured frigates *Invincible* and *Normandie* of wood, and the *Couronne* of iron. All were armed with the French 50-pounder smooth-bore of 16° centimètres calibre.* They were followed by two larger vessels, the *Magenta* and *Solferino*, laid down in 1859, and carrying fifty-four 16-centimètre guns in a two-decked battery. In them a spur for ramming appears for the first time. They again were followed in 1862 by ten frigates of very similar pattern to the *Gloire*, but carrying, instead of her 4·7 inch plates, armour 5·9 inches thick. They were also a trifle faster and more manageable. They were uniform in type, and this uniformity beyond doubt gave France an advantage which has in more recent years passed to England.

* The following are the English equivalents of the French calibres in centimètres ;

10 centimètres	=	3·9-inch.
14 centimètres	=	5·5-inch.
16 centimètres	=	6·3-inch.
19 centimètres	=	7·4-inch.
24 centimètres	=	9·4-inch.
27 centimètres	=	10·8-inch.
30 centimètres	=	11·8-inch.
32 centimètres	=	12·6-inch.
34 centimètres	=	13·4-inch.

They carried from 880 to 950 tons of plating each. Their successor was the small ironclad *Belliqueuse*, of 3750 tons, generally similar in design to our *Bellerophon*, though, of course, on a smaller scale. She was intended for cruising in distant waters and was of wood. Her battery consisted of four 19-centimètre, four 16-centimètre, and four 14-centimètre guns. In 1865 the *Alma* type was introduced, and seven vessels were built after it. Wood was abandoned for the upper works, but still retained for the hull of the ship below the water-line. There was an end-to-end belt, a central battery, and above this on either beam a barbette tower with fire ahead and astern. The barbettes were slightly sponsoned out from the sides, and each contained one 19-centimètre gun. In 1868 the *Océan*, a far more powerful ship, was launched, and in 1869 and 1870 she was followed by the sister ships *Marengo* and *Suffren*. The weight of armour carried rose to 1370 tons, and the thickness to 8·6 inches on the water-line. The hull was of wood, the upper works of iron. The battery was carried in a central armoured enclosure, and in four barbette towers, resting upon the armoured walls of the enclosure, amidships on either beam. The guns, as usual in the French type of tower, revolved on a turn-table inside a fixed armoured turret. The gunners were not adequately protected, but then on the other hand they could obtain a clear view of their enemy. In each tower was one 27 or one 24-centimètre gun, and in the central work four to six other heavy guns. Besides the heavy weapons an auxiliary armament of 12 and 14-centimètre guns was carried. The engines of the *Suffren* were compound.*

In 1868 the *Richelieu*, an improved *Océan*, was laid down. She had the four barbette towers of the earlier type, but a longer central battery. She carries in each tower a 24-centimètre gun; in her central battery are six guns of 27 centimètres, whilst one of 24 centimètres is placed forward under the forecastle. The armour is 8·6 inches thick. The speed on trial was 13·1 knots. The hull is of wood below the water-line; above it, outside the central battery, of iron. The weight of plating rises to 1690 tons. She was followed by three ships of similar type, which, however, differ slightly from her and from each other. The *Trident* has two barbette towers, and carries eight 27-centimètre guns and two 24-centimètre. The *Colbert* and *Friedland* carry, the former eight 27-centimètre and six

---

* Particulars of most of these vessels are given in Table X.

24-centimètre, the latter eight 27-centimètre guns, as their heavy armament. Their hulls are of wood, and their armour 8·6 inches at its thickest.

In 1872 an enormous advance was made. Wood was abandoned, the draught of the ships designed reduced; deck protection was introduced, and recessed ports adopted. The *Redoubtable* was the first battleship laid down which embodied these innovations. She is a central-battery and barbette ship, carrying in her central battery four 27-centimètre guns; in two barbettes above the casemate, one on each beam, two more 27-centimètre weapons, and aft a seventh gun of this calibre. The barbettes have no protection against artillery fire, but the central battery is completely enclosed by armour 10 inches thick. There is an end-to-end water-line belt, which amidships is 14 inches thick : 2502 tons of plating are carried. The ahead fire is delivered by four 27-centimètre guns, two in the barbettes and two in the central battery. The light or auxiliary battery is not forgotten, and eight 14-centimètre guns are disposed on the forecastle and quarter-deck. The speed on trial was 14·26 knots. The ship was originally fully rigged, but now only carries light military masts.

The *Dévastation* and *Courbet* followed the *Redoubtable*. The weight of armour is increased to 2700 tons, and the maximum thickness to 15 inches, but the belt is not end-to-end, the stern being left unarmoured. The general features of the *Redoubtable's* design are retained; there is the central battery carrying four guns, 34-centimètre in place of 27-centimètre, with fore-and-aft fire; there are the unarmoured barbettes above the central battery, carrying the 27-centimètre gun; but a heavy gun forward is added. Hydraulic gear of the Rendel pattern was fitted to the *Dévastation*, and subsequently to her sister, and to successive French ironclads. The *Dévastation* is perhaps the finest central battery ship that has ever been designed, and in all round fire was greatly superior to the English ironclads of her type and date. On trial she steamed 15·1 knots. She carries 900 tons of coal.

The *Amiral Duperré* was begun in 1876, some months after the *Dévastation*. In her the central battery completely disappears, and the barbette is triumphant. There are four barbette towers, two placed forward one on either bow, one amidships and one astern, at a height of 27 feet above the water. These barbettes are protected by 15-inch armour, and each contains one 34-centimètre 48-ton gun. They are, however, mere shallow trays of armour, resting

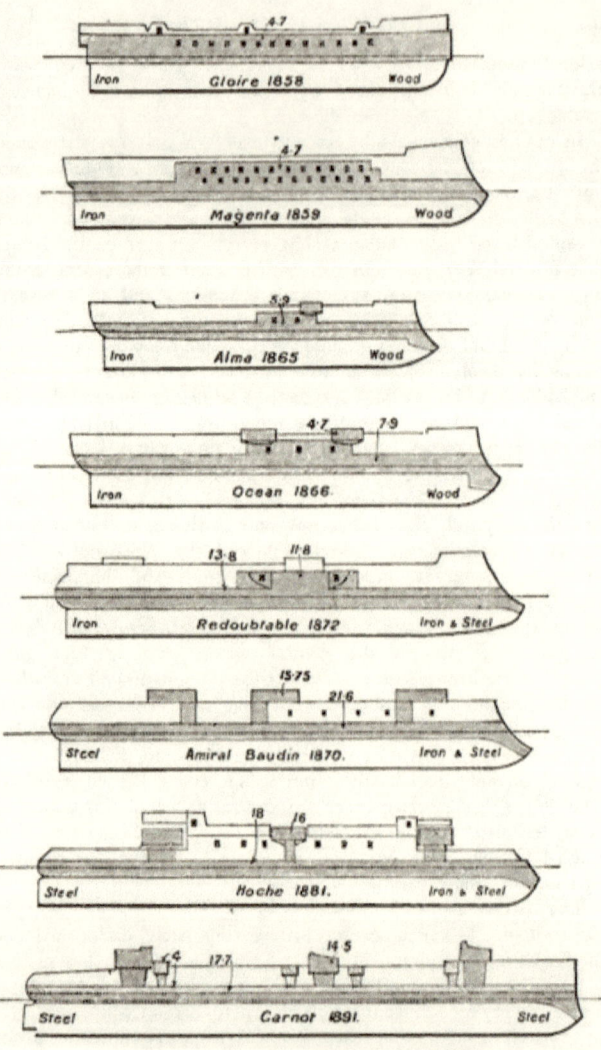

Types of French Ironclads. 1858-1891.
Material of Hull to Right.
" " Armour to Left.
Figures give thickness of Armour in inches.

Plate XLII.

upon the upper deck, with an armoured trunk running down to the protective deck; and they expose the gunners' heads and shoulders, whilst shells bursting underneath might bring them down through the ship's deck and bottom. There is a narrow end-to-end belt of armour $21\frac{1}{2}$ inches thick amidships. The weight of armour is 2900 tons. There are fourteen 14-centimètre guns mounted amidships. The chief defect of the ship is the great extent of unprotected side which she exposes to the enemy's fire. In this she resembles the *Inflexible*, though she differs widely from that ship in her high freeboard and end-to-end belt.

The *Amiral Baudin* and *Formidable*, which followed the *Duperré*, are generally similar to her. There are three, instead of four, barbettes, and all are placed on the centre-line. The armour is of steel, $16\frac{1}{2}$ inches thick on the barbettes and $21\frac{1}{2}$ inches amidships on the water-line, its weight reaching the very high figure of 4000 tons. The guns carried are three 75-ton weapons in the barbettes, and twelve 14-centimètre guns amidships. As in the *Duperré*, practically the whole of the ship's side is open to the smallest projectile, and only little patches and strips of very thick armour are carried. In 1880 were laid down three more barbette-ships, the *Marceau*, *Neptune*, and *Magenta*, and a fourth, barbette-ship and turret-ship combined, the *Hoche*. In these ships there are four heavy gun positions disposed lozenge-wise, one forward, one aft, and one on each beam. Thus, three guns can in most positions be brought to bear on an enemy. The 75-ton weapons of the *Baudin* give way to the long 34-centimètre gun in the first three. The armour is $17\frac{3}{4}$ inches thick on the narrow end-to-end belt; $13\frac{3}{4}$ inches on the barbette; and $3\frac{1}{2}$ inches on the deck. The *Hoche* differs from the others in having two turrets, instead of barbettes forward and aft, containing each one 34-centimètre gun; amidships she had two barbettes, each with one 27-centimètre gun. Her weight of armour is 3618 tons, and she is reported to be dangerously unstable; indeed, great fault has been found with all the four ships of this class. But if they are indifferently protected above the water-line, they carry very powerful armaments, as they have no less than seventeen 14-centimètre guns besides their main artillery.

The *Brennus* followed them after an interval of eight years during which France only laid down second-class ships. She carried two turrets, fore and aft; in the forward one are two 34-centimètre long guns of about 71 tons weight; in the after turret is one 34-centimètre gun. From end to end runs a 15¾-inch belt of compound armour, and above this, amidships, is a lightly plated citadel. On this citadel stand four small turrets, two on each beam, each carrying one 16-centimètre quick-firer. Six more of these weapons are mounted in the citadel and separated from each other by splinter-proof traverses. Thus ten 16-centimètre quick-firers are carried, of which five fire on either broadside and four ahead or astern. The ship has not ram bows but a perfectly straight stem. As originally equipped for sea she was so grievously overloaded that she lacked stability. Very considerable alterations are being made in her.

The Naval Defence Act, passed by the British Parliament in 1889, stimulated France to great exertions. In 1891 three first-class battleships were commenced—the *Charles Martel*, *Carnot*, and *Amiral Jauréguiberry*. A return was made to the lozenge-wise disposition of the heavy guns which had been abandoned in the *Brennus*. The armament consists of two 30-centimètre guns, one fore and one aft, and one 27-centimètre gun on either beam. The open barbette is abandoned and the turret adopted. An auxiliary armament of eight 14-centimètre quick-firers is carried; in the *Jauréguiberry* the quick-firers are placed in pairs in four lightly armoured turrets; on the other two each weapon has a separate armoured turret and ammunition hoist. Thus the guns are well separated, and all in armoured positions. The plating carried is thick on the heavy gun positions and water-line, where it varies between 17¾ and 10 inches. There is an end-to-end belt of this stout armour. Above this again is a belt of 4-inch armour about 4 feet deep, running from end to end, but carried up forward and aft in the line of the bow and stern waves. The small turrets all have 4-inch armour. There can be no doubt that these are extremely fine and powerful ships, carrying as they do over 4000 tons of armour, but they expose a very large unarmoured surface. Their speed is to reach eighteen knots.

Of closely similar design are the *Bouvet* and *Masséna*, in which the enormous weight of 4160 tons of armour is carried.* The *St. Louis*, *Charlemagne*, and *Gaulois*, however, are widely different. In them a return is made to the fore-and-aft system of mounting heavy guns, and the lozenge is abandoned. Four 30-centimètre guns are carried forward and aft, mounted in pairs in two turrets behind 15¾ inches

* Eight 10-centimètre quick-firers are carried on these two ships in addition to the 14-centimètre quick-firers.

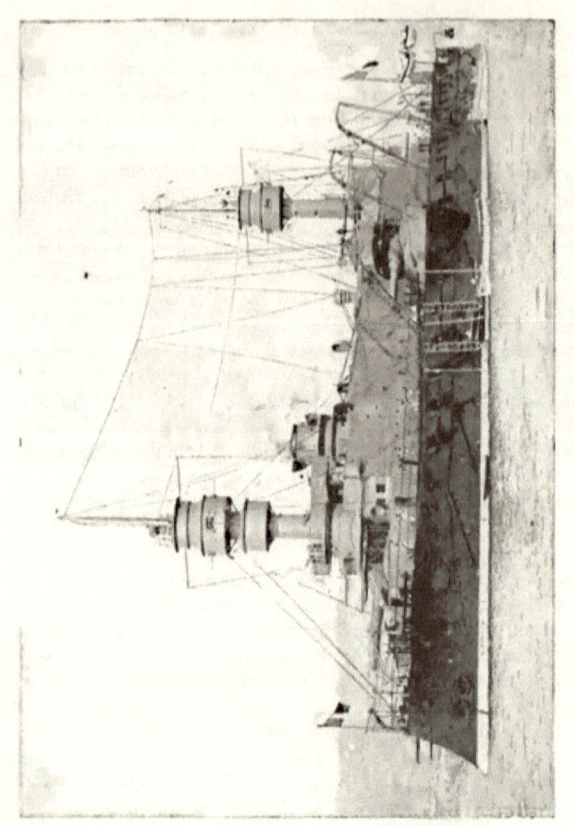

PLATE XLIII. THE FRENCH BATTLESHIP NEPTUNE.

APPENDIX I.

of armour. Eight 14-centimètre quick-firers are mounted on the main deck, four on each side, behind 3-inch hardened steel. Two more are on the upper deck, protected only by shields, whilst on the hurricane deck six 10-centimètre quick-firers are carried. There is an end-to-end water-line belt $15\frac{3}{4}$ inches thick amidships, tapering to the ends, and over this again a $3\frac{3}{4}$-inch belt. The armour-decks are two in number; one $3\frac{1}{2}$ inches thick at the upper level of the thick belt, the other $1\frac{1}{2}$ inches thick at its lower level. This gives these ships great protection against both ram and gun, as the belt is so strongly supported that it could scarcely be crushed in. The axial fire is very powerful. Six 14-centimètre, four 10-centimètre, and two 30-centimètre guns fire ahead or astern; on the broadside four 30-centimètre, five 14-centimètre, and three 10-centimètre. The guns carried are of enormous length and have very high muzzle velocities. The three ships and the *Masséna* all have the triple screw, which probably will add considerably to their manœuvring power if not to their speed.

No survey of the fighting strength of France would be complete which did not include her *garde-côtes cuirassés*, or armour-plated coast-defence ships. A few of these are good for nothing but harbour-work, but the great majority are capable of going to sea in moderate weather, and would certainly have to be reckoned with in the Mediterranean. The first of the class were the five armoured batteries built in 1855. The *Taureau*, a wooden ram with one 24-centimètre gun, mounted forward in a barbette, followed in 1863, and a little later four similar vessels carrying two 24-centimètre guns in a revolving turret forward. In 1864 a number of floating batteries for harbour defence were laid down. They were very inferior ships even at the date of their design, and were good for little work at sea. In 1872, after the war with Germany, a new type of vessel was introduced, similar in general design to our *Rupert* and *Glatton*. The vessels of this class are mastless monitors, carrying one turret and two guns each. Their names are the *Tempête*, *Tonnerre*, *Vengeur*, and *Fulminant*. They were succeeded by two vessels of somewhat different design, the *Tonnant* and *Furieux*. These have one barbette forward, and another aft, with one 34-centimètre gun mounted in each. There is a very thick end-to-end belt, but the freeboard is low, and the pair are not too seaworthy. Their design, however, was received with favour, and was repeated with improvements in the four ships, *Caiman*, *Indomptable*, *Requin*, and *Terrible*, which are larger, just as heavily armoured, and even more heavily armed. They carry each two 75-ton 42-centimètre guns, placed forward and aft, besides four 10-centimètre weapons. They were launched between 1881 and 1885. About the same date eight small armoured gunboats of very doubtful value were added to the fleet.

In the *Jemmapes* class the French coast-service vessel draws closely to the sea-going battleship. The *Jemmapes* and *Valmy* are of the *Caiman* pattern, with a 34-centimètre 70-ton gun fore and aft in closed turrets. Amidships is a high superstructure, at the angles of which are mounted four 10-centimètre guns. These are excellent little ships, very well armed, handy, fast; and even better are the *Bouvines* and *Tréhouart*, in which there is a high freeboard forward; in them the 34-centimètre guns are replaced by weapons of 30 centimètres, and the number of 10-centimètre quick-firers is increased to eight. The speed is also raised. An improved ship of this class, the *Henri IV.*, is to be laid down in 1896.

By purchase at the close of the American Civil War, France acquired the monitor *Onondaga*, and the large casemate-ship *Dunderberg*, which was renamed *Rochambeau*. Both soon disappeared from the French Navy List.

A third class of armourclad which has been built by France is the vessel for cruising on distant waters, or for encountering at home the cruisers which are now beginning to abound in all navies. The first ships of this class were the *Alma* and her sisters. These were followed some years later by the three small ironclads *Galissonière*, *Triomphante*, and *Victorieuse*. They are all three of wood, with a complete belt 6 inches thick, and $4\frac{3}{4}$ inches of armour on their battery or barbettes. They are merely weak and slow ironclads, and have no important advantages as cruisers. For fighting purposes they are about as bad ships as our *Nelson*. They were succeeded by four much better ships—the *Turenne*, *Bayard*, *Duguesclin*, and *Vauban*. The first two have hulls of wood, the last pair hulls of iron and steel. They are reduced copies of the *Duperré*, carrying four barbettes, arranged as hers are, one sponsoned out on each bow, one amidships in the centre line, and one astern. The thickness of armour and weight of guns are reduced by one half, and the heaviest weapon carried is the 24-centimètre breechloader. The speed of the four varies between fourteen and fourteen and a-half knots.

Three years passed between the launch of the *Duguesclin*, the most modern of the four, and the designing of the *Dupuy de Lôme*.

PLATE XLIV. THE FRENCH BATTLESHIP FORMIDABLE.
See p. 263.

In the meantime the importance of speed had been recognised, and the advent of high explosives had made armour more than ever necessary. The *Dupuy de Lôme* was commenced in 1886 and completed in 1894. She is a fast cruiser with a broad belt of hardened steel 4-inch armour reaching from some feet below the water-line to the level of her upper deck. Her armament consists of two 19-centimètre guns mounted amidships, one on either beam in sponsoned turrets, and six 16-centimètre quick-firers, each in a separate turret. Three of these guns are grouped forward, and three astern. End-on, two 19-centimètre and three 16-centimètre guns can be brought to bear; on the broadside, one 19-centimètre and four 16-centimètre. There is a double armour-deck as in the newest French battleships, and the speed is twenty knots, which has not been in practice attained. There are two military masts, and the funnels are of unequal size—a great disfigurement to the ship. But there is no denying her fighting value: she is well-gunned, fast, and well-protected. She was followed by a group of four similar but smaller cruisers. These have a wide belt of $3\frac{1}{4}$-inch steel, and $3\frac{1}{2}$-inch hardened-steel turrets containing their armament — six 14-centimètre and two 19-centimètre guns. As in the *Dupuy de Lôme*, the axial fire is very powerful. The speed is to be nineteen knots. In the *Latouche-Tréville* the turrets and ammunition hoists are operated by electricity. A fifth, a slightly larger and faster cruiser, the *Pothuau*, is also in hand, and a sixth is projected.

To unarmoured cruisers France has of recent years given great attention. Of the large first-class cruisers similar in design to the English *Blakes* and *Edgars*, she has only two constructed, the *Tage* and *Cécille*, but four more are being taken in hand.* In second-class cruisers she began with the *Sfax*, and the admirable *Isly*, *Alger*, and *Jean Bart*, which are fast and powerful vessels. In all her modern cruisers she has aimed at two things, speed and powerful axial fire. Ton for ton her cruisers are more heavily armed, than those of England. The best known types are the *Alger*, *Davout*, and *Pascal*. The first carries four 16-centimètre and six 14-centimètre guns; the second six 16-centimètre quick-firers and four of 10 centimètres; the last four 16-centimètre and ten 10-centimètre quick-firers. In smaller cruisers of the third class are the *Condor* and *Forbin* types, which are better suited for scouting than for fighting. They are by no means strong ships, and are too small to be of much use at sea.

* Including two which will be commenced in 1896.

In the construction of torpedo gunboats France led the way with her eight vessels of the *Bombe* class, which proved too weak for severe work at sea. The *Léger* and *Lévrier*, which followed, are larger, but larger still are the three fine vessels, *Casabianca*, *Cassini*, *D'Iberville*. The latter, which has been tested on the measured mile and at sea, is probably faster than any of our many torpedo gunboats; indeed she is perhaps the fastest torpedo gunboat in Europe.

With torpedo-boats France is very well supplied. From the first her sailors have attached great importance to torpedo warfare, and they are certainly second to none, whether in practical knowledge of their craft, or in the numerical strength of which they dispose. The French torpedo-boats fall into three classes. In the first are fifty-five large sea-going boats, over 125 feet in length. These might be able to accompany a squadron to sea even in the rough waters of the Atlantic. They regularly cruise with the French ironclads, but have not seldom, in bad weather, to make for port. Next come 173 boats of limited sea-going quality, some not good for much work at sea even in fine weather, others little inferior to the true sea-going craft. Finally, there are fourteen boats which are less than 86 feet in length, and which could be used only for harbour defence or attack. One has a hull of aluminium alloy, and did twenty-and-a-half knots on the measured mile. In her newest boats France has obtained very high speeds. Thus the sea-going *Chevalier*, on the measured mile, accomplished 27·2 knots in an hour. The *Forban* is expected to equal the *Sokul's* record with thirty knots.*

Of submarine boats France has four — the *Goubet*, *Gymnote*, *Gustave Zédé*, and *Morse*. The first two are of little serious value: the last two are larger, but perhaps not much better. The *Zédé* has a cigar-shaped hull, 131 feet long. Her displacement is 266 tons, and she carries a crew of eight men. Her motive force is electricity, stored in accumulators. The fumes from these have proved a source of great annoyance to her crew, and there have been extraordinary explosions on board when they were being charged. During her trials she descended forty and sixty feet, and moved about below the surface at a rate of six or eight knots, launching torpedoes. As, however, it is impossible to see at this depth she is a vessel of very doubtful value. The *Morse* is understood to be very similar to her in design.

* The *Sokul*, a Russian destroyer, built by Messrs. Yarrow, did 30 knots for a short time on her trial; the *Forban* has since done 31.

PLATE XLV.   THE FRENCH CRUISER ALGER.
See p. 107.

Considering the French fleet as a whole, there can be no doubt that it is a most formidable force. The first-class battleships are all extremely well protected on the water-line, have a good freeboard, and a great height of command for their guns. But in the older vessels very little protection is given to the guns, and practically none to the gunners. The large auxiliary batteries could not be used in action against a modern ship. Again, a very great extent of the side is left exposed to any projectile—a fault, however, which is equally shared by English battleships of their date. The French battleships of 1879—1886 have all one perilous weakness—that their heavy guns could be put out of action by bursting common shell underneath: this weakness occurs in our six "Admirals," but not in our turret-ships of the *Inflexible* type, nor in our battleships subsequent to the "Admirals." The French ships, too, are in many cases defective in the very important matter of stability. The *Hoche*, when three heavy guns are trained abeam, inclines fifteen degrees with quite a moderate helm. At this inclination her belt is submerged on the one side and her unarmoured bottom exposed on the other. Nor is the lozenge disposition of heavy guns without grave inconveniences. When the two heavy weapons amidships are trained axially, right ahead or right astern, their blast impinging upon the ship's works is liable to cause serious injuries to the structure, whilst it greatly interferes with the working of the auxiliary armament; a bugle has to sound for the gunners in the line of the blast to retire. The great difficulty and confusion which would result from this in battle is obvious. The French cruisers and battleships all exhibit this fault, and thus the axial fire, which is so formidable on paper, would dwindle very much in action. For example, the *Dupuy de Lôme* could never fight five guns end-on. Her 19-centimètre weapons firing past her 16-centimètre turrets would stun the men in them, besides blowing away the upper works of the ship. Of course the French understand this as well as their critics, and would never be likely to train the 19-centimètre guns axially. They have striven to give their guns the widest possible angles of fire, and might find these wide angles of great value, when, under a heavy fire, portions of the ship's armament had been disabled.

To compare the naval strength of France with the naval strength of England is a difficult matter. It is almost impossible to evaluate *personnel*, and it is not much easier to evaluate *matériel*. The ships which these two rival claimants to sea power have built in the past are so widely different that action alone can decide their respective merits, and it is just possible that the test of war might prove one or other type wholly unsuitable. Generally, English designers have striven to protect the men working the guns: to do this they have reduced the length of the ship's armour. French naval architects, on the other hand, have given little attention to the protection of the men, and a great deal to the protection of the water-line. In consequence they have reduced the breadth of the ship's armour. Perhaps the English ironclads of the 1875-1885 period are a little better than the French, though they are, where the muzzle-loader has been retained, worse armed. The *Inflexible* at least gives her gunners good shelter, and protects well the bases of the heavy gun positions. The *Duperré* and *Baudin* are defective in each of these points, and their barbettes with their thin gun-shields are mere shell-traps.

The test of a battleship's capacity to fight in line is age, and here it is a case of the younger the better. Indeed naval progress generally removes ships from the first-class in ten or twelve years. Not that they necessarily become valueless: a good and well-built ship can generally be re-armed and re-fitted. But after ten or twelve years the advance of artillery, of metallurgy, of boiler or engine-making, renders fresh combinations of the items which make up the compromise necessary. Quick-firers may lead to the substitution of a great extent of thin armour for a little thick armour, and automatic heavy guns firing large projectiles with great rapidity, may again compel a return to a limited extent of thick plating.

"Standard armoured ships," or first-class battleships, may then be defined to be ships not more than ten years old, reckoning from the date of their launch. It is usual to impose some limit of tonnage in addition, and as first-class battleships are generally reckoned, they must not displace less than 9000 tons. Yet small size is hardly such a serious disqualification as old age. In the so-called second-class battleships the armament is generally reduced, and there is less coal and ammunition, but there are certain instances where smaller ships are perfectly capable of work in line with their larger rivals. If recent battleships are to be subdivided, 8000 tons would appear to be a better point at which to draw the line. There will then be two classes of standard battleships, the one above 8000 and the other of and below 8000 tons, and in each case the ships must not have been launched more than ten years. The figures for England and

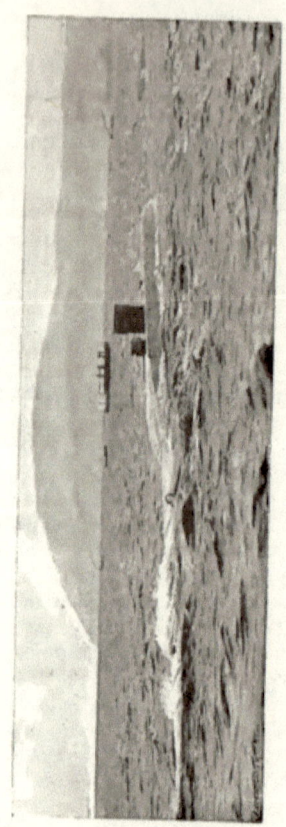

PLATE XLVI. THE FRENCH SUBMARINE-BOAT GUSTAVE ZÉDÉ.
See p. 268.

France, including in the class certain ships such as the *Conqueror*, *Baudin*, and three of the *Caiman* class, which the age limit would, if strictly enforced, exclude, are (in 1895):

		Over 8000.	Under 8000.	Total.
England	Building or completing......	8	0	8
	Ready for sea	20	2	22
France...	Building or completing......	8	0	8
	Ready for sea	7	8	15

This comparison is a little unjust to France as it excludes from the first class the *Courbet* and *Dévastation* which are exceedingly fine ships. On the other hand it does not show the homogeneity of the English fleet. Analysing the distinct types in either fleet and passing over small differences we get these figures:

English Battleships.
- 9 *Majestic* type. [Building.*]
- 1 *Renown*. [Building.]
- 2 *Centurion* type.
- 8 *Royal Sovereign* type.†
- 2 *Nile* type.
- 5 " Admirals."
- 1 *Sanspareil*.
- 2 *Hero* type.

French Battleships.
- 3 *Charlemagne* type. [Building.]
- 4 *Carnot* type. [Building.]
- 1 *Jauréguiberry*. [Building.]
- 1 *Brennus*.
- 3 *Magenta* type.
- 1 *Hoche*.
- 3 *Formidable* type.
- 4 *Jemmapes* type.
- 4 *Caiman* type.

Ships, 30. Types, 8. Average ships of one type, 3·75. Ships, 23. Types, 9. Average ships of one type, 2·55.

* Two are now ready for sea and are so reckoned in the first comparison.
† The *Hood* only differs in detail.

The nine *Majestics* and the eight *Royal Sovereigns* form two homogeneous squadrons, in which each individual ship is superior to any vessel that France possesses. Eight indifferent but closely similar ships would probably be found superior in war to eight ships, each of which was better than those of the homogeneous squadron, but each of which was at the same time different from the other. If ships are kept as far as possible identical—if a large number of each type are always laid down, the sailor who knows one knows all, and the training of the *personnel* is simplified. Similar and interchangeable armaments, similar turning-circles, and similar speed are factors of great value, and in our fleet they are combined with superior armament, coal endurance, and ammunition supply. The excess of tonnage which our newer ships exhibit when contrasted with French ships under construction is given mainly to these important elements in the compromise. On the other hand the French ships on the stocks are to be half a knot faster.

There can be no doubt that England is very much ahead in standard battleships. Taking into consideration the small size of eight of the French ships, England is to France here about as five to three. If we turn now to second-class armoured tonnage, launched between 1875 and 1885, and if this class, like the standard one, is subdivided, the line being drawn at 7000 instead of 8000 tons, the figures are these:

	Over 7000 tons.	Under 7000 tons.	Total.
England	14	3	17
France	5	8	13

Included in the English fourteen ships of over 7000 tons are the *Thunderer* and *Devastation*, which do not, strictly speaking, come within the age limit, but which have been completely re-armed, and are beyond question, fine fighting ships. The *Imperieuse* and *Warspite*, which verge upon the battleship, are also included. In the four smaller ships the *Belleisle*, *Orion*, and *Rupert* are reckoned. The last is an old ship, but, like the *Thunderer*, has been re-armed. These three are not capable of keeping the sea in all weather, but could probably be trusted for fighting in the Channel. Of the seventeen English ships only eight carry breech-loaders as their heavy armament; eight are still equipped with the muzzle-loader—a most serious weakness; and one has both breech and muzzle-loader. The French ships are made up as follows: The *Trident* and *Colbert*, both of wood, have been rejected; the *Friedland*, with hull of iron, though launched in 1873, has been included, and with her two of the more recent armoured cruisers of *Vauban* type, the *Duguesclin* and *Vauban*, and six *garde-côtes*. Of the total—thirteen—six are of questionable sea-keeping quality. The others are fine and strong ships; indeed the *Dévastation* is as good as, or better than, any ship which England possesses in this class. Taking muzzle-loaders into account, there is little to choose between England and France in this class.

There remain a certain number of ironclads which, though more than twenty years old, are capable of doing service in the line of battle or in reserve. In this category the figures are:

England, 17; France, 11.

The English figure includes the old ships of *Achilles'* type, which are never likely to do much fighting in line, the *Glatton* and the *Penelope*. All the French eleven are of wood, and six of them may be expected to disappear at a very early date.

APPENDIX I.

Of armoured ships, which it is difficult to class as battleships, England has five coast-defence ships, and France eight small armoured gunboats. England has further ten armoured cruisers, three of which, the *Shannon*, *Nelson*, and *Northampton*, are of no great value, whilst the other seven could not face battleships. France has six armoured cruisers of *Dupuy de Lôme* type, which are finer craft than the unprotected cruiser, but inferior to the battleship. England has no ships of similar type.

Merely considering figures and paying no attention to the duties which war would impose upon the respective fleets, there can be no doubt of the great preponderance of England. In standard ships her superiority is especially striking, but it is no less evident when we consider unarmoured ships. These may be divided into three classes: Over 6000 tons, over 2000 tons, and under 2000 tons. As the "protected cruiser," the fast unarmoured ship with horizontal deck-plating, is a modern conception, most are recent. One or two fast but old ships which lack the armour-deck, are included in figures given, but no ship appears which has not done more than the sixteen knots at her trial.

	Over 6000 tons.	Over 2000 tons.	Under 2000 tons.	Total.
England	17	68	17	102
France	4	22	12	38

Finally the torpedo flotillas of the two powers are as follows:

	Torpedo Gunboats.	Destroyers.	Sea-going Boats.	1st class Boats.	2nd class Boats.	Small.	Total is only for Boats.
England	31	62	43	26	4	93	166
France	13	8	47	53	84	50	234

As torpedo-boats are the weapon of the weaker power, England has given attention rather to vessels designed to combat them — to torpedo gunboats and destroyers. It is doubtful whether either of these types will prove able to endure much work at sea, and there are many who hold the third-class cruiser is the vessel to meet the torpedo-boat. But if they are useless for cruising, the new destroyers will at least be superior to any torpedo-boat for torpedo work.

Tables of naval strength, however, are only of value to give some faint adumbration of the truth: they cannot from their very nature deal with such vital points as organisation, training, discipline, and character. And here we have no means of comparison, no test except the stern trial of war. Yet the English sailor should be superior to the average French sailor, from the fact that the former is a long-service man, whilst the latter in many cases serves no more than four or five years in the fleet. Steadiness and discipline can only be assured by a long training, and there is little doubt that war would not find our sailors wanting in these essential qualities. On the other hand, with long service the provision of a trained reserve becomes difficult, and here France is better off than England.

Lastly, the numerous claims upon the British fleet in war are to be considered. It is not enough to be stronger than France; England must possess that degree of superiority which will enable her to confine her opponents' ironclads to their ports, and prevent the hostile commerce-destroyers from plundering her commerce. It is possible that two battleships to the French one would be necessary for a close blockade. It is certain that four cruisers to the French one will not be found any too many in war.

The probability of England having to confront an alliance, herself without allies, is one which politicians should consider. Russia and France will unquestionably be as strong or stronger in *matériel*, when compared with England, before the end of the century, unless England makes further and very determined efforts. On the other hand the single power has a great advantage against an alliance, and England holding the interior position could operate against the two allies, attacking each in detail.

In a final table the results already obtained are recapitulated, and the Russian fleet is included for purposes of comparison:

	England.	France.	Russia.	France & Russia.
Standard battleships	30	23	16	39
Second-class battleships	17	13	5†	18
Third-class battleships	20	11	6	17
Harbour service ironclads	5	8	13	21
Cruisers (armoured)	7*	6	4	10
,, (unarmoured) over 6000 tons	17	4	0	4
,, ,, over 2000 tons	68	22	3	25
,, ,, under 2000 tons	17	12	0	12
Torpedo gunboats	31	13	8	21
Destroyers	62	8	17	25
Sea-going torpedo-boats	43	47	55	102
Smaller torpedo-boats	123	187	116	303

* Excluding the *Shannon*, *Nelson*, and *Northampton*, which are reckoned as third-class battleships. † Including the three vessels of the *Admiral Orthakoff* type and the *Nachimoff*.

APPENDIX II.

* *Report of the French Committee in 1870 upon the practicability of attacking the Prussian Littoral.*

MEMBERS OF THE COMMITTEE : Rear-Admiral Dieudonné ; Duburquois, Chief of the Staff; Lacour, Colonel of Artillery Captains Quilio and Serras.

ALSEN. The depth of water will not permit an approach to this point within at most 3300 yards, at which distance attack would be useless owing to the plunging fire of the forts. Nothing can be done here without a force to land. It is further most probable that there is defence by submarine mines along the shore. These would have to be removed, and this could not be attempted until the squadron was supplied with the necessary apparatus.

DÜPPEL AND KAPPELN. Quite out of reach of the ships' guns. Too little water in the bays. We could get at them with armoured gunboats.

ECKERNFORD. The isolated works could easily be destroyed. They are, however, of no importance, and unless troops can be thrown on shore the reduction of the forts would be insignificant.

KIEL. It would be necessary to employ the whole strength of the squadron. The success of gun-fire is uncertain, on account of the height of the forts above the shore, and the losses which the assailants will certainly incur unless they can occupy the forts as they are silenced. The forts of Friedrichsort being destroyed, as the squadron would be unable to penetrate to the bottom of the bay within range of Kiel, owing to the obstructions, the torpedoes and all the means of defence which have there been accumulated, the French ships would soon be forced to retire without even knowing the result of their attack.

NEUSTADT. An open town without defence. The bay is so shallow that the French ships could not even reach with their projectiles the merchant ships, which are anchored some distance from the port, properly so called.

It is the same along the coast as far as

KOLBERG. A strong place, besieged in 1807, and attackable from the sea at 2400 yards. Before attacking there it will be necessary to make a reconnaissance to make certain that the houses along the shore, especially the Casino, do not mask fortifications which would compel changes in the plan of attack.

DANZIG. The fort at the entrance of the bay is within range of our upper-deck guns, but only at a distance of 4500 yards. The battery guns could not be used elsewhere with advantage.

CONCLUSION. Kolberg and Danzig alone can be attacked; but the small effect which will result from these two attempts will be of a nature to deprive the French squadron of the prestige of its force. In order to operate usefully, special vessels are required, and the prospect of forcing the enemy to assemble his troops on this part of the littoral. But this end is unattainable without a landing force.

* René de Pont Jest J.U.S.I., xxxiii., 230, and the original in the *Moniteur* (Tours).

APPENDIX III.

BRITISH IRONCLADS.

Sea-going Ships.

BROADSIDE.	CENTRAL BATTERY.	TURRET.	BARBETTE.	ARMOURED CRUISER.
{ Warrior *	{ Enterprise *	Monarch	Collingwood	Shannon
{ Black Prince	{ Favourite *	Captain †	{ Howe	{ Nelson
{ Defence *	Bellerophon	Devastation	{ Rodney	{ Northampton
{ Resistance *	Pallas * ‡	Thunderer	Benbow	{ Impérieuse
{ Hector *	Penelope	Dreadnought	{ Anson	{ Warspite
{ Valiant *	Hercules	Neptune	{ Camperdown	{ Aurora
Research * ‡	{ Invincible	Inflexible	{ Royal Sovereign	Australia
Achilles	{ Audacious	{ Ajax	{ Empress of India	Galatea
{ Royal Oak * ‡	{ Vanguard †	{ Agamemnon	Ramillies	Immortalité
{ Prince Consort * ‡	{ Iron Duke	{ Colossus	{ Repulse	Narcissus
{ Ocean * ‡	{ Swiftsure	{ Edinburgh	Revenge	Orlando
{ Caledonia * ‡	{ Triumph	{ Nile	Resolution	{ Undaunted
{ Royal Alfred * ‡	Sultan	{ Trafalgar	Royal Oak	
{ Lord Clyde * ‡	Alexandra	{ Victoria †	{ Centurion	
{ Lord Warden * ‡	Téméraire	{ Sanspareil	{ Barfleur	
{ Zealous * ‡	Superb	Hood	Renown	
{ Repulse * ‡	{ Belleisle		Majestic	
{ Minotaur	{ Orion		Magnificent	
{ Agincourt			Prince George	
{ Northumberland			Victorious	
			Jupiter	
			Mars	
			Caesar	
			Illustrious	
			Hannibal	

* Struck off Navy List.   † Lost at sea.   ‡ Wooden hull.

Sister ships are bracketed.

BRITISH IRONCLADS—*contd.*

Coast Service Ships.

⎧ Glatton **	⎧ Scorpion *	⎧ Cyclops
⎪ Meteor **	⎩ Wivern *	⎪ Gorgon
⎪ Thunder *†	Royal Sovereign *†	⎪ Hecate
⎩ Trusty **	Prince Albert *	⎩ Hydra
⎧ Ætna *	⎧ Viper *	Glatton
⎪ Terror *	⎩ Vixen *	Hotspur
⎪ Thunderbolt *	Waterwitch *	Rupert
⎩ Erebus *	⎧ Cerberus	⎧ Conqueror
	⎩ Magdala	⎩ Hero
	Abyssinia	

Total ironclad ships built since 1855 118
Lost at sea   ...   3
Struck off the list   ...   33
    Leaving eighty-two serviceable.

    * Struck off Navy List.   † Wooden hull.

    Sister ships are bracketed.

APPENDIX IV.

LEADING AUTHORITIES CONSULTED.

GENERAL NAVAL HISTORY.

*MAHAN. Influence of Sea Power upon History. London.
* ,, Influence of Sea Power upon the French Revolution. 2 vols. London.
JAMES. Naval History. 6 vols. London, 1886. (*Cited as* James.)
YONGE. History of the British Navy. 2 vols. London, 1863.
*COLOMB. Naval Warfare. London, 1895. (*Cited as* Colomb.)
* ,, Essays on Naval Defence. London. (*Cited as* Colomb's Essays.)
LAUGHTON. Letters and Despatches of Nelson. London, 1886.
,, Nelson. London, 1895.
CHABAUD-ARNAULT. Histoire des Flottes Militaires. Paris. 1889.

RECENT NAVAL ARCHITECTURE. ETC.

*Naval Annuals, 1886—1895. Portsmouth.
*EARDLEY WILMOT. Development of Navies. London, 1892.
LEDIEU ET CADIAT. Matériel Naval. 3 vols. Paris, 1889.
CRONEAU. Architecture Navale. 2 vols. Paris, 1894.
WHITE. Manual of Naval Architecture. London, 1894.
*PARIS. L'Art Navale. 2 vols. Paris, 1869.
HAUSER. Cours de Construction Navale. Paris, 1886.
DISLÈRE. La Guerre d'Escadre. Paris. 1876.
TROMP. Navires Cuirassés. Utrecht, 1880.
*BRASSEY. The British Navy. 5 vols. London, 1882.
*KING. Warships of the World. London. 1880.

REED. Our Ironclad Ships. London, 1869.
„ Modern Ships of War. London, 1888.
COLES. Letters, and the Opinion of the Press. London, 1866.
KRONENFELS. Das Schwimmende Flottenmaterial der Seemachte. Wien., 1880.
HUNIER [Pseud.]. Du Navire de Combat. Paris, 1892.
BALINCOURT. Étude sur les Navires d'Aujourd'hui. Paris, 1892.
LEROL. Les Armements Maritimes en Europe. Paris, 1889.
PÈNE-SIEFERT. Flottes Rivales. Paris, 1890.
WEYL. La Flotte de Guerre et les Arsenaux. Paris, 1894.
ROBINSON. The British Fleet. London, 1894.
The Technical Press, especially *Engineer; Engineering; Le Yacht; Rivista Marittima; Revue Maritime.*

ARMOUR, GUNNERY, TORPEDOES, &c.

HOLLEY. Ordnance and Armour. New York, 1865.
DAHLGREN. Shells and Shell Guns. Philadelphia, 1857.
HOWARD DOUGLAS. Naval Gunnery. London, 1861.
ORDE BROWNE. Armour and its attack by Artillery. London, 1887—1894.
COOKE. Naval Ordnance. New York, 1875.
VERY. Armour for Naval Use. Annapolis, 1883.
Treatise on Service Ordnance. London, 1893.
*LLOYD AND HADCOCK. Artillery and its Progress. Portsmouth, 1893.
Modern Naval Artillery. London.
DREDGE. Modern French Artillery. London, 1892.
SLEEMAN. Torpedoes. Portsmouth, 1889.
BARNES. Submarine Warfare. New York.
HOVGAARD. Submarine Boats. London, 1887.
BUCHARD. Torpilles et Torpilleurs. Paris, 1886.
BRANDT. Gunnery Catechism. New York, 1864.
NORMAND. Étude sur les Torpilleurs. Paris, 1885
*CLARKE. Fortification. London, 1890.

STRATEGY AND TACTICS.

CLERK. Naval Tactics. Edinburgh, 1804.
HOWARD DOUGLAS. Naval Warfare with Steam. London, 1864.
BUTAKOV. Nouvelles Bases de Tactique. Paris, 1864.
PHILLIMORE. Naval Tactics. London, 1859.
PENHOAT. Éléments de Tactique Navale. Paris, 1879.
FARRET. Études Comparatives de Tactique Navale. Paris, 1883.
   Études sur les Combats Livrés sur Mer de 1860—1880. Paris, 1881.
CHABAUD-ARNAULT. Essai Historique sur la Stratégie et la Tactique. Paris, 1889.
LULLIER. Tactique Navale. Paris, 1867.
ELLIOT. Treatise on Future Naval Battles. London, 1885.
BETHELL. Remarks on the Manœuvres of Two Vessels in Action. Portsmouth, 1881.
CAMPBELL. Naval Tactics. London, 1880.
*HOFF. Naval Tactics. Portsmouth, 1885.
 ,, Elementary Naval Tactics. New York, 1894.
NOEL. The Gun, Ram, and Torpedo. Portsmouth, 1885.
Z. AND MONTÉCHANT. Essai de Stratégie Navale. Paris, 1893.
 ,, ,, Guerres Navales de Demain. Paris, 1891.
Papers by *Sturdee, *Calthorpe, *Laird Clowes, Long, Fremantle, *Colomb, and others, in the *Journal of the United Service Institution.* London.
Papers by Wainwright, Alger, and others, in the *Proceedings of the United States Naval Institute.* Annapolis, U.S.A.
Anonymous Articles in the *Quarterly* and *Edinburgh Reviews.*

INTERNATIONAL LAW, COMMERCE-DESTRUCTION, &c.

DANSON. Our next War. London, 1894.
WARAKER. Naval Warfare of the Future. London, 1892.
NYS. Droit de la Guerre. Bruxelles, 1882.
LEMOINE. Précis de Droit Maritime. Paris, 1888.
HALL. International Law. Oxford, 1890.
COBBETT. Leading Cases in International Law. London, 1885.
WHEATON. International Law. London, 1889.

HOLLAND. Manual of Naval Prize Law. London, 1888.
TWISS. Continuous Voyages. London, 1877.
  „   Belligerent Right on High Seas. London, 1884.
FAUCHEVILLE. Du Blocus Maritime. Paris, 1882.
LAWRENCE. Essays on Disputed Questions in International Law. Cambridge, 1885.
PHILLIMORE. Commentaries upon International Law. 4 vols. London, 1882.

TECHNICAL PERIODICALS CONSULTED.

Journal of the United Service Institution. London. (*Cited as* J.U.S.I.)
Proceedings of United States' Naval Institute. Annapolis, U.S.A.
Information from Abroad. Navy Department. Office of Naval Intelligence. Washington, U.S.A.
Transactions of the Institution of Naval Architects. London.
United Service Magazine. London.
Army and Navy Gazette. London.
Broad Arrow. London.
Engineer. London.
Engineering. London.
Revue Maritime. Paris.
Le Yacht. Paris.
La Marine Française. Paris.
Rivista Marittima. Rome.
Mittheilungen des Seewesens. Pola.
Army and Navy Journal. New York.
Engineer. New York.
Mechanics' Magazine. London.
Marine Rundschau. Berlin.

KINBURN, BOMBARDMENT OF.

YONGE. History of the British Navy.
RUSSELL. The War. 2 vols. London, 1855.
Revue des Deux Mondes, vol. xiii. Paris.
Mechanics' Magazine. London.

GENERAL HISTORY OF AMERICAN CIVIL WAR.

*JOHNSON. Battles and Leaders of the Civil War. 4 vols. New York, 1887. (*Abbreviated to* Battles and Leaders.)
> A series of articles on the various actions by those who took part in the struggle: a storehouse of information, and well illustrated.

*SCHARF. History of the Confederate Navy. New York, 1887. (*Cited as* Scharf.)
> From the Confederate point of view.

PORTER. Naval History of the Civil War. London, 1887. (*Cited as* Porter.)
> Gives original documents, &c., but often inaccurate.

*MACLAY. History of the United States Navy. 2 vols. London, 1894. (*Cited as* Maclay.)
> A concise and accurate account of the U.S. Navy.

Reports of the Secretary for the Navy, 1862-5. Washington.
> Official documents, &c.

*The Navy in the Civil War. 3 vols. New York, 1883.
> Viz.: SOLEY. The Blockade and the Cruisers. (*Cited as* Soley.)
> AMMEN. The Atlantic Coast. (*Cited as* Ammen.)
> MAHAN. The Gulf and Inland Waters. (*Cited as* Mahan's Gulf.)
>> Deal fully with the strategic aspects of the war.

BOYNTON. History of the Navy during the Rebellion. 2 vols. New York, 1867.

STENZEL. The American Civil War. *United Service Magazine*, vol. cxxxi. London. (*Cited as* Stenzel.)

BADEAU. Military History of U. S. Grant. 3 vols. London, 1881.

BULLOCH. Secret Service of the Confederate States. 2 vols. London, 1883. (*Cited as* Bulloch.)

DAVIS. Rise and Fall of the Confederate Government. 2 vols. London, 1881.

GREELEY. American Conflict. Hartford, 1864.

Southern Historical Society's Papers. Richmond.

VON SCHELIHA. Coast Defence. London, 1868.

Daily Despatch. Richmond.

New York Herald. New York.

Report of the Secretary of the Navy in relation to Armoured Vessels. Washington, 1864.

BIGELOW. France and the Confederate States' Navy. New York, 1888.

SMITH. Confederate War Papers. New York, 1884.

The War of the Rebellion. Washington, 1880, &c.
    Official documents, Confederate and Federal.

NICOLAY and HAY. Abraham Lincoln. 10 vols. London, 1890.

WORKS BEARING UPON PARTICULAR INCIDENTS OF THE AMERICAN CIVIL WAR.
(IN ADDITION TO THE MORE GENERAL HISTORIES.)

Monitor.

 CHURCH. Life of Ericsson. 2 vols. London, 1890. (*Cited as* Church, Ericsson.)

 SWINTON. The Twelve Decisive Battles of the War. New York, 1867.

Mississippi, Opening of.

 DORSEY. Recollections of H. W. Allen. New York, 1866.

 MAHAN. Admiral Farragut. London, 1893. (*Cited as* Mahan, Farragut.)

Mobile.

 PARKER. Battle of Mobile Bay. Boston, 1878.

 Reports, &c., of [Confederate] Secretary of the Navy. Richmond.

The Cruisers and the Blockade.

 SEMMES. Service Afloat. London, 1887. (*Cited as* Semmes.)

 HAYWOOD. Cruise of the Alabama. Boston, 1886. (*Cited as* Haywood.)

 HOBART PASHA. Sketches from My Life. London, 1886.

  ,, ,, Never Caught. [Pseudonym, Captain Roberts.] London, 1867.

 ALABAMA CLAIMS. Case on the part of Her Majesty's Government. London.

 ALABAMA CLAIMS. Correspondence concerning Claims against Great Britain, transmitted to the Senate of the United States. 5 vols. Washington, 1869—1870.

THE AUSTRO-ITALIAN WAR OF 1866.

*Österreichs Kämpfe im Jahre 1866. Vol. v. Wien, 1899.
  (*Cited as* Staff History.)
    The official Austrian account, probably revised by Tegetthof.

CROUSSE. Bataille de Lissa. Brussels, 1891.
    A translation of the above.

*Rendiconti delle udienze pubbliche dell'Alta Corte, &c. Contro l'Ammiraglio Conte Carlo Pellion di Persano. Firenze, 1867. (*Cited as* Rendiconti.)
    The evidence, &c. given at Persano's trial.

L'Ammiraglio C. di Persano nella campagna navale dell' anno 1866. Torino, 1873. (*Cited as* Persano.)
    Persano's own story of the battle of Lissa.

AMICO. I Fatti di Lissa. 1868. (*Cited as* Amico.)

Processo del Conte C. Pellion di Persano. Milano, 1867.

Revue des Deux Mondes, November, 1866. Paris.

Revue Maritime, vols. xviii., xix. Paris.

*LAUGHTON. Studies in Naval History. London, 1887. (*Cited as* Laughton.)
    An admirable account of the war is given in Chapter V., "Tegetthof."

La Guerra in Italia nel 1866. Milano, 1867.

Times, Standard, Army and Navy Gazette.

SOUTH AMERICAN WARS, 1866—1870.

Revue Maritime, vol. xvii. Paris.
    Gives Commodore Rodgers' (U.S.N.) Report on the Bombardment of Callao.

Illustrated London News, 1866. London.
    Letters from British Officers at Valparaiso and Callao.

GARMENDIA. Recuerdos de la Guerra del Paraguay. Buenos Aires, 1889.

*THOMPSON. Paraguayan War. London, 1869.

*WASHBURN. Paraguay, 2 vols. Boston, 1871.

Revue des Deux Mondes, 1866—73. Paris.
    Numerous articles on the Paraguayan War; two on the Spanish War, vols. lii. and lxxvii.

CHABAUD ARNAULT. Histoire des Flottes Militaires.

KENNEDY. La Plata, Brazil, and Paraguay. London, 1869.

FRANCO-GERMAN WAR.

CHEVALIER. La Marine Française et la Marine Allemande pendant la Guerre de 1870—71. Paris, 1873. (*Cited as* Chevalier.)

BOUËT-WILLAUMEZ. Questions et Réponses au Sujet de nes Forces Navales. (*Cited as* Bouët.)

*RENÉ DE PONT-JEST. Campaign in the North Sea and Baltic. Journal United Service Institution. Vol. xxxiii.

CHABAUD ARNAULT. Histoire des Flottes Militaires.

F. JULIEN. L'Amiral Bouët-Willaumez. Paris, 1872.

The Franco-German War. Translated by Captain F. C. H. CLARKE. London, 1872, &c. (*Cited as* Staff History.)

RUSSO-TURKISH WAR.

*L'Année Maritime, 1878—1879. Paris.

*Journal United Service Institution, vol. xxii.—iii. London.

HOBART PASHA. *Blackwood's Magazine*, 1885. London.

ALEXANDRIA, BOMBARDMENT OF.

WALFORD. *Journal United Service Institution*, vol. xxvii. London.

*GOODRICH. Egyptian War. Two Parts. Washington, 1883.

L'Année Maritime, 1882. Paris.

Times, Standard, Army and Navy Gazette.

CAPTURE OF SFAX.

*Revue Maritime, vol. lxxvi. Paris.

L'Année Maritime, 1881. Paris.

CHILI AND PERU.

MARKHAM. War between Chili and Peru. London, 1882. (*Cited as* Markham.)

MACKENNA. Guerra del Pacifico. 4 vols. Santiago, 1880-2.

*MASON. War on the Pacific Coast. Washington.

BARROS ARANA. Histoire de la Guerre du Pacifique. Paris, 1881.

Journal United Service Institution, vol. xxv. London.

*Revue Maritime, vols. lxv—viii., lxxi. Paris.

L'Année Maritime. 1879—81. Paris.
   Gives official documents.

SHAH AND HUASCAR.

*Parliamentary Papers. 52. 1877. London.
Times, Standard, Army and Navy Gazette, Hansard Parliamentary Debates.

FRANCO-CHINESE WAR.

ROCHE AND COWEN. The French at Foochow. Shanghai, 1884.
CARRALL. Report of Imperial Chinese Customs.
*LOIR. L'Escadre de l'Amiral Courbet. Paris, 1894. (*Cited as* Loir.)
DE DONCOURT. Les Français dans l'Extrême Orient. Lille, 1884.

CHILIAN CIVIL WAR.

*LAIRD CLOWES. Naval Annual, 1892. Portsmouth. (*Cited as* Clowes.)
HERVEY. Dark Days in Chile. London, 1892. (*Cited as* Hervey.)
Times, Standard, Army and Navy Gazette.
Revue Maritime, vol. xcv., Paris.

BRAZILIAN CIVIL WAR.

*LAIRD CLOWES. Naval Annual, 1894. Portsmouth.
**Information from Abroad.* 1894. Washington.
**Revue Maritime.* March, 1895. Paris.
Times, Standard, Army and Navy Gazette, Le Yacht.

CHINO-JAPANESE WAR.

NORMAN. The Far East. London, 1895.
*JUKUCHI INOUYÉ. Japan-China War. Yokohama.
*LAIRD CLOWES. Naval Annual, 1895. Portsmouth.
      ,,   ,,   *Blackwood's Magazine*, 1895. London.

McGiffin. *Century Magazine*, 1895. New York. (*Cited as* McGiffin.)
> This has not been used in the text, as it did not appear till the book was in the press. It has been employed in the notes, and is by far the best Chinese account of the Yalu that has as yet appeared.

Broad Arrow, 1894. London.
> A series of articles on the Yalu, in which the hand of an eminent British strategist and tactician will be recognised.

Blackwood's Magazine, October and November, 1895.

Times, Standard, Pall Mall Gazette, Daily News, Army and Navy Gazette, United Service Gazette, Le Yacht, Marine Française, Le Moniteur de la Flotte, Marine Rundschau.

NAVAL CATASTROPHES.

Parliamentary Papers on sinking of *Captain*, *Vanguard* and *Victoria*.

Times, *Engineer*, and *Engineering* on the same, and on the loss of the *Grosser Kurfürst*.

*Thursfield. The Loss of H.M.S. *Victoria*. Naval Annual, 1894.

*Elgar. The Loss of the *Victoria*. *Nature*, vol xlix. (*Cited as* Elgar.)

IMAGINARY NAVAL WARS AND BATTLES.

Great Naval War of 1887. London, 1887.
Plus d'Angleterre. Paris, 1887.
Arnold-Forster. In a Conning Tower. London, 1891.
Nelson Seaforth. The Last Great Naval War. London, 1891.
Rope. Rome et Berlin. Paris, 1888.
The "Russia's Hope." London, 1888.
Eardley Wilmot. The Next Naval War. London, 1894.
Laird Clowes. The Captain of the *Mary Rose*. London, 1892.

APPENDIX V.—ILLUSTRATIONS.

Originally I had intended that authentic and accurate illustrations of the various incidents described should be included, but as there were difficulties in the way of procuring such, which would have added largely to the size and cost of the book, I decided that photographs of the leading English and French ships would be preferable, and probably as useful. These have been necessarily placed at intervals throughout the two volumes, and not opposite the matter which they illustrate. Captain Mahan's flagship, the *Chicago*, has been given as an example of a fine type of modern rigged cruiser with heavy armament and fair speed. A number of Chinese and Japanese ships are reproduced from Mr. Ogawa's photographs. The *Huascar* is from a print in the "Illustrated London News," and the battle of Lissa from an Austrian painting. For the 6- and 8-inch quick-firers I have to thank Sir W. Armstrong, Mitchell, & Co.; for the elevations of the *Chen Yuen* and *Naniwa*, which are from the "Naval Annual" of 1889, the Hon. T. A. Brassey; and for the "*Chen Yuen* in action," Mr. F. T. Jane. The elevations of English and French ironclads are all drawn to the same scale, and are compiled from the "Naval Annual," the "Engineer," Croneau's "Architecture Navale," and "Information from Abroad." The diagram of the *Victoria* sinking is from the Parliamentary Paper on the court martial. The "End of a Battleship" is reproduced from the "Cosmopolitan," and originally represented the destruction of the *Aquidaban* by a dynamite shell. The maps and plans are compiled by myself from various sources: the diagrams of the Yalu are necessarily to some extent conjectural. They are based upon the plans of Mr. Laird Clowes, Mr. Jukuchi Inouyé, and the "Revue Maritime" for January, 1895.

The illustration in the second volume, called "The Last of the *Victoria*," is from a photograph taken by Staff-Surgeon Collot, of H.M.S. *Collingwood*, and depicts H.M.S. *Victoria* as last seen off Tripoli, Syria, on the afternoon of the 22nd of June, 1893. It has been reproduced, by kind permission, from a print, the copyright of Mr. R. Ellis, of Valetta, Malta.

TABLE I.—UNITED STATES NAVAL ORDNANCE, 1861-5.

Name of Gun.	Description.	Weight of Gun. lbs.	Weight in lbs. of Shot.	Weight in lbs. of Shell.	Charge. lbs. Maximum.	Charge. lbs. Ordinary.	Maximum Range. Yds.	Elevation to give Range. Degrees
15-inch	Smooth-bore	42,000	440	352	50	35	2100	7°
14-inch	Smooth-bore	15,700	196	135	30	15	3440	15°
10-inch	Smooth-bore	{ 16,000 / 12,000 }	124	101	{ 40 / 15 }	{ 25 / 12½ }	3000	11°
9-inch	Smooth-bore	9000	90	73	13	10	3300	15°
8-inch	Smooth-bore	{ 10,000 / 5500 }	65	52	{ 20 / 7 }	{ 16 / 9 }	2600 / 1600	10° / 5°
64-pounder	Smooth-bore	10cwt.	64	52	16	8	...	...
32-pounder	Smooth-bore	{ 57cwt. / 27cwt. }	32	26	{ 9 / 4 }	{ 6½ / 3⅓ }	2731	10°
8-inch	Parrot Rifle	...	{ 135 / 154 }	135	16	...	2100	5°
6'4-inch	Parrot Rifle	...	{ 70 / 100 }	80 / 100	10	...	8453	35°
5'8-inch	Parrot Rifle	...	60	50	6	...	...	...
4'2-inch	Parrot Rifle	...	30	29	3½	...	6700	25°
3'6-inch	Parrot Rifle	...	20	18	2	...	4400	15°

Where there are two or more guns of the same calibre, but of different weights and sizes, and only one figure is given for range and elevation, that figure refers to the most powerful pattern.

TABLE II.—UNION FLEET AT NEW ORLEANS.

	13-in. Mort.	11-in. S.B.	10-in. S.B.	9-in. S.B.	8-in. S.B.	32-pr. S.B.	100-pr. R.	80-pr. R.	50-pr. R.	30-pr. R.	20-pr. R.	Total Guns.	Howitzers. 24-pr.	Howitzers. 12-pr.	Howitzers. Total	Grand Total	Tons.
Hartford				24							2	26		2	2	28	1900
Brooklyn				22								24		2	2	26	2070
Richmond		1	1	20			1					22				22	1929
Pensacola		2		20			1					23	2		2	25	2155
Mississippi					19	4						21		1	1	22	1692
Oneida					8	4				1		9	2		2	10	1032
Varuna		2										10				11	1300
Iroquois		2			1				1			10	3		3	11	1016
Cayuga		1								1		2	2		2	4	507
Itasca						4				1		4	2		2	4	507
Katahdin						4				1		2	2		2	4	507
Kennebec		1								1		2	2		2	4	507
Kineo		1								1		3	2		2	5	507
Pinola										1		2	2		2	4	507
Sciota		1								1		2	2		2	4	507
Winona		1								1		2	2		2	4	507
Wissahickon		1								1		2	2		2	4	507
Total Fleet		13		88	27	10	3	1		8	12	166	17	9	26	192	
Mortar Division	19	1		9	20	46	1			2	2	101	8	1	9	110	
	19	14	2	97	47	56	4	1	1	9	14	267	25	10	35	302	

CONFEDERATE FORTS AND SQUADRON.

	7-in. R.	32-pr. R.	10-in. S.B.	9-in. S.B.	8-in. S.B.	42-pr. S.B.	32-pr. S.B.	24-pr. S.B.	16-pr. S.B.	13-in. Mort.	10-in. Mort.	8-in. Mort.	8-in. How.	7-in. How.	24-pr. How.	12-pr. How.	Grand Total
Ft. Jackson (Proper)	2	2	2		3	6	15	25				2		1	10	1	74
Water Battery					2			1									
Ft. St. Philip (Proper) Outer Batteries	1	1			4	6	9	20			1		1		1		52
" " Outer Batteries	2		1	3	4			6			4		1				16
Louisiana				3			2										8
McRae				1			2	2									2
Jackson								2									2
Manassas						1											1
Gov. Moore							2										2
General Quitman							2										2
Warrior							1	1									1
Stonewall Jackson							1	1									1
Defiance							1	1									1
Resolute						1											1
General Lovell						1											1
Breckinridge							1	1									1
Two launches																2	2
	4	3	3	4	14	13	33	56	4		6	3	1	1	10	3	156

K. = Rifle. S.B. = Smooth-bore. How. = Howitzer. Mort. = Mortar.

TABLE III.

UNION FLEET'S ATTACK ON FORT SUMTER.
April 7th, 1863.—Details:

	15-in. S.B.	11-in. S.B.	150-pr. R.	Shots Fired.	Times Hit.	Tons.	Thickest Armour. Inches.
Weehawken	1	1	—	26	53	844	11
Passaic	1	1	—	13	35	844	11
Montauk	1	1	—	27	14	844	11
Patapsco	—	1	1	10	47	844	11
Catskill	1	1	—	22	20	844	11
Nantucket	1	1	—	15	51	844	11
Nahant	1	1	—	14	36	844	11
Keokuk	—	2	—	3	90	677	3
New Ironsides	2	14	2	9	65?	3486	4½
	7	22	3	139	520		

CONFEDERATE FORTS OPPOSING ATTACK.

	10-in. S.B.	9-in. S.B.	8-in. R.M.L.	8-in. S.B.	42-pr. R.	32-pr. R.	32-pr. S.B.	10-in. Mort.	Total.
Fort Johnson	—	—	—	—	—	1	—	1	—
" Sumter	4	2	2	8	7	5	13	2	44
" Moultrie	5	—	—	9	—	1	5	7	31
Battery Bee	—	—	—	1	—	—	—	—	6
" Beauregard	—	1	—	—	—	1	—	—	2
" Gregg	—	—	—	—	—	1	—	—	2
" Wagner	1	—	—	1	—	—	—	—	2
	10	3	2	19	7	8	18	10	77
Rounds fired	385	80	86	336	140	366	343	93	2220

TABLE IV.—UNION FLEET AT MOBILE.

	15-in. S.B.	11-in. S.B.	10-in. S.B.	9-in. S.B.	32-pr. S.B.	150-pr. R.M.L.	100-pr. R.M.L.	60-pr. R.M.L.	50-pr. R.M.L.	30-pr. R.M.L.	20-pr. R.M.L.	Total Guns.	Tons.	Thickest Armour. Inches.
Tecumseh	2	—	—	—	—	Single-turret	—	Monitor	—	—	—	2	1034	10
Manhattan	2	—	—	—	—	Single-turret	—	Monitor	—	—	—	2	1034	10
Winnebago	—	4	—	—	—	Double-turret	—	Monitor	—	—	—	4	970	8½
Chickasaw	—	4	—	—	—	Double-turret	—	Monitor	—	—	—	4	970	8½
Hartford	—	—	—	18	—	—	2	—	—	—	—	21	1900	—
Brooklyn	—	—	2	20	—	—	2	—	1	—	—	24	2070	—
Richmond	—	—	—	18	—	—	2	—	1	—	—	20	1029	—
Lackawanna	—	2	—	4	—	—	1	—	—	3	—	8	1533	—
Monongahela	—	2	—	—	6	—	1	—	—	3	—	11	1378	—
Ossipee	—	—	—	4	—	—	1	—	—	—	—	9	1240	—
Oneida	—	2	—	—	4	—	—	—	—	3	—	10	1032	—
Galena	—	—	—	8	—	—	—	—	2	—	—	8	738	3
Seminole	—	1	—	1	6	—	—	—	—	—	—	6	805	—
Port Royal	—	—	—	4	6	—	—	—	—	—	—	6	974	—
Metacomet	—	—	—	3	—	—	—	—	—	—	2	6	829	—
Itasca	—	—	—	—	3	—	—	—	—	—	2	5	507	—
Kennebec	—	—	—	—	3	—	—	—	—	—	—	3	507	—
	4	18	1	77	27	2	11	—	3	10	4	159		

TABLE V.—UNION FLEET AT THE SECOND BOMBARDMENT OF FORT FISHER, WITH ITS ARMAMENT.

	15" S.B.	12" S.B.	11" S.B.	10" S.B.	9" S.B.	8" S.B.	150-pr. R.	100-pr. R.	60-pr. R.	30-pr. R., 32-pr. S.B.	20-pr.	Shots Fired.	Killed.	Wounded.	Missing.
Line No. 1.															
Brooklyn	...	...	...	...	20	...	...	2	2	...	...	...	3	12	0
Mohican	...	...	...	...	6	...	...	1	...	2	...	436	12	0	0
Tacony	...	2	...	...	4	...	...	...	...	...	...	620	0	0	0
Kansas	...	...	...	...	2	...	...	1	1	...	...	485	0	1	0
Yantic	...	...	...	...	2	...	...	1	1	...	...	235	2	2	0
Unadilla	...	...	1	...	...	...	...	...	...	...	1	403	0	0	0
Huron	...	...	1	...	...	...	...	...	...	1	...	300	0	5	0
Maumee	...	...	...	...	...	...	...	1	...	{ 2 S.B. } { 1 }	...	340	0	0	0
Pequot	...	...	...	...	...	...	1	...	...	{ 6 S.B. } { 1 }	...	498	3	5	0
Pawtucket	...	...	1	...	4	...	...	1	...	...	...	463	0	0	0
Seneca	...	...	1	...	...	...	...	...	...	...	1	252	0	0	0
Pontoosuc	...	...	...	...	4	...	...	2	...	...	2	318	0	7	0
Nereus	...	...	...	...	...	...	...	...	1	{ 2 } { 5 S.B. }	...	540	3	3	0
Line No. 2.															
Minnesota	...	...	1	...	42	...	1	4	...	...	...	1054	13	23	0
Colorado	...	...	1	...	40	...	1	...	...	...	...	786	3	14	0
Wabash	...	...	...	...	42	...	1	...	...	...	...	1035	0	12	0
Susquehanna	...	...	...	...	12	...	2	...	...	...	...	836	3	15	8
Powhattan	...	...	1	...	14	...	...	3	...	...	...	...	3	19	7
Juniata	...	...	...	...	...	6	...	1	...	2	...	1003	5	10	0
Shenandoah	...	...	2	...	...	...	1	...	...	1	...	317	6	0	5
Ticonderoga	...	...	...	...	12	...	...	...	...	1	...	552	1	6	0
Vanderbilt	...	...	...	...	12	...	...	2	...	2	...	170	0	0	0
Mackinaw	...	...	1	...	6	...	...	...	...	...	...	939	0	2	0
Tuscarora	...	...	...	...	...	6	...	1	...	2	...	214	3	12	0
Line No. 3.															
Sant. de Cuba	...	...	...	...	...	...	...	...	...	{ 3 } { 5 S.B. }	...	199	1	9	0
Ft. Jackson	...	...	...	...	8	...	...	1	...	2	...	...	1	10	0
Osceola	...	...	1	...	4	...	...	1	...	...	...	319	0	0	0
Sassacus	...	...	...	...	4	...	...	2	...	...	2	262	0	0	0
Chippewa	...	...	...	...	1	...	...	...	...	...	1	74	0	0	0
R. R. Cuyler	...	...	...	...	...	...	...	...	...	{ 2 } { 2 S.B. }	...	49	0	0	0
Maratanza	...	...	...	...	1	...	...	1	...	...	...	...	0	0	0
Rhode Island	...	...	...	...	1	5	...	...	...	2	...	299	8	2	0
Monticello	...	...	...	...	2	...	...	1	...	3	...	262	4	4	0
Alabama	...	...	...	...	1	...	...	...	...	{ 2 } { 6 S.B. }	...	...	0	0	0
Montgomery	...	...	...	1	4	...	...	...	...	1	...	580	2	4	0
Iosco	...	...	...	...	4	...	...	2	...	...	...	558	2	12	0
Ironclads.															
New Ironsides	...	...	14	...	...	...	2	...	2	...	...	971	0	0	0
Monadnock	4	...	...	...	...	...	...	...	...	...	...	441	0	0	0
Canonicus	2	...	...	...	...	...	...	...	...	...	...	207	0	3	0
Mahopac	2	...	...	...	...	...	...	...	...	...	...	153	0	0	0
Saugus	2	...	...	...	...	...	...	...	...	...	...	212	0	1	0
Flagship.															
Malvern	...	...	...	...	...	...	...	...	...	...	...	...	3	1	0

R. = Rifled.   S.B. = Smooth-bore.

TABLE VI.—THE SOUTHERN COMMERCE-DESTROYERS AND THEIR PRIZES.*

Steamers.	Vessels whose fate is not stated.	Vessels destroyed.	Vessels on which Cargo only was destroyed.	Vessels Bonded or Sold.	Vessels Released or Recaptured.	Vessels used as Cruisers.	Total Prizes.
Alabama	...	53†	2	11	2	1	69
Sumter	...	7	...	2	9	...	18
Florida	...	28	4	4	...	1	37
Georgia	...	5	...	4	...	...	9
Shenandoah	...	31	...	5	...	...	36
Tallahassee	...	22	...	5	2	...	29
Olustee (ex-Tallahassee)	...	4	2	...	...	...	6
Nashville	...	1	1	...	...	...	2
Calhoun	3	...	...	...	...	...	3
Chickamauga	1	2	1	...	...	...	4
Boston	1	...	1	...	...	...	2
Sailing Vessels.							
York	...	...	...	...	1	...	1
Tacony	8	1	1	4	1	...	15
Clarence	7	...	...	1	...	...	8
Jeff. Davis	7	...	...	...	1	...	8
Winslow	5	...	...	...	...	...	5
Retribution	1	1	1	...	...	...	3
Echo	2	...	...	...	...	...	2
Chickamauga	1	2	1	...	...	...	4
							261

* The figures given in this table are based upon the returns of ships captured, in Scharf, p. 807-816, and differ slightly from those given in the text in some instances. There are the same differences and discrepancies in Scharf.
† Includes the *Hatteras*.

TABLE VII.—ITALIAN FLEET AT LISSA.

	Name of Ship.	Class of Ship.	Thickest Armour. In.	Men.	Tonnage.	Horse-powr.	Guns.	Name of Captain	Shots Fired.	Rifled Guns. Armstrong			Smooth Bores.			
										25-cm.	20-cm. long	16-cm.	20-cm.	16-cm.	12-cm.	8-cm.
1	*Re d'Italia*†	1st class Frigate	5	600	5700	800	36	Capt. Faa di Bruno	240	2	2	30	4			
2	*Re di Portogallo*	,, ,,	5	550	5700	800	28	Capt. Ribotti	294	2	2	26				
3	*Principe di Carignano*	2nd class ,,	4½	440	4086	600	22	R.-Adm. Vacca Capt. Saudi				18	4			
4	*Maria Pia*	,, ,,	4½	484	4250	700	26	Capt. Del Carretto	189			22	4			
5	*Castelfidardo*	,, ,,	4½	484	4250	700	27	Capt. Cacace	384			23	4			
6	*Ancona*	,, ,,	4½	484	4250	700	27	Capt. Piola	238			23	4			
7	*San Martino*	,, ,,	4½	484	4250	700	27	Capt. Roberti	198			22	4			
8	*Affondatore*†	Turret Ram	5	390	4070	700	7	{Adm. Persano Capt. Martini}		2						
9	*Formidabile*	Central Battery	4½	356	2700	390	20	Capt. Saint Bon				16	4			
10	*Terribile*	,, ,,	4½	356	2700	450	20	Capt. De Cosa				16	4			
11	*Palestro*†	Armd. Gun-boat	4½	250	2000	300	5	Com. Cappellini			2		2			
12	*Varese*	,, ,,	4½	250	2000	300	5	Com. Fincati			2		2			
13	*Carlo Alberto*	1st class Frigate	None	580	3200	400	50	Com. Pucci					7	10	32	
14	*Duke of Genoa*	,, ,,	,,	580	3315	450	54	Capt.Di Chiavenna					8	10	32	
15	*Gaeta*	,, ,,	,,	580	3030	450	54	Capt. Cerruti					8	12	34	
16	*Garibaldi*	,, ,,	,,	580	3086	450	54	Capt. Vinegliano					8	12	34	
17	*Maria Adelaide*†	,, ,,	,,	550	3459	600	52	{V. Adm. Albini Capt. Di Monale}				9	23			
18	*Principe Umberto*	,, ,,	,,	580	3691	600	50	Capt. Acton					8	10	32	
19	*Victor Emmanuel*	,, ,,	,,	580	3415	580	50	Capt. Imbert				8	8	32		
20	*San Giovanni*	Corvette	,,	345	1790	210	20	Com. Burcone				6	10	24		
21	*Governolo*	,,	,,	260	1700	450	6	Com. Gogola						6		
22	*Guiscardo*	,,	,,	190	1400	300	6	Com. Pepi						6		
23	*Ettore Fieramosca*	Despatch Boat	,,	190	1400	380	2	Com. Baldisarotto				2	2			
24	*Messaggiere*	,, ,,	,,	128	1000	380	2	Com. Giribaldi				2	2			
25	*Esploratore*	,, ,,	,,	108	1000	380	2	Com. Orengo				2	2			
26	*Giglio*	,, ,,	,,	29	250	60	1	Lieut. De Negri								
27,28,29	Three Gunboats, each			63	262		4									
30	Washington	Hospital Ship	,,	98	1400	250	2	Lieut. Zuccavo								2
	Ships 30			10,572	84,422		636			4	6	206	130	228	17	2

Also two transports, *Independenza* and *Piemonte*, and two unarmed merchant steamers, *Flavio Gioja* and *Stella d'Italia*.

† Ironclads in italics. ‡ Ships sunk.

Approximate weight of projectiles: 25-cm., 500lbs.; 20-cm., 300lbs.; 16-cm., 100lb. Smooth Bore: 16-cm., 72lbs.; 20-cm., 68lbs.

Rifled: 12-cm., 50lbs.; 8-cm., 15lbs.

TABLES. 297

TABLE VIII.—AUSTRIAN FLEET AT LISSA.

	Class of Ship.	Thickest Armour Inches.	Men.	Tonnage. Builders Measurements.	Horse Power, Nominal.	Guns.	Shots fired.	Description of Guns. 60-pr. R.	48-pr. S.B.	30-pr. S.B.	12-pr. R.	Killed.	Severe.	Slight.	Total.	Captain.
1 *Archduke Ferdinand Maximilian*	1st Class Frigate	5	499	5130	800	18	176	18						5	8	Rear Adm. Tegetthof. Capt. Baron Sterneck.
2 *Hapsburg*	" " "	4½	428	5130	800	18	170		16			1			1	Capt. Faber.
3 *Kaiser Maximilian*	2nd Class Frigate	4½	386	3588	640	30	217		16							" Groller.
4 *Prinz Eugen*	" " "	4½	386	3588	650	30	234		16					3	3	" Barry.
5 *Don Juan*	" " "	4½	386	3588	650	26	277		14				1	4	5	" Wiplinger.
6 *Drache*	3rd Class Frigate	4½	343	3064	500	26	121		10			1	2	4	7	" Baron Moll.
7 *Salamander*	" " "	4½	343	3064	500	26	211		10					2	2	" Kern.
8 *Kaiser*	Line of Battle Ship	None	904	5194	800	92	850	10		74		24	37	38	99	Commodore Petz.
9 *Novara*	Frigate	" "	552	2497	500	52	343	4		44	4	7	3	17	27	Capt. Klint.
10 *Schwarzenberg*	"	" "	547	2614	400	46	286	4		36	4					" Millosich.
11 *Radetzky*	"	" "	395	2198	300	33	289	3		24	4					" Aurnhammer.
12 *Adria*	"	" "	395	2198	300	33	221	3		24	4					Com. Daufalik.
13 *Donau*	"	" "	395	2198	300	31	326	3		24	2					" Pittner.
14 *Archduke Frederick*	Corvette	" "	284	1474	230	22	255	2		16	2					" Florio.
15-21 Seven gunboats each	Gunboat	" "	139	852-860	230	4	Total, 446		2							
22-23 Two sloops each	Sloop	" "	100	501	99	6	Total, 33									
24 *Empress Elizabeth*	Imp. Yacht	" "	186	1450	350	4	71	1			3					Capt. Oesterreicher.
25 *Andreas Hofer*	Tender	" "	186	770	180	2	51									Lieut. Lund.
26 *Greif*	Imp. Yacht	" "	103	1200	300	2	5									Com. Kronowetter.
Ships 26			7529	55,944	10,060	532	4450	42	120	252	6	38	50	29	179	

Also the unarmed merchant steamer *Stadium*. Ironclads in Italics. S.B. = Smooth-bore. R. = Rifle.
Approximate weight of projectiles, English lbs.: 60-pr. = 65lb.; 48-pr. = 48lb.; 40-pr. = 36lb.; 30-pr. = 30lb.; 24-pr. = 25lb.; 12-pr. = 15lb.

TABLE IX.—COMPARISON OF FLEETS AT LISSA.*

	Ironclads.		Wooden Ships.			Small Craft.			*Non-combatant Ships.	Total.		All Ships are Included.				Weight of Shot fired in one round.†				Loss.		
	No.	Guns.	No.	Guns.		No.	Guns.			Ships.	Guns.	No. of Men.	Tonnage.	Guns. Rifled.	S.B.	Rifles. lbs.	Sm. Bores. lbs.	Total. lbs.	Killed.	Wounded.	Total.
Austria	7	176	7	304		9	40		4	27	532	7871	57,344	121	412	7130	16,400	23,530	38	138	176
Italy	12	343	11	382		3	11		8	34	645	10,866	86,022	276	369	28,700	24,536	53,236	606‡	39	639‡

* Including all Avisos, Hospital Ships, Merchant Steamers, &c. † Excluding Austrian 12-pounder rifles, and Italian of 12 and 8-c/m., particulars of which are not to hand. ■ Slight differences will be found in the totals here given as compared with the either tables.
‡ Includes those lost on board the *Palestro* and *Re d'Italia*; number not accurately known. The transport and merchant vessels are excluded in the other tables, since all the ships with each fleet are included here, whereas the transport and merchant vessels are excluded in the other tables.

TABLE X.—TYPES OF FRENCH IRONCLADS IN 1870.

	Tonnage	Length. Ft.	Beam. Ft.	Draught. Ft.	Armour on water-line. Inches.	Horse-Power.	Speed. Knots.	Rifled Breechloader Armament.	
Magnanime	6613	259	55·9	27·10	5·9	3452	14	VIII 24-c/m., IV 19-c/m.	Frigate; broadside battery.
†*Couronne*	5700	260	54·3	25	4·7	2000	12	VIII 24-c/m., IV 19-c/m.	,, ,,
**Magenta*	6600	284	57·3	27·10	4·9	3500	13	X 24-c/m., IV 19-c/m.	,, ,,
Solferino	6680	284	57·3	27·10	4·7	3500	13	X 24-c/m., IV 19-c/m.	,, barbette and broadside.
§*Gloire*	5600	252	55·9	26	4·7	2500	13	XXXVI 16-c/m.	,, broadside.
Alma	3600	239	46·3	23	6	1800	12	VI 19-c/m.	Corvette; barbette and central battery.
Cerbère	3600	216	59·6	19	9	1400	11	II 24-c/m.	Coast defence turret-ship.
Taureau	2400	216	52·6	19	9	1400	13	II 24-c/m.	,, barbette-ship.
Océan	7350	283	57·7	27	8	3700	13	IV 27-c/m., IV 24-c/m., II 16-c/m.	Frigate; barbette and central battery.

Guns: 27-c/m. = 10·8-inch, shot 176lb.; 24-c/m. = 9·45-inch, shot 176lb.; 19-c/m. = 7·6-inch, shot 165lb.; 16-c/m. = 6·3-inch, shot 99lb.

GERMAN IRONCLADS.

	Tonnage	Length. Ft.	Beam. Ft.	Draught. Ft.	Armour on water-line. Inches.	Horse-Power.	Speed. Knots.	Rifled Breechloader Armament.	
König Wilhelm	9602	359	60	26	8	7000	14·7	XVIII 24-c/m., V 21-c/m.	Frigate; broadside battery.
Friedrich Karl	5600	290	54·6	24	5	3450	13·5	XVI 21-c/m.	,, ,,
Kronprinz	4738	286	...	24	5	4700	14·3	XVI 21-c/m.	,, ,,
Pr. Adalbert	1450	160	30	16	4·7	1184	9·5	I 21-c/m., II 17-c/m.	Coast defence turret-ship.
Arminius	1550	197	...	13	4·7	1184	10·5	IV 21-c/m.	,, ,,

Guns: 24-c/m. = 9·45-inch, shot 176lb.; 21-c/m. = 8·24-inch, shot 174lb.; 17-c/m. = 6·8-inch, shot 111lb.

TABLE XI.—FLEETS OF CHILI AND PERU, 1878.

CHILI.

Description.	Name of Ship.	Tonnage.	Horse-Power.	Nom. Speed.	Launched. Date.	Thick Armour.	Armament.
Cent. Battery Battleship	Blanco Encalada	3450	3000	13	'75	9	VI 9-in. 12 ton M., I 20-pr., I 9-pr., 1 7-pr. B., I 7-pr. B., II 1-in. Nords., I Machine.
" "	Almirante Cochrane	3370	2920	13	'74	9	VI 9-in. 12 ton M., I 20-pr., I 9-pr. B., I 7-pr. B., II 1-in. Nord., I Machine.
Corvette	O'Higgins	1650	1100	13	'66	...	III 7-in. M., II 70-pr. M., IV 40-pr. M.
"	Chacabuco	1100	1100?	10?	...	...	III 7-in. M., II 70-pr. M., IV 40-pr. M.
Sloop	Abtao	1050	1000	9	'64	...	III 7-in. M., III 30-pr.
"	Esmeralda	850	200	7	...	...	XIV 40-pr.
"	Magallanes	772	1230	11	'74	...	I 7-in. M., I 60-pr. M.
"	Covadonga	412	140	8	...	...	II 70-pr. M.

Armed Merchant Steamers or Transports, Consiño, Loa.

PERU.

Description.	Name of Ship.	Tonnage.	Horse-Power.	Nom. Speed.	Launched. Date.	Thick Armour.	Armament.
Coast Defence Monitor	Huascar	1130	1200	12	'65	5½	II 10-in. 12 ton M., II 40-pr. M., I 12-pr. M., I Machine.
Broadside Ironclad	Independencia	2004	2200	12	'64	4½	I 8-in. M., III 7-in. M., XII 70-pr. M., IV 30-pr.
Coast Defence Monitor	Manco Capac	1034	1300	6	Compltd. '66	10	II 15-in. 440-pr. Smooth-bores.
" "	Atahualpa	1034	1300	6	Compltd. '66	10	II 15-in. 440-pr. Smooth-bores.
Corvette	Union	1500	1200	13	'64	...	XII 70-pr. [Also given as II 100-pr., and I heavy French gun.]
Gunboat	Pilcomayo	600	1060	9	'74	...	II 70-pr., IV 40-pr.

TABLES.

TABLE XII.—SHIPS WHICH TOOK PART IN THE BOMBARDMENT OF ALEXANDRIA.

Name.	Displacement.	Horse-power.	Draught of Water.	Date.	Armour. Side.	Armour. Bulkhead.	Turret or Batt.	Armament. All heavy guns Muzzle-loaders.	Crew.	System.
	Tons.									
Alexandra (2–1½" deck)	9490	8610	26' 6"	'77	14"–12"	8"	9⅜"–6"	II 25-ton, X 18-ton, XXX light	650	Central Battery.
Inflexible (3" deck)	11,880	8000	25' 5"	'81	24"–16"	22"–14"	18"	IV 80-ton, VIII 20-pr. R., XXIII light	484	Turret.
Sultan	9290	7720	26' 6"	'71	9"–6"	6"	8"	VIII 18-ton, IV 12-ton, VII 20-pr. R., XIX light	490	Central Battery.
Superb	9170	6580	26' 3"	'80	12"–7"	6"	10"	XVI 18-ton, VI 20-pr. R., XXV light	620	Central Battery.
Temeraire (1½"–1" deck)	8540	7500	27' 2"	'77	11"–8"	5"	10" & 8"	N 25-ton, IV 18-ton, XXXI light	524	Cent. Batt. and Barbette.
Invincible	6010	4910	22' 9"	'70	8"–6"	5"–4"	10"–9"	X 12-ton, IV 6-ton, II 7-inch MH light	450	Central Battery.
Monarch	8320	4750	26' 9"	'69	7"–4½"	5"	10"–9"	IV 25-ton, III 64-pr. MH light	325	Turret.
Penelope	4430	4720	17' 6"	'68	6"–4½"	4½"		VII 9-ton, III 40-pr. R., II 20-pr. R., XIII light	273	Broadside.
Bacon	603	380						I 18-ton, II 64-pr., II 20-pr. R.	75	
Bittern	585	850	10' 1"	'69				I 18-ton, II 64-pr. R., III light	90	Gunboats.
Condor	780	770	13' 2"	'76				I 18-ton, II 64-pr., III light	100	
Cygnet	440							II 64-pr., II 20-pr. R.	60	
Decoy	440							II 64-pr., II 20-pr. R.	60	
Helicon	1000	1080	10' 5"	'65				II 20-pr. R.	90	Despatch Boat.

E. = Breechloader.

TABLE XIII.—ARMAMENT OF THE ALEXANDRIA FORTS (GUNS MOUNTED ONLY).

	Rifled						Smooth Bore.				Mortars.			Total.			Grand Total.
	Muzzle-loaders.					40-pr. B.L.	15"	10"	8"	6"'s	20"	18"	13"	11"	Rifled.	S.B. and Mortars.	
	10"	9"	8"	7"													
Fort Silsileh	1							3							1	4	6
Fort Pharos	1	3	3					6	31			4			8	41	49
Fort Ada	1	3	1	3		2		14	11			5			9	40	49
Ras-el-Tin Lines	1	3	2	2		1	4	15	28			6			9	53	62
Fort Saleh Aga							2	5	6			1	2		4	16	20
Battery								4	10								
Fort Oom-el-Kubebe			1					6	3			3	2	2	4	18	22
Fort Kamaria		1				2		4	5							6	
Fort Mex								4	4								
Mex Lines			3				4	13	9			3	1		7	27	34
Fort Marsa			2	3		2		4	3			1	1		7	14	21
Fort Marabout																4	4
								9	16			2	2		2	37	39
	5	18	14	4		3	10	84	117	1	24	4	9	44	249	293

TABLE XIV.—SHOT AND SHELL EXPENDED AT ALEXANDRIA BY THE BRITISH FLEET.

	Guns.							Total Heavy Guns.	Guns.				Grand Total.	Martini-Henry.	Nordenfelt.	Gatling.	Rockets.
	16″ 80-ton.	12″ 25-ton.	11″ 25-ton.	10″ 18-ton.	9″ 12-ton.	8″ 9-ton.	7″ 6½-ton.		64-pr.	40-pr.	20-pr.	9 & 7-pr.					
Alexandra	...	...	48	221	...	...	...	269	...	...	138	12	407	...	4000	340	...
Sultan	...	...	...	132	50	...	...	187	...	...	139	12	338	...	1800	2000	...
Superb	...	...	...	310	...	...	...	310	...	...	60	41	411	...	1161	880	...
Penelope	117	...	...	...	48	331	...	331	...	96	23	...	360	...	1672	...	11
Monarch	...	117	...	...	48	...	21	186	...	...	...	181	367	...	340	2860	...
Tenerraire	88	...	130	84	...	...	...	220	...	...	8	...	225	5000	160	1000	...
Invincible	88	...	130	84	126	...	...	126	100	...	18	...	250	1850	2000	...	21
Inflexible							16	88	22	...	120	10	268	...	2000	...	...
Beacon	...	...	...	...	...	...	16	16	32	...	53	8	101	320	...	...	3
Condor	...	...	...	...	...	...	65	65	128	96	...	...	201	1000	...	200	13
Bittern	...	...	...	...	...	...	33	33	...	...	...	...	89	...	...	...	...
Cygnet	...	...	...	...	...	...	...	...	101	...	42	...	143	...	...	...	...
Decoy	...	...	...	...	...	...	...	...	49	...	10	...	69	40	...	...	...
Helicon	...	...	...	...	...	...	...	...	6	...	...	...	6	...	...	...	...
	88	117	184	752	224	331	135	1731	412	192	621	282	3198	10,160	16,333	7100	37

{Gunboats: Beacon, Condor, Bittern, Cygnet, Decoy, Helicon}

TABLE XV.—FRENCH AND CHINESE SHIPS ENGAGED OFF FOOCHOW.

FRENCH.

	Speed. Knots.		Tonnage.	Horse-Power.	Crew.	Guns.	Damage, &c.
Triomphante ...	13	Armoured corvette 6-4¾" plating	4727	2400	410	VI 24-c/m, III 19-c/m., VI 14-c/m.	None.
Duguay Trouin	10	Composite cruiser	3189	3740	300	V 19-c/m., V 14-c/m.	Hammock nettings amidships, starboard side, carried away.
Villars	14	Cruiser, Wooden	2268	2790	250	XV 14-c/m.	Struck a little forward of 3rd gun-port.
D'Estaing	15	Cruiser, Wooden	2236	2790	250	XV 14-c/m.	None.
Volta Courbet	13	Sloop, Wooden ...	1300	1000	160	III 14-c/m., III 10-c/m.	Shot-hole a little above W.L. amidships to starboard.
Lynx	10	Gunboat	452	450	120	III 14-c/m., II 10-c/m.	None.
Aspic	10	Gunboat	471	427	120	II 14-c/m., II 10-c/m.	None.
Vipère	10	Gunboat	471	427	120	II 14-c/m., II 10-c/m.	None.

Two Torpedo Launches, Nos. 45 and 46, 16 knots speed, spar-torpedoes, crews of 10 men each.

CHINESE.

	Speed. Knots.		Tonnage.	Horse-Power.	Crew.	Guns.	Damage, &c.
Yang Woo ...	13	Composite cruiser	1400	1250	270	VIII 3½ ton M.L.R., I 6 ton M.L.R.	Torpedoed, set on fire, and sunk.
Foo Poo	10	Sloop, Wooden ...	1258	610	70	VI 45-pr., I 18 ton M.L.R.	Ran away, sunk, and back broken.
Chi-an	10	Sloop, Wooden ...	1258	610	150	I 6-in. VI 45-pr.	Burnt and sunk.
Fei Yuen	10	Sloop, Wooden ...	1258	610	150	VI 45-pr.	Burnt and sunk.
Ching Wei	10	Sloop, Wooden ...	578	480	100	I 3½ ton M.L.R. IV. 45-pr.	Sunk.
Foo Sing	8	Sloop, Wooden ...	558	410	90	I 3½ ton M.L.R. II 45-pr., II 40-pr.	Sunk.
Yu Sing	9½	Sloop, Wooden ...	260	...	30	III Small	Sunk.
Yang Pao } Chun Hing }	9	Transports	{ 1450 } { 1450 }	610	150	None	{ Sunk. { Burnt and sunk.
Chen Sing } Fuh Sing }	8	Rendel Gunboats	{ 250 } { 250 }	389	30	I 16 ton, 10-in. M.L.R.	Both sunk.

Eleven war junks ; seven launches fitted with spar-torpedoes.

TABLE XVI.—CONGRESSIONAL AND BALMACEDIST SQUADRONS, 1891.

CONGRESSIONAL.

Description.	Name of Ship.	Tonnage.	Horse-Power.	Nom. Speed.	Launched. Date.	Thickness of Armour.	Torp. Tubes.	Armament.
Cent. Battery Battleship	Blanco Encalada	3450	3000	13	'75	9	1	VI 8-in. 14 ton B., IV 6-pr. Q.F., IV 1-in. Nords., II Mach.
"	Almirante Cochrane	3370	2920	13	'74	9	3	VI 8-in. 14 ton B., IV 6-pr. Q.F., IV 1-in. Nords., II Mach.
Coast Defence Monitor	Huascar	1130	1200	12	'65	5½	...	II 8-in. B., II 40-pr. B., II Mach.
Cruiser	Esmeralda	2840	6500	18	'84	Deck 1	3	II 10-in. B., VI 6-in. B., II 6-pr. Q.F., VI Mach.
Corvette	O'Higgins	1670	1100	10	'66	...	...	III 7-in. B., IV 40-pr. B.
Sloop	Abtao	1050	1000	9	'64	...	...	I 70-pr. B., IV 40-pr. B.
"	Magallanes	772	1150	11	'74	...	...	I 7-in. B., I 64-pr. M., III 20-pr. B.

Armed Merchant Steamers and Transports: Acomcagua (4112 tons, 12 knots, II 5-in. B., I 40-pr. B., VI small); Maipo (2199); Amazonas (1970); Dithmarschen (962); Isidora de Cousino (995); Biobio (343); Copiapo (1337); Itata (1760); Limari (669); Cachapoal (2160 tons); Charaponi; Carlos Roberto (644); Valparaiso, Cachapoalete, Bismarck, Tolten, Maule, Trufnao, Minero, Miraflores, Condor, Huemul.

BALMACEDIST.

Description.	Name of Ship.	Tonnage.	Horse-Power.	Nom. Speed.	Launched. Date.	Thickness of Armour.	Torp. Tubes.	Armament.
Torpedo Gunboat	Almirante Lynch	750	4500	21	'90	...	4	III 14-pr. Q.F., IV 3-pr. Q.F., II 1-in. Nords.
"	Almirante Condell	750	4500	21	'90	...	4	III 14-pr. Q.F., IV 3-pr. Q.F., II 1-in. Nords.
Armed Steamer	Imperial	2362	3000	15	...	...	4	I 6-in. B., VIII Small.
Torpedo Boats	Sargento Aldea	70	700	19	'86	...	4	II Q.F.
"	Geale	35	480	18	'81	...	...	...

And eight other torpedo boats.

Abbreviations: B. = Breech-loader. M. = Muzzle-loader. Q.F. = Quick-firer.

TALBE XVII.—BRAZILIAN CIVIL WAR.

Melloist Fleet.

Class of Ship.	Name.	Tons.	I.H.P.	Date.	Thickest Armour.	Speed.	Torpedo Tubes.	Armament.
2nd Class Battleship	Aquidaban	5000	6200	'85	in. 11½	Knots 15	5	IV 9'2-in. 21-ton B., IV 5'7-in. B., VI 4'7-in. Q.F., II 2-in., XI 1-in., and V-45-in. Nordenfelts
Coast defence monitor	Javary	3640	2500	'75	13	11	..	IV 10-in. M.L., IV 1-in. Nords., II Machine
River monitor	Alagoas	340	180	'86	4½ Deck	7	..	I 70-pr. M.L., II 1-in. Nords.
Cruiser	Almirante Tamandare	4465	7500	'90	1½	17	6	N 6-in. Q.F., II 4'7-in. Q.F., V 1-in. Nords.
"	Republica	1300	3300	'92	2	17	4	VI 4'7-in. Q.F., VI 6-pr. Q.F., VI Machine
"	Guanabara Hulk	2200	3000	'77		14	..	IX 5'2-in. M.L., IV machine or Nords.
"	Trajano	1400	2400	'73		13	..	VI 4-in. M.L., II 1-in. Nords., II machine
Gunboat	Marajo	450	400	'85	No Armour.	10	..	II 6-in. B., II 6-pr. Q.F., II 1-in. Nords.
" "	Liberdade	250	280	'64		7'5	..	IV 12-pr. B., IV 1-pr. Q.F., IV 1-in. Nords.
Transport	Madeira, paddle	1400	1200	'73		12	..	II 6-pr. B.
"	Purus, paddle	1355	1200	'74		12'3	..	II 12-pr. S.B.

Armed Merchant Steamers: Mercurio (1120 tons), Jupiter (1124), Urano (1119), Venus (1171), Meteoro (1062), Marte (1121), Pallas (845), Esperança (823), Vieira da Cunho.

Torpedo Boats: Iguatemy, Marcilio Diaz, Araguary, each 150 tons, 1550 H.P., 25 knots, 4 torpedo tubes, 2 6-pr. Q.F.; four torpedo-boats (probably) 52 tons, 600 H.P., 13 knots, 2 torpedo-tubes, 2 1-in. Nords.

Peixotoist Fleet.

Class of Ship.	Name.	Tons.	I.H.P.	Date.	Thickest Armour.	Speed.	Torpedo Tubes.	Armament.
Coast defence monitor	Bahia	1000	1640	'65	in. 5½	Knots 10	..	II 7-in. M., II 1-in. Nords.
Gunboat	Tiradentes	800	1200	'92		14'5	2	IV 4'7-in. Q.F., III 6-pr. Q.F., IV 1-in. Nords.
" "	Primeiro de Março	780	750	'81		10	..	VII 4'5-in. B., IV 1-in. Nords.
" "	Bracunnot	160	160	'72		8	..	}
" "	Paranahyba	840	900	'78		12	..	I 6-in. M., II 32-pr. B., II 1-in. Nords.
" "	Cabedello	210	200	'86		9	..	II 4'5-in. B., IV 1-in. Nords.
" "	Inicladora	260	260	'83		9	..	II 4'5-in. B., IV 3-pr. Q.F., IV 1-in. Nords.
Torpedo Gunboat	Gustavo Sampaio	480	2300	'93		18	..	II 20-pr. Q.F., IV 3-pr. Q.F.

Armed Merchant Steamers: Nictheroy, Rio de Janeiro, Itaipu, America, Advance, Finance, Allianca, Seguranpa, Vigilanpia.

Torpedo Boats: Piratiny; 12 of various types, Yarrow and Schichau built; 2 American built.

B. = Breech-loader. Q.F. = Quick-firer. M. = Muzzle-loader. Nord. = Nordenfeldt.

TABLE XVIII.—CHINESE FLEET AT THE YALU.

	Tonnage.	Inches of Armour on Belt.	Inches of Armour on Glacis.	Inches of Armour on Deck.	Max. Speed Trial. K'ts.	Date of Launch.	Where Built.	Torpedo Tubes.	Military Masts.	Armament: 10½ & 9½	Armament: 8½	Armament: 5.9	Armament: 5.1 & 4.7	Armament: Q.F. 4.7	Small Q.F. and Machine.	Projectiles over 10 lbs. fired in 10 minutes.* No.	Projectiles over 10 lbs. fired in 10 minutes.* Total Weight. lb.	Captain, &c.
Line of Battle.																		
Ting Yuen	7430	14	12	3	14.5	'81	Stettin	3	2	4					12	31	10,940	Admiral Ting. Commodore Lin.
Chen Yuen	7430	14	12	3	14.5	'82	Stettin	3	2	4		2			12	32	10,940	Commodore Lin.
King Yuen	2850	9.4	7/8	3	16.5	'87	Stettin	4	2			2	2		13	20	3390	
Lai Yuen	2850	9.4	7/8	3	16.5	'87	Stettin	4	2			2	2		13	20	3390	
Tsi Yuen	3355		9/8	3.9	15	'83	Stettin	4	2			2	2		10	20	3390	
Chi Yuen	2300			3.9	18	'83	Elswick	4	1		2	2			16	15	4370	
Ching Yuen	2300			3.9	18	'86	Elswick	4	1		2	2			16	25	4370	Capt. Fong.
Chao Yung	1350			3.P	16.5	'81	Low Walker	3	2		2				7	38	4940	Capt. Tang.
Yang Wei	1350			3.P	16	'81	Low Walker	3	2		2				7	38	4940	
Kwang Kai	1290			1.2	15.?	'91	China	4				2		1.P	8	6½	2240	
Ping Yuen	2100	8	5	2	10.5	'90	China	2			1.P		1.P	1.P	8	14	2770	
Kwang Ping	1000			1	15.?	'91	China	4				2		1.P	8	6½	2240	
Inshore.																		
Torpedo Boats		Fu Lung (28 tons, 15 knots sea-speed, III tubes, Schichau-built); Cho oi-Ti (99 tons, 15 knots sea-speed, III tubes)																
13 Ships	35,415							44		8	5	12	15	10	2	130	304	58,610

For Notes see page 307.

TABLE XIX.—JAPANESE FLEET AT THE YALU.

		Tonnage.	Inches of Armour on Belt.	Inches of Armour on Guns.	Inches of Armour on Deck.	Max. Speed Trial.	Date of Launch.	Where Built.	Torpedo Tubes.	Military Masts.	12·6	10·2	9·4	6·8, 6, and 5·9	Q.F. 6	Q.F. 4·7	Light and Machine.	Projectiles over 10lb. fired in 10 minutes.* No.	Total Weight. lb.	Captain, &c.
Main Squadron	Matsushima	4277	...	11·8	1·5	K'ts. 17·5	'90	La Seyne	4	1	1*	...	...	...	...	12*	22	393	16,470	V.-Admiral Ito. Capt. Omoto.
	Itsukushima	4277	...	11·8	1·5	17·5	'89	La Seyne	4	1	1*	...	...	...	...	11*	22	393	16,470	Capt. Yoko-o.
	Hashidate	4277	...	11·8	1·5	17·5	'91	Yokosuka	4	1	1*	...	...	...	...	11*	22	393	16,470	Capt. Hidaka.
	Chiyoda	2450	4·6	...	1	19·5	'89	Glasgow	3	3	...	...	...	...	...	10*	17	390	13,500	Capt. Uchida.
	Fuso	3718	9	8	...	13·2	'77	Thames I.W.	...	...	...	...	4*	2*	...	...	9	22	4700	Capt. Arai.
	Hiyei	2200	4·5	...	...	13	'78	Milford H.	...	...	...	...	...	3* { 6*	...	...	4	41	4720	Capt. Sakurai.
Flying Squadron	Yoshino	4150	...	4·5	...	23	'92	Elswick	5	2	...	...	...	...	4	8*	22	290	18,000	R.-Admiral Tsuboi.
	Takachiho	3650	...	3	...	18·7	'85	Elswick	4	2	...	2*	...	6*	...	...	12	38	7000	Capt. Kawara.
	Naniwa	3650	...	3	...	18·7	'85	Elswick	4	2	...	2*	...	6*	...	...	12	33	7000	Capt. Nomura.
	Akitsusu	3150	...	2·5	...	19	'92	Yokosuka	4	2	...	...	...	4*	...	6*	10	210	12,750	Capt. Togo.
	Akagi	615	...	...	...	12	'88	Onohama	...	...	...	...	...	...	...	1*	11	35	2730	Capt. Kamimura. Comm. Sakamoto.
Not in Line.	Saikio	2913	...	...	...	14	Armed steamer of Nippon Yusen Kaisha.	...	...	...	...	...	...	...	...	...	...	...	V.-Adm. Count Kabayama. Comm. Kano.
12 Ships		39,297							32	14	3	4	4	27	8	59	154	1863	119,790	

For Notes see page 307.

TABLES.

TABLE XX.—COMPARISON OF TWO FLEETS, EXCLUDING GUNBOATS, TORPEDO BOATS, AND VESSELS NOT BUILT FOR WAR PURPOSES.

	No. of Ships.	Average Tonnage of each Ship.	Average Age.	Average Speed.	Speed of Slowest Ship.	Armament. Over 6in.	Armament. 6in. and over.	Armament. Large Q.F.	Armament. Small.	Armament. Total.	Period of 10 mins. Shots.	Period of 10 mins. Average of Weight of.	Average Weight of each Shot.
			Years.	Knots.								lb.	lb.
China	12	2951	8·5	15·75	10·5	13	27	2	130	172	13	4885	145
Japan	10	3575	7·2	17·70	13	11	23	66	154	254	185	11,700	63

CHINESE GUNS.

Reference Letter.	Calibre of Gun.	Weight of Gun.	Weight of Project.	Rate of Fire Assumed Shots in 10 minutes.
	inch.	tons.	lb.	
1	12	35·4	255	3
2	9·8	26	400?	4
3	10·2	21·7	412	4
4	8·2	9·5	217	5
5	5·9	4	112	10
6	5·1	3	98	15
7	4·7	1·3	33	15
8	4·7	2	45	50

JAPANESE GUNS.

Reference Letter.	Calibre of Gun.	Weight of Gun.	Weight of Shot.	No. of Shots Fired in 10 minutes.
	inch.	tons.	lb.	
a	12·6	66	990	1
b	10·2	21·7	412	4
c	9·4	14·6	396	4
d	6·0	5·5	117	7
e	5·9	4	112	10
f	5·9	4	112	10
g	6	7	100	30
h	4·7	2	45	50

NOTES TO TABLES XVIII.–XX.

* The rate of fire assumed for the various calibres of guns will be found in the last column of the tables of guns. The figures do not, of course, represent the actual number of shots fired per 10 minutes by each gun during the action, but only give some approximation to what that rate might have been, for purposes of comparison.

† The name and all the particulars of this ship are doubtful. Capt. McGiffin gives her three 6-inch and four 5-inch guns.

‡ It is not certain whether this was a quick-firer. It is counted as a slow-firer. (6 in Chinese table of guns.) Capt. McGiffin gives her three 4·7-inch guns.

‡ All guns breechloaders.   Q.F. = Quick-firer.

Ships in italics were lost.   P. = Partial deck.

TABLE XXI.—DETAILS OF JAPANESE LOSSES AT THE YALU.

Name of Ship.	Killed.		Wounded.		Died of Wounds.		Position of Ship according to her loss.
	Officers.	Men.	Officers.	Men.	Officers.	Men.	
Matsushima	2	33	5	71	1	21	1
Hiyei	3	10	3	34	0	4	2
Itsukushima	...	13	1	17	0	1	3
Akagi	2	9	2	15	0	0	4
Akitsusu	1	4	0	10	0	0	5
Fusoo	0	2	2	10	1	2	6
Yoshino	0	1	2	9	1	0	7
Hashidate	2	1	0	9	0	0	8
Saikio	0	0	1	10	1	0	9
Takachiho	0	1	0	2	0	0	10
Naniwa	0	0	0	1	0	0	11
Chiyoda	0	0	0	0	0	0	12
	10	80	16	188	4	28	
			294				

TABLE XXII.—LEADING TYPES OF ENGLISH BATTLESHIPS.

1. Progress in Speed, Dimensions, and Armour.

Date of Completion.	Name.	Length.	Ratio of length to breadth.	Tonnage.	Horse-power.	Speed.	Material of Armour.	Armour Deck.	Water-line Belt.		Inches, armour.		Total weight of Armour, tons.	
									Length.	Thickness.	Primary Battery.	Secondary Battery.		
		Ft.				Knots		in.	Feet.	in.				
1860	Warrior	380	6·5	9210	5270	14·3	Iron	...	250	4½	4½	...	925	High freeboard
'64	Achilles	380	6·5	9820	5700	14·4	Iron	...	Whole length	4½	4½	...	1200	High freeboard
'68	Hercules	325	5·5	8680	6750	14	Iron	...	Whole length	9	9	...	1240	High freeboard
'71	Sultan	325	5·5	9290	8000	14·1	Iron	...	Whole length	9	9	...	1800	High freeboard
'73	Devastation	285	4·6	9330	6650	13·8	Iron	3·2	Whole length	12	14	...	2900	Low freeboard
'77	Alexandra	325	5·1	9490	8600	15	Iron	1½	Whole length	8	12	...	2350	Low freeboard
'81	Inflexible	320	4·7	11,860	8010	13·8	Iron and compound	3	110	24	18	...	3540	Low freeboard
'86	Colossus	325	4·8	9420	7500	16·75	Compound	3	123	18	10	1	2360	Low freeboard
'86	Collingwood	325	4·8	9500	9570	16·8	Compound	3	140	18	18	3	2280	Low freeboard
'89	Victoria	340	4·8	10,470	14,000	16·7	Compound	2½	163	18	18	4	1845*	Low freeboard
'90	Nile	345	4·7	12,590	12,000	16·7	Compound and nickel-steel	3	230	20	18	5	4200	Low freeboard
'92	Royal Sovereign	380	5·0	14,150	14,000	18	Compound and nickel-steel	3	250	18	17	6	4400	High freeboard
'93	Centurion	360	5·1	10,500	13,000	18¾	Compound and nickel-steel	2½	200	12	9	7	...	
'96	Renown	380	5·2	12,350	12,000	18	Harveyed steel	3	200	8	10	6	4300	High freeboard
'96	Majestic	390	5·2	15,000	12,000	17½	Harveyed steel	4	220	9	14	6	4300	High freeboard

* Weight of vertical armour only.

Thickness of armour given is in most cases the maximum thickness.

TABLE XXII.—LEADING TYPES OF ENGLISH BATTLESHIPS.—Continued.

2. PROGRESS IN ARMAMENT.

Name of Ship.	Primary Armament.	Secondary Armament.	Anti-Torpedo Armament.	Guns of and above 20-pounder						Wt. of Guns. Tons.	Weight of Shell			Muzzle Perforat. thro' Wgt. Iron.	
				Broadside		Ahead Fire		Stern Fire			Heaviest Gun.	Aux. Guns.	Hvy. Guns.	Aux. Guns.	
				Guns.	Wt. in lbs.	Guns.	Wt. in lbs.	Guns.	Wt. in lbs.				Ins.	Ins.	
Warrior	{ II 100-pr. B. { IV 40-pr. B.* { XXXIV 68-pr. S.B.	......	......	11	1474	...	...	1	100	110	800	...	Ins. 6	...	
Achilles	{ IV 112-ton M. { XX 68-pr. S.B.	......	......	12	1192	...	...	1	256	143	256	...	11	...	
Hercules	{ VIII 18-ton M. { II 12-in. M. { VI 64-ton M.	......	......	9	2494	1	256	1	256	207	410	...	13	...	
Sultan	{ VIII 18-ton M. { IV 12-ton M.	......	......	6	2152	2	512	2	512	192	410	...	13	...	
Devastation	{ II 35-ton M.	......	......	4	2856	2	1428	2	1428	140	714	...	15½	...	
Alexandra	{ X 18-ton M.	......	......	6	2592	4	1904	2	820	230	548	...	13½	...	
Inflexible	{ IV 16-ton M.	......	XXXIII.	8	6880	4	6800	2	3400	323	1700	10	24	13	
Colossus	{ IV 45-ton M.	V 6-in. B.	XXX.	4	2856	4	2856	2	1428	205	714	100	22½	10⅝	
Collingwood	{ IV 45-ton M.	VI 6-in. B.		7	3156	4	3156	2	1428	210	714	100	22½	10⅝	
Victoria	{ II 110-ton B. { I 29-ton B.	XII 6-in. B.	XXXI.	9	4500	2	3600	3	700	305	1800	100	37½	10½	
Nile	IV 67-ton B.	VI 4½-in. Q.F.	XXI.	7	5135	2	2500	4	2500	280	1250	45	34	13	
Royal Sovereign	IV 67-ton B.	X 6-in. Q.F.	XXXV.	5	5500	4	2700	4	2700	338	1250	100	34	16	
Centurion	IV 29-ton B.	X 4½-in. Q.F.	XXIX.	9	2225	4	1090	4	1090	136	500	45	25	12	
Renown	IV 29-ton B.	X 6-in. Q.F.	XXIX.	9	2225	4	1200	4	1200	196	500	100	25	16	
Majestic	IV 46-ton B.	XII 6-in. Q.F.	XXXVI.	10	4000	2	2100	4	2100	268	850	100	38½	16	

*As designed. Later changed to XXVII 68-pounder S.B.   X 110-pounder B.

B., Breech-loader Rifled.   M., Muzzle-loader Rifled.   Q.F., Quick-firer.   S.B., Smooth-bore.

TABLE XXIII.—PROGRESS IN CRUISERS.

Launched.	Name.	Tonnage.	Length. Ft.	Beam. Ft.	Ratio of Length to Breadth.	Armour on Deck. Inch.	Armour on	Guns. Inch.	Armament.	Guns over 30-pr. Weight of Armament. Tns	Weight of Broadside. Lbs	Speed Max. Trial. Kts	*Coal Endurance. Miles.
1866	Inconstant	5780	337	50	6·4	None	...	None	X 12 ton M., VI 6½ ton M.	159	1736	16·2	2780
1869	Active	3080	270	42	6·7	None	...	None	XVIII 64-pr. M.	58	576	15·1	2000
1876	Bacchante	4130	280	45	6·1	None	...	None	XIV 7-in. M., II 64-pr. M.	69	926	15·0	4070
1881	Canada	2380	215	43	5·0	1½-in. over centre of ship	...	None	II 7-in. M., XII 64-pr. M.	48	492	13·0	5400
1883	Amphion	3750	300	46	6·5	1½-in. over centre of ship	...	None	X 6-in. B., XVI Light	40	500	17·0	10,000
1885	Mersey	4050	300	46	6·5	3·2	...	None	VI 8-in. B., X 6-in. B., XIV Light	70	920	18·0	7400
1888	Medea	2800	265	41	6·4	1½	...	None	VIII 4·7-in. Q.F., XII Light	16	180	20·0	8000
1890	Pearl	2575	265	41	6·4	1½	...	None	VIII 4·7-in. Q.F., II 6-in. B., XIII Light	22	335	20·0	10,000††
1891	Apollo	3400	300	43	6·9	2·1	...	None	II 6-in. Q.F., VIII 4·7-in. Q.F., XIII Light	36	335	20·0	10,000††
1892	Æolus	3600	300	43½	6·9	2·1	...	None	II 6-in. Q.F., VIII 4·7-in. Q.F., XIII Light	30	380	20·0	10,000††
1893	Astræa	4360	320	49	6·5	2·1	...	None	II 6-in. Q.F., VIII 4·7-in. Q.F., XIII Light	30	380	20·0	15,000††
1894	Eclipse	5600	350	53	6·6	2¼ Deck.	...	None	II 6-in. Q.F., VIII 4·7-in. Q.F., XIII Light	47	435	19·5	15,000††
1876	Nelson	7930	280	60	4·6	3	Belt. 9	9?	IV 18 ton M., VIII 12 ton M.	166	1825	14·5	5300
1883	Imperieuse	8400	315	62	5·0	2	10	...	IV 9·2-in. B., X 6-in. B., XIX Light	125	1640	16·7	7000
1886	Aurora	5600	300	56	5·3	3	10	9?	II 9·2-in. B., X 6-in. B., XXV Light	94	1260	19·0	8000
1889	Blake	9200	375	65	5·7	6·3	None	6-in. casem.	II 9·2-in. B., X 6-in. B., XXIII Light	94	1260	22·0	15,000
1890	Edgar	7350	360	60	6·0	5·1	None	6-in. casem.	II 9·2-in. B., X 6-in. Q.F., XXIV Light	114	1260	20·0	10,000
1895	Terrible	14150	500	71	6·9	4·2½	None	6-in. casem.	II 9·2-in. B., XII 6-in. Q.F., XXXVII Light	128	1360	21·9	25·000

M. = Muzzle-loader. B. = Breech-loader. Q.F. = Quick-firer.

* Nominal endurance ; actual endurance from one-half to two-thirds this.
† Calculated endurance with bunkers full. Figures doubtful.

P. = Bulkhead only.

TABLE XXIV.—ENGLISH HEAVY GUNS.

	Length of Bore in Calibres.	Weight of Gun in Tons.	Weight of Projectile.	Weight of Charge.	Muzzle Velocity. Foot-seconds.	Muzzle Energy. Foot-tons.	Perforation of Wrought Iron at Muzzle.	Energy of Gun at Muzzle per Ton of Weight.
A. Muzzle-loaders.			lbs.	lbs.			inches.	
7-inch	15·8	4½	114	22	1325	1395	7·9	310
8-inch	14·75	9	170	35	1390	2398	9·7	266
9-inch	13·9	12	250	50	1440	3681	11·4	307
10-inch	14·5	18	410	70	1370	5408	13·1	300
11-inch	13·1	25	548	85	1360	7028	14·2	281
12-inch	12	25	614	85	1292	7190	13·4	288
12-inch	13·6	35	714	140	1390	9566	15·9	273
12½-inch	15·84	38	818	210	1575	14,070	19·3	371
16-inch	18	80	1700	450	1540	27,960	24·3	350
B. Breechloaders.								
7-inch*	14·2	4½	108	11	1100	909	5	222
6-inch	26	5	100	48	1960	2665	12·4	533
8-inch	29·6	15	210	118	2150	6730	17·1	449
9·2-inch	31·5	22	380	166	2030	10,920	20·3	496
10-inch	32	29	500	252	2040	14,430	22·3	498
12-inch	25·25	45	714	250	1914	18,130	22·6	403
12-inch†	35·43	46	850	...	2400	33,940	38·5	738
13½-inch	30	67–69	1250	630	2016	35,230	34·2	526
16¼-inch	30	110½	1800	960	2087	54,390	37·5	492
C. Quick-firers.								
4·7-inch	40	2	45	12	2188	1494	12	711
6-inch	40	7	100	30	2200	3356	16·1	479
8-inch‡	44·6	20	250	55	2500	10,830	26·1	541

* Old-type screw breech.
† New type wire gun (*Majestic* class).
‡ Not as yet mounted in the English fleet. An Armstrong gun.

TABLE XXV.—SUMMARY OF TORPEDO OPERATIONS EXCLUDING STATIONARY TORPEDOES.

Date.	Place.	Assailant Party.	Assailed Party.	Means Employed.	Type of Torpedo employed.	Number of Torpedoes Fired.	Minimum Distance between Assailant and Assailed.	Name of ship Attacked.	State of ship when Attacked.	Time and State of Weather when Attack made.	Result of Attack to Assailed.	Damage to Assailants.	Page.
April 9, '63	Charleston	Confeds.	Federals	Special launch	Spar	Seen before tor. could be used		*Ironsides*	?	?	No damage to either		1
Oct. 5, '63	Charleston	Confeds.	Federals	Special launch	Spar	1.	Contact	*Ironsides*	At anchor	Night	No damage; torpedo exploded against the side-armour	13 bullet holes in boat; two men captured	103
Feb. 17, '64	Charleston	Confeds.	Federals	Submarine boat	Spar	1.	Contact	*Housatonic*	At anchor	Night	Sunk	Boat sunk and all drowned	103
March 6, '64	N. Edisto River	Confeds.	Federals	Special launch	Spar	Tor. could not be used		*Memphis*	Slipped her cables, and in motion	Night	If try fire opened, but boat got away		104
April 9, '64	Hampton Roads	Confeds.	Federals	Launch	Spar	1.	Contact	*Minnesota*	At anchor	Night	Severe damage	Escaped without injury	104
April 19, '64	Charleston	Confeds.	Federals	Special launch	Spar	Seen before tor. could be used		*Wabash*	Slipped her cables and in motion	Night	Sunk the boat before she was hit	Boat sunk	103
Oct. 27, '64	Albemarle Sound	Confeds.	Federals	Launch	Spar	1.	Contact	*Albemarle*	Moored to shore, log boom round	Dark & rainy night	Sunk	Boat sunk; two drowned, 29 taken prisoners	111-3
May 29, '77	Ylo, Peru	English	Peruvian	*Shah's* launch carriage	Whitehead	1.		*Huascar*	In motion at sea	Day	Torpedo missed	Nil	309
May 15, '77	Batum	Russians	Turks	4 launches	Towing	1.	which did not explode	?	At anchor	Night	Torpedo fouled Turkish ship but failed to explode	Nil	291
May 26, '77	Braïlov	Russians	Turks	4 launches	Spar	II.	Contact	*Seifi*	Stationary in River Danube	Night, dark and rainy	*Seifi* sunk; both torpedoes exploded under her	Nil	290-2
June 11, '77	Sulina	Russians	Turks	6 launches	Spar	II.	Contact?	*Idjlalieh*	At anchor, had nets out	Night	Both torpedoes probably struck booms; no result	One launch sunk	293-4
June 23, '77	Danube, Month of Aiuta	Russians	Turks	2 launches	Spar	1.	Non-e fired	?	In motion	Day	Attack easily repulsed	Nil	294
Dec. 27, '77	Batum	Russians	Turks	4 launches	Whitehead	II.		*Mahmudieh* ?	At anchor	Night	Both torpedoes missed; no harm done to Turk	Nil	301-2
Aug. 23, '77	Sokhum Kalè	Russians	Turks	5 launches	Towing	1.		*Assar i Chevket*	At anchor, boats, &c., round her	Night, moon eclipsed	One torpedo exploded without effect; a second fouled *Assar*, but did not explode	Nil	{296-300}
Jan. 25, '78	Batum	Russians	Turks	2 launches	Whitehead	II.	70-90yds.	?	At anchor	Night, foggy	Ship sunk; both torpedoes probably hit	Nil	302-3
Aug. 27, '79	Antofagasta	Peruvians	Chilians	*Huascar's* dirigible torpedo		1.		*Stsas* and *Magelanes*	At anchor	Day	Torpedoes ran back on *Huascar*, and endangered ship	Nil	322
April 19, '80	Callao	Chilians	Peruvians	Jacques-Ozide, torpedo boats	Spar	1.	Contact	*Union*	At anchor behind boom	Night	Torpedo exploded against boom; no damage done	Nil	333
Feb. 14, 85	Sheipoo	French	Chinese	2 launches	Spar	II.	Contact	*Fogen*	At anchor	Dark night	Chinese ship sunk	Slight injuries to one launch	ii 13-5

Continued on next page.

TABLE XXV.—SUMMARY OF TORPEDO OPERATIONS EXCLUDING STATIONARY TORPEDOES.—Continued.

Date	Place	Assailant Party	Assailed Party	Means Employed	Type of Torpedo Employed	Number of Torpedoes Fired	Minimum Distance between Assailant and Assailed	Name of ship Attacked	State of ship when Attacked	Time and State of Weather when Attack made	Result of Attack to Assailed	Damage to Assailants	Page
Aug. 23, '84	Foochow	French	Chinese	3 launches	Spar	III.	Contact	Yang-Foing	At anchor	Day, dur'g Fr'it attack	Both Chinese ships sank	One launch disabled, one man killed	8
Jan. 27, '91	Valparaiso	Congressionalists	Balmacedists	Blanco's launch	White-head	I.	?	Imperial	At anchor		Missed		21
April 23, '91	Caldera Bay	Balmacedists	Congressionalists	Lynch, Condell, tor. gun boats	White-head	V.	100–250yds.	Blanco Encalada	At anchor	Just before dawn, cloudy	Blanco Encalada sank; one hit	Lynch hit four times; no damage	22–9
April 15, '93	Sta. Catherina	Peixotoists	Mellonist	Sampaio, tor. gun boat, and 3 torpedo boats	White-head	IV.	160yds.	Iquidaban	At anchor	Night, very dark	Iquidaban sank; one hit	Sampaio 25 hits; no one hurt; no damage; other boats one hit	43–9
Sept. 17, '94	Off the Yalu	Chinese	Japanese	Torpedo boat	White-head	II.		Hiyei	In motion	D'y, dur'g bat.	No hits; ship not injured	Boat not injured	93
Sept. 17, '94	Off the Yalu	Chinese	Japanese	Torpedo boat	White-head	III.	50yds.	Saikio	In motion	D'y, dur'g bat.	All torpedoes missed	No injury to boat	93
Feb. 2, '95	Wei-hai-wei	Japanese	Chinese	Several torpedo boats	None used			Chinese fleet	At anchor	Night	Seen and fired upon by Chinese; attack aband'd		129
Feb. 5, '95	Wei-hai-wei	Japanese	Chinese	10 torpedo boats	White-head	VI.	150yds.	Ting Yuen, Lai Yuen ?	At anchor	Dark night, 10° below zero	Ting Yuen struck and sunk	Only one boat uninjured; one sank; one ran ashore; 10 or 13 killed	130–2
Feb. 6, '95	Wei-hai-wei	Japanese	Chinese	6 torpedo boats	White-head	?	?	Lai Yuen,† Wei Yuen, Ching Yuen?	At anchor	Dark night	Lai Yuen capsized	No lives in boat	133

* A previous attack had been made upon her, details uncertain.   † Three ships are allowed for in the summary for both attacks.
‡ Both boats were afterwards re-floated.   § Doubtful

SUMMARY.

Type of Torpedo	Total Number of Attempts	Against Ships in motion		Against Ships at rest				Loss to Assailant Boat
		Number of Attacks	Successful	Number of Attacks	Some success in.	Ships sunk.	Ships damaged	
Spar	13	3	0	10	8	6	—	4
Towing	2	0	0	2	0	0	0	0
Whitehead	11	3	0	8	5	6	0	2
Lay	1	0	0	1	0	0	0	0
	27	6	0	21	13	12	1	6

I.—INDEX OF ACTIONS.

I. FLEET ACTIONS.

Lissa, i, 230-248.
Yalu or Haiyang, ii, 82-103.

II. ACTION BETWEEN SINGLE SHIPS OR SMALL SQUADRONS.

Aconcagua and torpedo-gunboats, ii, 29-30.
Alabama and *Hatteras*, i, 154-5.
„ „ *Kearsarge*, i, 157-164.
Albemarle, and Northern gunboats, i, 108, 109-110.
Angamos, Battle of, i, 323-332.
Arkansas and *Carondelet*, i, 71-2.
Asan, Action off, ii, 67-71.
Assar-i-Chevket and *Vesta*, i, 304-5.
Atlanta and *Weehawken*, i, 97-100.
Charleston, Action off, i, 87-9.
Covadonga and *Esmeralda*, i, 253.
Foochow, Battle of, ii, 5-10.
Fort Pillow, „ „ i, 67-8.
Heligoland, Action off, i, 226.
Iquique, Battle of, i, 315-321.
Memphis, Battle of, i, 68.
Merrimac, *Congress*, and *Cumberland*, i, 14-20.
Merrimac and *Monitor*, i, 25-32.
Meteor and *Bouvet*, i, 279.
Mobile, Battle of, i, 114-134.
Riachuelo, Battle of, i, 259-260.
Shah and *Huascar*, i, 308-9.
Selma and *Metacomet*, i, 127-8.

III. ACTIONS BETWEEN SHIPS AND FORTS.

Off Alexandria, i, 337-357.
„ Callao, i, 254-256.
„ Charleston, i, 92-96, 101-2.
„ Fort Fisher, i, 137-140, 141-2.
„ „ Donelson, i, 64-5.
„ „ Henry, i, 63-4.
„ Grand Gulf, i, 79-80.
„ Lissa, i, 220—224.
„ Min River, ii, 11-12.
„ Mobile, i, 124-7.
„ New Orleans, i, 45-57.
„ Port Hudson, i, 74-77.
„ Rio de Janeiro, ii, 38.
„ Sfax, ii, 1-4.
„ Vicksburg, i, 70, 78-9.
„ Wei-hai-Wei, ii, 133.

IV. TORPEDO ACTIONS.

[*See also* Table XXV.]

Albemarle, Sinking of, i, 110-113.
Aquidaban, Sinking of, ii, 43-49.
Batum, at, i, 298, 302, 303-4.
Braila, at, i, 290-2.
Caldera Bay, ii, 22-30.
Sheipoo, ii, 13-15.
Sukhum Kalé, i, 298-310.
Wei-hai-wei, ii, 129-132.

II.—INDEX OF NAMES.

ABBREVIATIONS.—des. = described; m. = mentioned; M. = Map or Plan; n. = note; Pl. = Plate; q. = quoted; Tab. = Table. A few other obvious abbreviations have been employed.

A.

Abtao, Tab. xi, xvi, xxv; des. i, 313; at Antofagasta, i, 322.

Aconcagua Tab. xvi; des. ii, 18; engages *Lynch* and *Condell*, ii, 29-30, 27.

Achilles, Tab. xxii, des. ii, 220; at Alexandria, i, 343, 349; with Channel fleet, ii, 189; now of little use, ii, 272.

Active, Tab. xxiii, ii, 255.

Adalbert Prince, commands German squad., i, 273, 275.

Adalbert, see *Prinz Adalbert*.

Adler, engaged off Heligoland, i, 226; at Kiel, 1870, i, 278.

Admiral Ortshakoff, ii, 145.

"Admiral" class consists of six barbette ships, named after famous admirals, ii, 230-1; bases of their barbettes unprotected, ii, 164, 269; patches of armour on, ii, 240; ii, 271.

Adria, Tab. viii.

Adriatic, M. xi, i, 216; Austro-Italian war in, i, 209-250.

Advance, improvised warship, ii, 41.

Æolus, Tab. xxiii, ii, 257.

Ætna, at Kinburn, i, xxxiii.

Affondatore, Tab. vii; des. i, 212-3; Persano's faith in her, 215-6; tels. for her, 217; waits for her, 218; she arrives at Lissa, 222; place in line, 232; unmanageable, 233; Persano moves to her, 233; part in battle, 238, 240, 242, 243, 244, 250; sinks at Ancona, 245; mentioned by Persano, 249.

Affonso Pedro, des. ii, 43; torpedoes *Aquidaban*, 46.

Africa, *Alabama* off coast of, i, 156, 172.

Agamemnon, des. ii, 229; m. 228.

Agamemnon v. *Melpomène*, ii, 138.

Agincourt, des. ii, 221; in Channel squad., 185.

Aguilar, on *Blanco Encalada*, ii, 25; drowned, 27.

Aguirre, on *Huascar*, i, 328; killed 329.

Ajax, des. ii, 228-9; m. 335.

Akagi, Tab. xix; loss at Yalu, xxi; with J. Fleet, ii, 84; position in line, 88; hotly engaged, 91; retires, 91; re-enters battle, 93; loss, 105; damage, 110; speed, 112; loses mast, 167; devolution of command on, 181; m. 60, 61, 101.

Akerman, i, 286.

Akitsusu or Akitsushima, Tab. xix; loss at Yalu, xxi; des. ii, 60; off Asan, 67; with Flying Squad. at Yalu, 84; position, 88; off Port Arthur, 93; loss, 105; off Wei-hai-wei, 128, 130; m. 71, 96.

Alabama. State of, i, 64; coast blockaded, 181.

Alabama, Confederate cruiser, des. i, 152; off Azores, 153; on the Banks, 153; at Martinique, 153; sinks *Hatteras*, 154-5; in central Atlantic, 156; at Pulo Condor, 157; reaches Cherbourg, 157; action with *Kearsarge*, 158-163; sinks, 163; her crew saved, 164; gunnery, 164-5; her crew, 153, 159 n.; prizes taken, 157; effects of her cruise, 153 n., 169 n.; measures which might have been taken against her, 169-173; m. 171.

Alabama, U.S.N., Tab. v.

Alabama claims, i, 174 n.

Alagoas, Tab. xvii; at Curupaity, i, 263; m, ii, 25.

Albatross, at Port Hudson, i, 74; destroys stores, 77.

Albemarle, Tab. xxv; construction, i, 106-7; des. 107; actions with Federal gbs., 108-110; first torpedo attack on, 110; second, 111-113; sunk and raised, 113.

Albemarle Sound, i, 106, 111, 180.

Albini commands wooden squadron, i, 209; against attack on Lissa, 220; conduct, 221-2; insubordinate, 223; ordered to land men, 224; position of, when Austrians appeared, 225; Persano signals to him, 232; conduct during battle of Lissa, 234-5, 240, 246, 250; disgraced, 251; m. 248.

Albrecht, on *Ting Yuen*, ii, 87; saves her, 97.

Alderney, i, 210.

Alexandra, English, elevation, Pl. xxxvii, ii, 220; Tab. xii, xxi; Pl. xviii, i, 350; des. i, 338, ii, 223; at Alexandria, i, 338, 342, 343; opens fire, 344; anchors, 346; loss, 349; damage, 350; has armour deck, ii, 227; compound engines, 254.

Alexandra, Austrian, at Lissa, i, 236.

Alexandria, U.S.A., i, 80.

Alexandria, Egypt, M. xviii, i, 340; riots at, i, 336; E. squad. off, 336; forts at, 340-1; general order of E., 342-3; bombardment, 344-8 damage to forts, 351-2; Egypt. account, 356-7; E. sailors landed, 349; m. ii, 137.

Alger, ii, 267; Pl. xlv, ii, 268.

Algiers. Line Toulon-Algiers in war of 1870, i, 275, 280; m. ii, 1.

Alliança, ii. 41.

Alma, Tab. x; elevation, Pl. xlii, ii. 262; des. ii, 261; at Sfax, 2-3; m. i, 267, ii, 191.

Almanza. in Pacific, i, 252; bombards Callao, 255-6.

Almirante Cochrane, Tab. xi, xvi; cf. elevation of sister ship, *Blanco Encalada*, Pl. xxi, ii, 28; des. i, 313; cleaned, 322; faster than *Huascar*, 322; sights her, 324; comes up fast, 325; action opened with *Huascar*, 325; attempts to ram, 329, 331; shells Arica, 334; explosion on board, 335; declares against Balmaceda, ii, 16, 17; m. i, 321, ii, 22.

Almirante Condell, Tab. xvi, xxv; des. ii, 16-7; arrives from Europe, 21; sinks *Blanco Encalada*, 23-28; engages *Aconcagua*, 29-30; at Caldera, 31.

Almirante Lynch, Tab. xvi, xxv. References as to *Almirante Condell*.

INDEX OF NAMES.

Almirante Tamandaré, hit, i, 262-3; explosions, ii, 181.
Almirante Tamandaré, Tab. xvii; des. ii, 36; in Melloist fleet, *ib.*; fires ballistite, 39.
Alsen, ii, 275.
Althea, i, 134.
Amazonas at Riachuelo, i, 260.
America, South, *Alabama* off, i, 156, 170; wars in, 252-264, 306-335; ii, 16-50.
America, ii, 41, 42.
American Civil War. *See* United States, Confederates; Index i, actions; Index iii, Blockade, Commerce Destroyers, International Law.
Amethyst, des. i, 307; seeks *Huascar*, *ib.*; action with *Huascar*, 308-9; not hit, 310.
Amezaga, q. ii, 117.
Amiral Baudin, elevation, Pl. xlii, ii, 262; des. ii, 263; defects, 147, 270.
Amiral Duperré, des. ii, 262-3; defects, 164, 270.
Amiral Jauréguiberry, des. ii, 264; m. 271.
Amoy, ii, 65.
Amphion, Tab. xxiii; des. ii, 255; in Mediterranean, 196.
Ancona, It. fleet at, i, 215; Austrians off, 216-7; Persano delays at, 218; leaves, 220; *Affondatore* sinks at, 229; m. 221.
Ancona, Tab. vii; des. i, 213; unready, 216; under Vacca, 219; at Lissa, 225; position in line, 232; gap after, 234, 236; collision with *Varese*, 240; badly handled, 250; m. 247.
Andes, i, 314.
Andrada, *see* America.
Angamos, Battle of, M. xvii, i, 326; des. 322-331.

Angamos bombards Callao, i, 333; des. 332.
Angioletti, Minister of marine, i, 215.
Angostura, i, 264.
Anson, similar to *Collingwood*; elevation Pl. xxxix, ii, 232; des. ii, 231. *See also* "Admirals."
Antilles, i, 269.
Antofagasta, i, 322; ii, 18.
Apollo, Tab. xxiii; ii, 257.
Aquidaban, Tab. xvii, xxv; elevation forward, M. xxiv, ii, 46; hits on, 37, 39; torpedo attack on, and sinking of, 43-8; no large q.-f., 49; raised and repaired, 50; name changed to *24 de Maio*, *ib.*; mainstay of Mello, 49; m. 135, 137.
Aquidaban, River, i, 264.
Arabi Pasha, i, 336, 349.
Araguay, i, 260.
Archduchess Frederick, Tab. viii, i, 230.
Archer, English, ii, 256-7.
Archer, Confederate, i, 149.
Arcona, i, 270, 278.
Arcona, ii, 46.
Arens, i, 294.
Argentine Confederation, war with Paraguay, i, 257-9.
Argonaut, i, 286 n.
Arica, i, 307, 315, 322; bombarded, 334.
Ariel captured by *Alabama*, i, 154, 157, 168.
Arkansas, State, i, 37.
Arkansas, i, 71-2.
Armide, Tab. x, i, 267; in Baltic, 277.
Arminius, Tab. x; des. i, 269; breaks blockade, 277, 281; in Elbe, 278.

INDEX OF NAMES.

Armstrong, Sir W., Mitchell, & Co., ii, 36, 245, 248, 250, 255; *v*. also Elswick.

Arrogante, i, 208.

Asan, C. land at, ii, 51; action off, ii, 67-71; sinking of *Kow-shing* off, 72-75; m. 80.

Aspic, Tab. xv.

Assar-i-Chevket, Tab. xxv; des. i, 287; at Sulina, 289; torpedo attack on, 299-300; action with *Vesta*, 304-5; m. ii, 137.

Assar-i-Tewfik, i, 287.

Astrea, Tab. xxiii, ii, 141, 257.

Asuncion, i, 263.

Atahualpa, Tab. xi; des. i, 312; m. 333.

Atalante, Tab. x, i, 267; in North Sea, 275.

Athens, Policy of, in war, i, 184 n.

Atlanta, i, 179.

Atlanta, ex *Fingal*, details of, i, 97-99; defeat and capture by *Weehawken*, 99-100.

Atlanta, see *Tallahassee*.

Atlantic, Russ. fleet in, 1877, i, 286.

Audacious, elevation, Pl. xxxvii, ii, 220; des. ii, 222.

Augusta, breaks blockade, i, 278, 281; captures three French ships, 278, 280.

Aurora, Tab. xxiii, ii, 256-7.

Aurora, see *Gustavo Sampaio*.

Australia, ii, 208.

Austria. War with Italy, i, 211-251; fleet off Ancona, 216; Tegetthof, commander, 225; state of fleet, 226-7; compared with It., 227, want of guns, 226, 228; *personnel* ill-affected, 227; ill-trained, 228; Lissa bombarded, 221—225; fleet puts to sea, 229; battle of Lissa, 230-248; war with Denmark, i, 226.

Avni-Allah, i, 287.

Azazieh, i. 287.

Azores, *Alabama* at, i. 152-3; strat. importance of, 170, 172; *Augusta* blockade at, 278.

B.

Bacchante, Tab. xxiii, ii, 255.

Bahamas and blockade-runners, i, 86, 147, 185-187.

Bahia, seizure of *Florida* at, i, 150; importance of, 170.

Bahia, Tab. xvii, i, 259, 261, 263.

Bailey, i, 45.

Baldwin, i, 110.

Bali, i, 290.

Balkans, i, 288.

Ballistite, a smokeless powder used by *Tamandaré*, ii. 39.

Balmaceda, President, revolt against, ii, 16-17; downfall of, 32-4; m. 79.

Baltic, Campaign in 1870, i, 274, 276-7; French strategy in, 271-281; difficulty of landing, 281; Russian fleet in 1877, 286; canal to North Sea, 285.

Bangkok, i, 169.

Banjo, ii, 84.

Banks, General, i, 75, 80-1, 154.

Barbadoes, i, 149.

Barfleur, sister to *Centurion*, Tab. xxii; elevation, Pl. xxxix, ii, 232; des. ii, 236.

Barham, ii, 196-7.

Barnaby, i, 228.

Barros, i, 263.

Barroso, i, 263-264; hits on, 262.

Barroso, i, 258, 260-1.

Basilisk, i, 226.

Basques, m. i, 252.

Bat, i, 188.

Baton Rouge, i, 61, 69, 72.
Battenberg, Pr. Louis of, ii, 152.
Battery Bee, i, 87.
„ Gregg, i, 87.
„ Marion, i, 87.
Batum, M. xvi, i, 298; boom at, i, 297; t. attacks on Turks at, 298, 301—3.
Batoushka, i, 286.
Baudin, see Amiral Baudin.
Bayard, in China Sea, ii, 12; at Sheipoo 13; des. ii, 266.
Bazaine, Marshal, i, 251.
Bazin, Ship, ii, 258.
Bazley, i, 113.
Beacon, Tab. xii, i, 340.
Beaufort, i, 184.
Beaufort, i, 15, 19.
Beauregard, General, q. i, 93, 97.
Bedford, Lieutenant, i, 348.
Behring's Sea, i, 167.
Béja, ii, 1.
Belfast, i, 170, 172.
Bell, Commander, i, 45.
Bella Vista, i, 261.
Belleisle, purchased from Turkey, i, 224, 272.
Bellerophon, elevation Pl. xxxvii. ii, 220; des. ii, 221; improvements in, 254; m. 185, 261.
Belliqueuse, des. ii, 261; m. i, 267, 275.
Belmont, Battle of, i, 63.
Belmonte, i, 260.
Benbow, des. ii, 231; weakness of, 164. See "Admirals."
Benjamin Constant, ii, 36.
Benton, des. i, 62; at Vicksburg, 78; at Grand Gulf, 79.
Berenguela, in Pacific, i, 252; at Valparaiso, 253-4; at Callao, 255-6; damaged, 255.

Beresford, Lord Chas., conduct at Alexandria, i, 346-7.
Berkvirdelen, i, 289.
Berlin Decree, i, 197.
Bermuda, case of the, i, 200.
Bermuda and the blockade, i, 86, 98, 168, 186, 193; *Florida* coals at, 149.
Bertin, ii, 58.
Biberibé, i, 260.
Bilboa, i, 199.
Biobio, Tab. xvi; at Caldera, ii, 23, 26.
Birkenhead, i, 152, 168, 307.
Biscay, Bay of, i, 278; ii, 184-6.
Bittern, Tab. xii; at Alexandria, i, 340, 348.
Bizerta, ii, 1.
Black Prince, sister to *Warrior*, Tab. xxii; elevation, Pl. xxxvii, ii, 220; des. 250.
Black Sea, Russian fleet in, i, 134, 286; Turkish in, 287.
Blake, i, 155.
Blake, sister to *Blenheim*, Tab. xxiii; Pl. ix, i, 174; des. ii, 256; m., i, 311.
Blakely guns; early rifled, i, 147, 152.
Blanca in Pacific, i, 252; at Valparaiso, 253; at Callao, 255-6.
Blanche, ii, 257.
Blanco Encalada, Tab. xi, xvi, xxv; elevation, Pl. xxi, ii, 28; des., i, 313; searches for *Huascar*, 322; sights her, 323; pursues, 324; enters action, 328; damage, 330-1; blockades Callao, 333. Revolts against Balmaceda, ii, 16-17; under fire at Valparaiso, 19-20; launch attacks *Imperial*, 21, at Caldera, 22; torpedoed and sunk, 22-29; attempts to raise, 33; m, 135-137.
Blanquilla, i, 153.

Blenheim, Tab. xxiii, Pl. ix. i, 174; des. ii, 256.

Blitz, i, 226.

Boggio, lt. deputy, on *Re d'Italia*, i, 218; at Lissa, 221; does not change ship, 233; drowned, 242; complains of lt. gunnery, 247.

Boghaz Pass, i, 343.

Bolivia. War with Chili, i, 312; army, 332.

Boltun, i, 286.

Bombe, ii, 257, 268.

Bordeaux, i, 289.

Borneo, i, 157.

Boston, i, 184, 202.

Bouët Willaumez, commands Baltic squad., i, 272; puts to sea, *id.*; instructions, *id.* n.; off Jahde, 273; fresh orders, 273-4; in Baltic, 276-7; strictures on, 281-2.

Bouledogue, i, 267.

Bourbaki, General, i, 272.

Bourbon, i, 269.

Bourke, Hon. M., Captain of *Victoria*, ii, 197; misgivings, 197-8; character of Tryon, 198; jockeys with screws, 199; goes below, 202; acquitted of all blame, 205-6.

Bouvet, gunboat, des. i, 279; action with *Meteor*, 279.

Bouvet, battleship, similar to *Carnot*; elevation, Pl. xlii, ii, 262; des. ii, 264.

Bowling Green, i, 64, 65.

Braila, i, 288, 290-2.

Brandenburg, compared with *Majestic*, &c., ii, 242-3.

Bratec, i, 286, n.

Brazil, Outrage on neutrality of, i, 150-1, 156; war with Paraguay, 257-8; fleet, 259; *personnel*, 259; battle of Riachuelo, 260-1; Humaita, 262-3; boarding attacks, 264; defeat of Lopez, 264. Revolt of Mello, ii, 35; Peixoto's fleet, 36; fighting at Rio, 37-40; Peixoto's acquired fleet, 40-41; collapse of Melloists at Rio, 42; torpedoing of *Aquidaban*, 43-49; lessons of war, 49-50.

Brazil, i, 264.

Brazos Island, i, 185.

Breckinridge, Tab. ii.

Brennan Torpedo, a controllable torpedo for coast defence, ii, 259.

Brennus, Pl. viii, i, 160; des. ii, 263-4; water-tube boilers, 254; m. 116, 135, 271.

Brooke, designs *Merrimac*, i, 4, 5; m. 167.

Brooklyn Navy Yard, i, 168.

Brooklyn, Tab. ii, iv; at New Orleans, i, 45; collides with *Kineo*, 48; receives a hot fire, 50; supports *Hartford*, 51; at Vicksburg, 70-1; at Port Hudson, 74-5; at Mobile, 120; stops under Fort Morgan, 122-3; passes the fort, 125-6; loss, 132; at Fort Fisher, 137; blockades Mississippi, 144-5; off Galveston, 154-5.

Brown, ii, 34.

Bruat, i, xxxiii, 211.

Brunel, on the turret, i, 220.

Buchanan, commands *Merrimac*, i, 6; ill, 15; wounded 19; commands *Tennessee*, 117; his attempts to ram, 125-6; tactics, 127; action with Federal fleet, 130-1; wounded, 131; m. 25, 129.

Buenos Aires, i, 259, 261; ii, 21.

Bulk Light, i, 270.

Bulloch, q. i, 81 n., 149 n., 160 n., 167 n., 170 n., 171 n., 173 n., 175 n., 199 n.

Burgoyne, commands *Captain*, ii, 184; drowned, 188; m. 185, 187.
Butler, General, i, 138-9.
Byng, i, 224, 251.

C.

Cabral, i, 259, 203; boarding attack on, 264.
Cadiz, i, 146.
Cæsar, *Majestic*, ii, 237.
Caïman, des. ii, 265-6; m. 271.
Cairo, i, 38, 62.
Cairo, des. i, 62 n.; sunk by mines, 73, 84.
Caldera Bay, M. xxi, ii, 21; torpedo affair in, ii, 22-27; action with *Aconcagua* off, 29; visited by torpedo craft, 31.
Caldwell, i, 44.
Caleb Cushing, i, 149.
Caledonia, ii, 221.
Callao, bombarded, i, 255-6; guns mounted, 255; torpedo affair off, 332; blockade of, 333-4; long-range bombardment, 333; m. 208, 253, 315.
Cambrian, ii, 213.
Camperdown, Loss at, ii, 106; frigates at, 139.
Camperdown see *Collingwood*, "Admirals"; in Mediterranean fleet, ii, 196; turning circle, 197; turns towards *Victoria*, 199; collides with her, 200-1; precautions on board, 201; damage to, 205, 160; telegraph fails, 168, 201; force of blow, 207.
Canada, Tab. xxiii; ii, 255.
Canaries, i, 156; strategical importance, 170, 171, 172.
Canet guns on Japanese ships, ii, 58; hits at Yalu, 111; power of, 179, 250.

Canseco, i, 322.
Canton, i, 169; squad. ii. 62.
Cape, The, i, 167, 170-1.
Cape Blanco, i, 156.
Cape Comorin, i, 157.
Cape Fear River, i, 135, 186.
Cape San Roque, i, 156, 170.
Capetown, i, 156.
Cape Verde, i, 171.
Capitan Prat, ii, 17.
Cappellini, on *Palestro*, i, 241.
Captain, des. ii, 183; stability, 184; favourable opinions of, 185; in B. of Biscay, *id.*; rolls heavily, 185-6; vanishes, 186; last moments, 187-8; verdict of court-martial, 189; m. 207, 225, 239.
Carbajal, i, 327.
Carignano v. *Principe di C.*
Carlo Alberto, Tab. vii, i, 222, 242.
Carlson, i, 104.
Carnot, elevation, Pl. xlii, ii, 262; des. ii, 264; m. 154, 271.
Carolina, North, i, 106, 181.
Carolina, South, i, 78, 86, 177, 181.
Caronelet, des. i, 62; at Fort Henry, 63; at Fort Donelson, 64-5; passes Id. No. 10, 67; at Fort Pillow, 68; engages *Arkansas*, 71; passes Vicksburg, 78; at Grand Gulf, 79.
Casabianca, ii, 268.
Cassini, ii, 268.
Castelfidardo, Tab. vii; des. i, 213; short of petty officers, 215; at Ancona, 216; at Lissa, 219, 222, 225; position in line, 232; engages *Kaiser*, 239; damage, 245.
Catskill, Tab. iii; at Charlestown, i, 101; hits on, ii, 246.

Y 2

Cattegat, i, 284.
Caucasus, i, 287.
Cayuga, Tab. ii; at New Orleans, i, 45; passes forts, 46-7; engages Confederate gunboats, 54.
Cécille, ii, 267.
Centurion, Tab. xxii; elevation, Pl. xxxix, ii, 232; des. ii, 236; nickel-steel armour, 253; m. 234, 237, 271.
Cerbère, Tab. x, i, 267.
Chacabuco, Tab. xi; i, 313.
Chacal at Sfax, ii, 2.
Chalmers, ii, 254.
Chalmette, i, 56.
Champion, ii, 6, 27.
Channel Isles, i, 210.
Channel Squadron, French, in 1870, i, 267, 278.
Chao Yong or *Yung, see Tshao Yong*.
Charlemagne, des. ii, 264-5; compared with *Majestic*, &c., 242-5; m. 271.
Charles Martel, ii. 264.
Charleston, M. v, i, 92; North anxious to reduce, i, 86; defences of, 86-7; action off, 87-9; Dupont ordered to attack, 91; the attack, 92-5; Beauregard on, 97; Dahlgren attacks, 101-2; torpedo affairs off, 103-4; fall of, 137, 185; blockade of, 183; m. 2, 64, 121, 135, 165, 183, 190, 194, 208; ii, 168.
Château Renault, ii, 12.
Chattanooga, i, 179.
Chemulpho, Japs. land at, ii. 51; m. 80.
Chen Sing, Tab. xv; ii, 5.
Chen Yuen, Tab. xviii; elevation and deck plan, Pl. xxiii, ii, 62; in action, Pl. xxix, ii, 100; side after battle, Pl. xxx, ii, 110; cf. also *Ting Yuen*, sister ship, Pl. xxxii, ii, 122; des. ii, 62-3; defects, 63-4; reported off Asan, 70, 72, 78; preparations on board, 79; in G. of Korea, 83; place in line, 86, 88; decks drenched, 87; in battle of Yalu, 91-4, 97-8; shots fired, 109; hits on, 111; speed, 112; on fire, 97, 113; torpedoes of, 114; value of armour, 121; projectiles on board, 125; at Wei-hai-wei, 126, 128; surrendered, 133.
Cherbourg, i, 134; action off, 157-164; near Alderney, 210; Baltic fleet fits out at, 271-2; m. 275, 276; harbour enclosed, ii, 134.
Cherub, i, 150.
Chesapeake Bay, i, 179, 185.
Chestakoff, i, 290-2.
Chevalier, ii, 257, 268.
Chi-an, Tab. xv; ii, 5; position, 6; sunk, 7.
Chicago, Pl. vi, i, 96.
Chickasaw, Tab. iv; des. i, 119; at Mobile, 119; position, 120; under Ft. Morgan, 123; attacks *Tennessee*, 131; bombards Ft. Powell, 134.
Chicora, blockade-runner, i, 194.
Chicora, Confederate; action off Charleston, i, 87-8.
Chih Yuen, Tab. xviii; Pl. xxxi, ii, 114; des. ii, 65; in G. of Korea, 83; position in line, 88; attacks Akagi, 91, 95; engaged by *Yoshino*, 92; attempts to ram and is sunk, 92, 98, 101; cause of loss, 98, 111; deck did not save her, 120; m. 102, 114, 119, 159.
Childers, ii, 184.
Chili. War with Spain, i, 253; issues letters of marque, *id.*; Valparaiso bombarded, 253-4; captures *Covadonga, id.*; re-

INDEX OF NAMES. 325

quested to seize *Huascar*, 206. War with Peru and Bolivia, 312; fleet, 313-4, Tab. xi; configuration, 314, ii, 18; good gunnery, i, 318-9; loss of *Esmeralda*, 315—320; search for *Huascar*, 322; disposition of fleet, 322-3; capture of *Huascar*, 330; gains command of sea, 332. Congressional revolt, ii, 16-7; desultory warfare, 19-21; torpedoing of *Blanco Encalada*, 22-30; downfall of Balmaceda, 32; *Itata* affair, 33-4.

Chillicothe, i. 73.

Chiltern, i, 337-8.

China. War with France, ii, 4; squadron on the Min, 5-6; destroyed, 7-11; French pass Min forts, 12; torpedoing of *Yu-yen* at Sheipoo, 13-15; rice, 15. Quarrel with Japan, 51; C. troops land at Asan, *id.*; government of China, 54; C. navy 56, 62-65; docks, 65-6; Europeans in fleet, 66; action of *Tsi Yuen* and *Yoshino*, 67-71; *Kowshing* sunk, 72-5; breaches of international law, 76-78; C. fleet at sea, 79-80; orders of Li Hung Chang, 81; fleet with convoy leaves Taku, 83; battle of Yalu, 87-101; ships sunk, 102 n.; cowardice of C., 104; gunnery, *ib.*, 108; loss, 105-6; guns, 108-9; damage, 110-1; ammunition, 125, n.; fleet at Port Arthur, 126; retires to Wei-hai-wei, 126; Port Arthur captured, 127; torpedo attacks at Wei-hai-wei, 128-132; capture of Wei-hai-wei, 133.

China Sea, i, 170.

Ching Yuen, sister of *Chih Yuen*, Pl. xxxi, ii, 114; Tab. xviii, xxv; des. ii, 65; in G. of Korea, 83; place in line at Yalu, 88; part in battle, 92-3, 99; loss, 105; hits on, 111; fires, 99, 113; torpedoes of, 114; at Wei-hai-wei, 126; torpedoed, 132; sunk, 133; m. 119.

Chin Wei, Tab. xv, ii, 5-6.

Chippewa, Tab. v, i, 137.

Chiyoda, Tab. xix; losses, xxi; des. ii, 58; at Yalu, 84; position, 88; torpedoes *Yang Wei*, 93; no loss, 105.

Chokai, ii, 84.

Choutka, i, 294.

Chun Hing, Tab. xv.

Cincinnati, i, 38.

Cincinnati, des. i, 62 n; at Fort Henry, 63; rammed, 68; sunk, 80-1, 84.

City Point, i, 105.

Clarence, i, 149.

Clowes, W. Laird, q., i, 22; ii, 49, 67, 159.

Clydebank, ii, 208.

Cochrane, see *Almirante Cochrane*.

Colbert off Sfax, ii, 2-3; des. 261; m. 272.

Coles' turret design, i, 8, 33, 306, ii, 224-5, 220; designs *Captain*, 183-4; drowned on her, 185-6.

Collingwood, Tab. xxii; elevation, Pl. xxxix, ii, 232; of "Admiral" class; des. ii, 230-1; defects, 233; m. 196.

Collins, i, 150-1.

Colomb, q. ii, 138, 169.

Colombo, i, 259, 263.

Colonel Lovell, i, 67.

Colorado, Tab. v, i, 137.

Colossus, Tab. xxii; elevation, Pl. xxxix, ii, 232; des. ii, 229; defects, 240; m. 231, 239.

Columbiad, a heavy smooth-bore gun firing shell, i, 1.

Columbus, i, 61, 64-5.

Comisa, i, 220, 222, 224.
Comus, sister to *Canada*. Tab. xxiii, ii, 227.
Concon, ii, 32.
Condor, English, Tab. xii; at Alexandria, i, 340; well handled by Lord C. Beresford, 346-47, 348.
Condor, French, ii, 267.
Conestoga, des. i. 62; m. 63-4.
Confederates, Southerners, Secessionists, Rebels, Citizens of the Confederacy of eleven States which seceded from the United States in 1861; naval resources, i, 1-2, 106-7, 117-18; cotton, 177, 196; configuration, 179, 204-5; population, 179-80; food supply, 37, 82-3; manufactures, 178, 193-4; importance of Mississippi to, 37, 83; rise of prices, 83, 195-6; artillery, 40. 136; torpedo department, 102-5, 115; warships, 1-6, 41-2, 87, 98, 106-7, 116-17; cruisers, 144-169.
Congress, at Hampton roads, i, 14; attacked by *Merrimac*, 15; tries to escape, 18; burnt, 19; m., 34.
Congréve, i, xxxii.
Connyngham, i, 173.
Conqueror, elevation, Pl., xxxix, ii. 232; des., ii, 229; m., 232.
Conrad or *Tuscaloosa*, i, 156.
Constantine, Grand Duke, des. i, 292; off Sulina, 293; off Sukhum Kalé, 298-9, off Batum, 301-3; m., 286.
Constantinople, i, 287.
Cooke, i, 106-7.
Copiapo, ii, 24.
Coquette, i, 188, 195.
Coquimbo, i, 253.
Cordite, a smokeless powder, ii, 112.
Corrientes, i, 261.
Corvette Pass, i, 342.

Courbet at Foochow, ii, 4; destroys Chinese squadron, 7—12; passes Min forts, 12; at Sheipoo, 13-14.
Courbet. des. ii. 262.
Couronne, Tab. x.; des. ii, 260; m. i, 150, 267, 275.
Couting Island, ii, 12.
Covadonga, Tab. xi; captured from Spain by Ch., i, 252-3; des. 313-14; left at Iquique, 315; action with *Independencia*, 316-19; escapes, 320; damage to, 321; sunk, 334; m., 322, 323, 329.
Craven, T.A.M., drowned on *Tecumseh*, i, 124.
Craven, T.T., i, 49-51; 311 n.
Crimean War. Bombardment of Kinburn, i, xxxiii-vi. landing on Crimea, 281; high-angle fire, 355; m., 207. ii, 218, 220, 245.
Cuba, i, 147, 187.
Cumberland, Hampton Roads, i, 14; attacked by *Merrimac*, 16-17; heroism of her crew, 17; rammed 16; value of her resistance, 18; m, ii, 160.
Cumberland River, i, 61, 63.
Curupaity, i, 262, 263.
Curuzu, i, 262.
Cushing. Torpedo attack on *Albemarle*, i, 111-3.
Custozza, i, 217.
Cuyler, R.E., Tab. v, i, 137, 148.
Cyclops, des. ii, 226.
Cygnet, Tab. xii; i, 340.
Czarevitch, i, 290-2.
Czarevna, i, 290-2.

D.

Da Gama, ii, 39, 42.
Dahlgren, in command at Charleston, i, 100; unsuccessful attacks, 101; m. 121, 194.

INDEX OF NAMES.

Dahlgren guns, heavy smooth-bores designed by the above, i, 1, 5, 10, 26.
Dalmatians in Austrian fleet, i, 227.
D'Amico, i, 220-1.
Danube, torpedo actions on, i, 289-295; Sulina attacked, 295-7; m. 287, 288.
Danzig, i, 278; ii, 276.
Daring, Pl. x, i, 208; des. ii, 258.
Dauphin Island, i, 114.
"David," i, 103, 208; ii, 259.
Davis, i, 67, 69, 74.
Davout, ii, 267.
Dawkins, ii, 191-2.
Decoy, Tab. xii, i, 340.
Deer, i, 188.
Deerhound, i, 160, 163.
Defence, ii, 220 n.
De Grasse, i, 38.
De Gueydon, i, 277.
De Horsey, commander on Pacific station, i, 306; attacks *Huascar*, 308-310.
De Kalb, ex *St. Louis*, des. i, 62; at Ft. Henry, 63; at Ft. Donelson, 64-5; on Yazoo, 80; sunk, 84.
Delaware, River, i, 154.
Delaware, State, i, 149, 179.
Denmark, orders turret-ship, i, 8, 33; war of 1864, 33, 226; France hopes for her alliance, 271; m., 273, 276, 280-81.
Depretis succeeds Angioletti, i, 215; despatch to Persano, 217-8; responsible for attack on Lissa, 220.
Desaix, ii, 2.
D'Estaing, Tab. xv; at Foochow, ii, 4, 6-9; descends Min, 11.
Desterro, ii, 43, 45.

"Destroyer," des. ii, 257-8; m. ii, 148; i, 208.
Destroyer, see *Piratiny*.
Devastation, English, Tab. xxii; elevation, Pl. xxxvii, ii, 220; des. ii, 225-26; crew, 213; deck, 227; low freeboard, 239; hand-worked guns, 247; m. 233, 272.
Dévastation. French floating battery, i, xxxii-vi.
Dévastation, French battleship, des. ii, 262; hydraulic machinery, 247; m. 274.
Diaz, Marcilio, Tab. xvii, ii, 37.
D'Iberville, ii, 268.
Dieppe, i, 273.
Dieudonné, i, 267, 272, ii. 275.
Dilaver Pasha, i, 289.
Dixon, i, 104.
Djigit, i, 290-2.
Docka, i, 286 n.
Doctor Batey, i, 74.
Dog river, i, 117.
Donau, Tab. viii, i, 227.
Don Juan of Austria, Tab. viii; des., i, 226-7; position in line at Lissa, 230; part in battle, 243.
Doubasoff, i, 290-1.
"Double-enders," des., i, 182 n.
Douglas, Gen. Sir H., q., i, 4, 34.
Dover, i, 157. ii, 135.
Drache, Tab. viii; des., i, 227; position at Lissa, 230; engages *Palestro*, 235; part in battle, 243.
Dragon, i, 29.
Dreadnought, des., ii, 226; m., 196, 202, 239.
Dryad, ii, 213.
Duboc, ii, 13-14.
Duburquois, ii, 275.
Duckworth, Admiral, i, 38, ii, 50.

INDEX OF NAMES.

Duguay Trouin, Tab. xv; at Foochow, ii, 4, 6-9; descends Min, 11-12.
Duguesclin, ii, 266.
Duke of Genoa, Tab. vii.
Dumbarton, i, 165.
Duncan, General, i, 42, 44, 58.
Dunkirk, i, 276, 278.
Duperré, *see Amiral Duperré*.
Dupont or *Du Pont*, captures Port Royal, i, 184, 281; in command before Charlestown, 89; attacks unsuccessfully, 92-95; by order, 91; refuses to renew attack, 95 n.; recalled, 100; m. 97, 121, 194.
Duppel, ii, 275.
Dupuy de Lôme, French naval architect, ii, 219.
Dupuy de Lôme, Pl. xv, i, 311; des., ii, 266, 257; end-on fire, 269; fit for line, 143; m. 273.
Dursternbrock, i, 270.

E.

Eads, J. B., naval architect and engineer, designs Mississippi gunboats, i, 62, 73; monitors, 119; introduces armour-deck, ii, 227.
East Gulf Squadron, i, 185.
Eastport, i, 64.
Echo, i, 143.
Eckernford, ii, 275.
Eckernsünde, i, 33.
Ec aireur, ii, 13.
Eclipse, Tab. xxiii; m. ii, 140-142.
Edgar, Tab. xxiii, ii, 256; m. i, 311, ii, 155, 196.
Edward's Ferry, i, 106.
Egypt. Riots at Alexandria, i, 336; English fleet before, 336-340; Egyptian forts, 340-1; English ultimatum, 342-3; bombardment, 344-348; damage to forts, 351-2.

Eider Canal, i, 278.
Elbe, i, 226, 274, 277, 278.
Electric Spark, captured by Florida, i, 150, 168.
Elgar, Professor, q. ii, 174, 207.
Elisabeth, Empress, Tab. viii; at Lissa, i, 237-8.
Elisabeth i, 277-8.
Ellerbeck, i, 270.
Ellet, i, 68.
Elswick, ii, 36, 59. *See also* Armstrong.
Enterprise, English, ii, 221.
Enterprise, U.S.N., ii, 6-7.
Era No. 5, i, 73.
Erebus, i, xxxiii.
Ericsson, J., designer of *Monitor*, i, 6-7; letter to Navy Department, 6-7 n.; introduces turret, 8-9; his critics, 11-2; ships cannot fight forts, 91; night attacks, 97; invulnerability of his ships, i, 266, ii, 246; later designs, i, 89; the *Destroyer*, ii, 40; *Stockton*, ii, 211; and the Stevens' battery, ii, 218; armour deck, 227; and Napoleon III., i, 8, ii, 218.
Esmeralda, wooden cruiser, Tab. xi, des. i, 313; captures *Covadonga*, 253; left at Iquique, 315; action with *Huascar*, 316-319; sinks, 319; m. 331. Steel cruiser, Tab. xvi; des. ii, 17-8, 255; aids insurgents, 16; m. 22, 29, 22.
Essex, frigate, i, 38, 133, 150.
Essex, gunboat, des. i, 62; at Ft. Henry, 63; attacks Arkansas, 72.
Esperança, ii, 39.
Esploratore, Tab. vii, i, 216, 217, 220, 225.
Europeans in Chinese Navy, ii, 66, 87.
Excellent, i, 152.

F.

Faa di Bruno, at Lissa, i, 236-7.
Farragut, D. G. life, i, 38-9; at New Orleans, 45; his conduct, 48-9; difficulties, 58-9; at Vicksburg, 69-70; attacks *Arkansas*, 72; at Port Hudson, 74-6; destroys stores, 77; leaves Mississippi, 80; at Mobile, 118, asks for ironclads, 119; his dispositions, 120; letter to his wife, 120-1; on position of admiral, 45, 120, ii, 151-2, n.; climbs rigging, 1, 121; takes the lead, 124-5; passes the fort, 127; rams the *Tennessee*, 130; narrow escape, 130; on his crew, 133; use of chain armour, 160; compared with Tegetthof, 226, 228; with Bouet Willaumez, 281-2; on armour, ii, 120-1; m. 50, 144, 151, 153.
Fasana Canal, i, 228, 229, 248.
Favourite, ii, 221.
Fayal, i, 153, 273.
Fearless, ii, 196-7.
Federals, or Northerners, the inhabitants of those States which were faithful to the Union in 1861. *See* United States.
Feiseen, renamed *Inhanduay*, ii, 41.
Fei Yuen, Tab. xv; at Foochow, ii, 5, 6-9.
Ferdinand Maximilian, Archduke, abbreviated to *Ferdinand Max* or *Max*, Tab. viii; des. i, 226-7; flagship, 229; place in line at Lissa, 230; rams *Re d'Italia*, 236-7; rams unknown vessels, 238; share in the battle, 242; m. 243; ii, 160.
Fernandina, i, 184.
Fernando de Noronha Island. Neutrality infringed, i, 151, 156; strateg. importance of, 170-1.
Ferré Diego, killed on *Huascar*, i, 327.

Feth-i-Bulend, i, 287, 293.
Feth-ul-Islam, i, 289, 290.
Fieramosca, Ettore, Tab. vii, i, 212.
Finance, ii, 41.
Fingal v. *Atalanta*.
Finisterre, ii, 185, 186.
First of June, battle of, ii, 106, 115.
Fishbourne, q. i, 34.
Flandre, des. i. 267, and Tab. x. 272-277.
Flavio Gioja, i, 219.
Florida, i, 177, 181, 184, 187.
Florida, ex *Oreto*, des. i, 146-7; at Nassau, 147; runs into Mobile, 148; runs out, *id.* destroys commerce, 148-9; prizes captured, 149; on Northern coast, 150; seized at Bahia, 150-1; sunk, 151.
Florida Straits, i, 170.
Flusser, C. W., killed, i, 108.
Flying Squadron, Japanese, off Asan, ii, 67; attacks *Tsi Yuen*, 67-71; at Yalu, 84, 88, 124.
Fong commands *Tsi Yuen*, ii, 67-8; sentenced to death, 71; carelessness, 72; misconduct at Yalu, 99, 100.
Foo Ching, ii, 66.
Foochow, destruction of Chinese squad. off, ii, 4-11; dock, ii, 65; squadron, 62, 134.
Foo Poo, Tab. xv; at Foo-chow, ii, 5, 9.
Foo Sing, Tab. xv, xxv; at Foochow, ii, 5, 6-7, 65, 126.
Foote, in command on upper Mississippi, i, 28, 63; wounded, 67.
Forbach, i, 275.
Forban, ii, 268.
Forbin, ii, 207, 216.
Fo est Queen, i, 78.

Formidabile, Tab. vii; des., i, 212-13; at bombardment of Lissa, 222-23; position during battle of Lissa, 225; steams off, 232; damage during bombardment, 245; Pl. xliv, ii, 266.

Formidable, sister of *Amiral Baudin*, which see, ii, 263, 147.

Fort Ada, Tab. xiii; des. i, 340; guns disabled, 351; m., 342, 347, 349, 355, 356.

Fort Ajemi, Tab. xiii; des. i, 340; m., 356.

Fort Andes, ii, 10.

Fort Beauregard, i, 87, Tab. iii.

Fort Buchanan, i, 137.

Fort Bueros, ii, 19.

Fort Charles, i, 60.

Fort Constantine, i, xxxi.

Fort De Russy, i, 80.

Fort Donelson, capture of, i, 64-5.

Fort Fisher, its importance, i, 135; des. 135-7; first naval expedition against, 137; the powder boat, 138; bombardment 139-40; second expedition, 140; bombardment, 141; capture, 142.

Fort Gaines, des, i, 115; capture, 134.

Fort Henry, capture, i, 63-4.

Fort Hindman, i, 73.

Fort Hospital, i, 347, 352, Tab. xiii.

Fort Jackson, des. i, 40; bombarded by fleet, 43-4; passed by fleet, 45, 54; garrison, 57; m., 39, 56, 59, 127, Tab. ii.

Fort Kamaria, i, 340, Tab. xiii.

Fort Lage, ii, 38.

Fort La Mercede, i, 256.

Fort Lighthouse, 342-43, 346; ceases fire, 347.

Fort Marabout, Tab. xiii, i, 340; engaged by *Condor*, 346-47.

Fort Marsa, Tab. xiii, i, 340, 346.

Fort Martello, i, 351.

Fort McAllister, i, 90.

Fort Monroe, i, 35.

Fort Morgan, des. i, 114-15; passed by the fleet, 122-127; shots fired, 128; captured, 134.

Fort Moultrie, i, 92.

Fort Nikolaiev, i, xxxiv.

Fort Oom-el-Kubebe, Tab. xiii; des. i, 340; guns disabled, 352; m., 343.

Fort Pharos, Tab. xiii; des. i, 340; evacuated, 347; guns disabled, 351; m. 343, 349, 355, 356.

Fort Pillow, i, 61; battle of, 67-8; captured, 68.

Fort Powell, i, 114, 134.

Fort Ras-el-Tin, Tab. xiii; des. i, 340; Moncrieff gun, 341; guns disabled, 352; m. 337, 342, 346.

Fort St. Philip, Tab. ii; des. i, 40; passed by fleet, 46-54; fall of, 56; m. 45-9.

Fort Saleh Aga, Tab. xiii, i, 340; guns disabled, 351.

Fort Santa Cruz, ii, 37-8.

Fort São João, ii, 38, 39.

Fort Silsileh, Tab. xiii; guns mounted at, i, 337; position, 340; m. 342, 351.

Fort Sumter, Tab. iii; position, i, 86-7; attacked by Dupont, 92-95; by Dahlgren, 101-102; Ericsson urges night attack on, 97.

Fort Valdivia, ii, 19.

Fort Villegagnon, ii, 37, 38, 42.

Fort Wagner, Tab. iii, i, 87, 92, 101.

Foster, General, i, 107.

Foudroyante, i, xxxii.

Fourichon commands French North Sea squad., i, 267, 275-277.

Fox, G. V., i, 38, 91, 184.
France. Unprepared state of navy in 1870, i, 265; ships, 266-268; expedition to Baltic, 271; dispatch of a squad. 272-3; Fourichon leaves Mediterranean, 275; difficulty of coaling, 276; gunners landed, 276; blockade, 278; and China, *see* China. Declares rice contraband, ii, 15; navy, ii, 260-266; compared with English, 269-274.
Franquet, i, 279.
Fraser, Trenholm & Co., i, 194.
Fratesti, i, 289.
Frederick, Archduke, Tab. viii, i, 238.
Friedland, ii, 261, 272.
Friedrich, Karl. See *Prinz*, &c.
Friedrichsort, i, 270.
Frisbee, i, 53.
Fuentes, ii, 22, 30.
Fuh Sing, Tab. xv, ii, 5.
Fulminant, ii, 265.
Fu Lung, Tab. xviii.
Funk, i, 167.
Furieux, ii, 265.
Fusoo, Tab. xix; loss, xxi; des. ii, 57; at Yalu, 84; position, 88; hotly engaged, 91; very slow, 103; loss, 105; kept station, 113; m. 124.

G.

Gaeta, Tab. vii.
Gaines, i, 118, 128.
Galatz, i, 288.
Galena, Tab. iv, i, 9, 120.
Galissonière, des. ii, 266; at Sfax, ii, 2-3; in the East, 12.
Galsworthy, Captain, ii, 73-75.
Galvez, i, 256.

Garezon, i, 329.
Garibaldi, i, 211.
Garibaldi, Tab. vii, i, 221.
Gauloise, i, 267, 272.
General Beauregard, i, 67, 69, 85.
General Bragg, i, 67-8.
General Jeff. Thompson, i, 67.
General Lovell, Tab. ii, i, 68-9, 85.
General Quitman, Tab. ii, i, 42.
General Stirling Price, i, 67, 68, 78, 85.
General Sumter, i, 67, 68.
General Van Dorn, i, 67.
Genesee, i, 74-5.
Geneva Arbitration, i, 149, 174.
Genoa, i, 213.
Georgia, State of, i, 177, 178, 181.
Georgia, ex *Japan*, des, i, 165-6; m, 170, 174.
Germany, coast of, i, 268; fleet, 269, 270; measures for coast defence, 273; dispositions during war of 1870, 278; *Arminius* and *Elisabeth*, 277; *Meteor* and *Bouvet*, 279; well prepared, 283.
Gibraltar, i, 146; ii, 134, 217.
Gibson, Milner, q. i, 168.
Giglio, Tab. vii.
Giraffe, i, 188.
Gladiateur, ii, 2.
Glasgow, i, 165.
Glassell, i, 103.
Glatton, floating battery, i, xxxiii.
Glatton, turret-ship, ii, 226, 272.
Gloire, Tab. x; des. ii, 218-9, 260; m, i, 7, 267; ii, 220.
Gobernador Island, ii, 35.
Golden Rocket, i, 145.
Goni, ii, 24, 27.

Gonzalez, ii, 25.
Goodrich, q. i, 354.
Gorgon, ii, 226.
Goubet, ii, 259, 268.
Gourdon, ii, 13-4.
Gourko, General, i, 287.
Governolo, Tab. vii, i, 240.
Governor Moore, Tab. ii; des. i, 41-2; engaged, 55; sunk, 56.
Goya, i, 261.
Grand Duke Constantine, see Constantine.
Grand Gulf attacked, i, 70; falls, 80; m. 70, 77.
Grant, General, i, 63, 64; at Shiloh, 65; at Vicksburg, 73, 77-8; testimony to services of fleet, 82, 83; m. 24, 80, 81, 84.
Gravelotte, i, 281.
Great Belt, i, 277.
Great Eastern, ii, 241.
Greene, S. D., on *Monitor*, i. 14, 26, 30-1.
Greville, i, 315.
Griffiths, i, 328.
Grille, i, 278.
Grosser Kurfürst, des. ii, 190; with German squad., 192; rammed by *König Wilhelm*, 192-3; sinks, 193-4.
Guacoldo, i, 332, 333.
Guale, Tab. xvi; ii, 30-1.
Guanabara, Tab. xvii; ii, 36.
Guerrière, ii, 174.
Guiscardo, Tab. vii; i, 220, 221.
Gustave Zédé, ii, 268; Pl. xlvi, ii, 270.
Gustavo, Sampaio, ex *Aurora*, des. ii, 36; arrives, 39; torpedoes *Aquidaban*, 44-49; damage to, 46; m. 42, 43.
Guyenne, i, 267, 272.
Guzman, ii, 27.
Gymnote, ii, 268.

H.

Habana, see *Sumter*.
Habsburg, Tab. viii; des. i, 226; position at Lissa, 230; part in battle, 242.
Haines Bluff, i, 80.
Haiyang, Battle of. See Yalu.
Haiyang-tao Island, ii, 84, 85.
Halifax, i, 168.
Halpine torpedo, ii, 258.
Hampton Roads, i, 14, 22, 24, 104, 151.
Hanneken, Major Von, strategical adviser to Admiral Ting, ii, 66; on the *Kowshing*, 72-3, 75; on *Ting Yuen*, 87; wounded, 96.
Hannibal, ii, 237. See also *Majestic*.
Harding, L., i, 346.
Harriet Lane, i, 183.
Hartford, flagship of Farragut; off Mississippi, i, 37; position at battle of New Orleans, 45-6; in battle, 48-52; at Vicksburg, 70-1; at Port Hudson, 75-6; destroys stores, 77; at Mobile, 120, 121-3; engages *Tennessee*, 130; loss, 132, Tab. ii, iv; m. ii, 144.
Harvey process, des. ii, 253.
Hashidate, Tab. xix; losses, xxi; Pl. xxii, ii, 58; des. ii, 58-9; at Yalu, 84; place in line, 88; hoists Ito's flag, 92; loss, 105; hits, 110, 167.
Hatteras Island, i, 180, 184.
Hatteras, sunk by *Alabama*, i, 154-5, 157.
Havana, i, 56, 144, 171, 187, 195, 279.
Havre, i, 278.
Hayti, i, 153.
Hecate, ii, 226.
Heckmann, Herr, ii, 87.

Hector, ii, 189.
Heimdal, i, 226 n.
Helicon, Tab. xii; i, 340.
Heligoland, Action off, i, 226; French off, i, 275.
Henri Grace à Dieu, ii, 212.
Henry Clay, i, 78.
Herbal, i, 263-4.
Hercules, Tab. xxii; des. ii, 221-2; m. 185; elevation, Pl. xxxiii, ii, 220.
Hero, ii, 229, 271.
Heroine, i, 267, 275.
Hertha, i, 270, 278.
Hervey, M., ii, 21, 28, 31.
Hickley, ii, 190.
Higgins, Col., i, 44, 52, 58.
Hirst, i, 298.
Hisber i, 289.
Hiyei, Tab. xix; losses, xxi; des. ii, 58; mythical battle, 79; at Yalu, 84; place in line, 88; cut off from line, 91; passes between battleships, 91; detached for repairs, 93; part in battle, 94-5; loss, 104-5; hits on her, 109; exterior appearance, 110; slow, 112; m. 101, 126.
Hobart Pasha, runs American blockade, i, 192, 195; with Turkish fleet, 287; q. 293, 297, 301.
Hobby, Engineer, i, 110.
Hoche, elevation Pl. xlii, ii, 262; des. ii, 263; m. 269, 271.
Hoel, i, 67.
Hoffman, Herr, ii, 87, 100.
Holyhead, i, 170, 172.
Hong Kong, ii, 61.
Hood, ii, 235, 240, 271.
Hoste, Sir W., i, 219.
Hotchkiss gun, ii, 250.

Hotspur, ii, 226.
Hotham, i, 20.
Housatonic, Tab. xxv; i. 80; torpedoed, 103-4, 208.
Howell torpedo, ii, 40.
Hsutan, ii, 72.
Huascar, Tab. xi, xxv; Pl. xvi, i, 328; des. i, 306; crew mutiny, 306; molests Eng. ships, 306; attacked by *Shah*, 308-9; escapes, 309; damage, 309-10; compared with her opponents, 310-1; speed in 1879, 314; action with *Esmeralda*, 315-9; rams and sinks *Esmeralda*, 319; bad gunnery, 317-8; crew demoralised, 318; damage 320; attacks *Magal'anes*, 321; harries coast, 321; captures *Rimac*, 322; sighted by Chilians, 323; battle of Angamos, 324-331; crew demoralised, 327; attempts to ram, 328; surrenders, 330; repaired, 332. Under Chilian flag, Tab. xvi; blockades Callao, 333; declares for Congress, ii, 16-7; at Caldera, 22.
Huasco, ii, 22.
Humaita, Attacks on, i, 262-264; abandoned, 264.
Humber, i, 343.
Hunt-Grubbe, i, 343.
Huron, Tab. v, i, 137.
Hüsum or Büsum, i, 278.
Hydra, ii, 226.
Hyène, ii, 2.

I.

Ida, i, 134.
Idjilalieh, Tab. xxv, i, 287; torpedo attack on, 293-4.
Ignacio, Admiral, i, 263.
Iguatemi, Paraguayan, i, 260.
Illinois, State of, i, 61.
Illustrious, see Majestic, ii, 237.

Iltis, ii, 75.

Imperial, improvised war ship, ii, 16; sent to Caldera, 20-22, 30; stratagem, 32, Tab. xvi.

Imperieuse, Tab. xxiii; elevation, Pl. xxxix, ii, 232; des. ii, 256; m. 272.

Inconstant, Tab. xxiii, ii, 185-6, 189, 254.

Independencia, Tab. xi; des. i, 312; action with *Covadonga*, 316-319; attempts to ram, 319; strikes rock and is abandoned, 319-20; m. 314-5, ii, 177.

Independenza, Tab. vii, i, 240.

Indianola, des. i, 73; rammed and sunk, 74, 84.

Indomptable, ii, 265.

Inflexible, Tab. xii, xxii; Pl. xvii, i, 338; elevation, Pl. xxxix, ii, 232; des. i, 338, ii, 228; at Alexandria, 338, 343; part in the bombardment, 344-7; little ammunition left, 348; loss, 349; damage, 350; effect of her fire, 352-3; m. 342; tactical influence of her design, ii, 154; denuded of armour, 240; compound armour, 253; m. 196, 203, 270.

Ingles, on the Japanese, ii, 53, 103.

Intrepide, ii, 2.

Invincible, English, *see* sister ship *Audacious*, Tab. xii; des. i, 339; ii, 222; at Alexandria, i, 337; flagship, 336; manœuvres, 347; loss, 350; damage, *id.*; m. 342, 343, 344.

Invincible, French, i, 267; in North Sea, 275; fuel runs short, 276.

Iosco, Tab. v, i, 137.

Iowa compared with *Majestic*, &c., ii, 242-5.

Ipiranga, Paraguayan, i, 260.

Iquique, Attack on *Magallanes* at, i, 321; action off (*Esmeralda*), 315-320; m. 307; ii, 28.

Iris, ii, 255.

Iron Duke, with Channel squad., ii, 189; rams *Vanguard*, 190; her ram not damaged, 192; court-martial, 192; m. 58, 160, 190, 191.

Ironsides, New; des. i, 90; solid armour, 9; at Charleston, 93-4; hits, 101; ii, 246; torpedo attack on, i, 103; at Fort Fisher, 137, 139, 140; Tab. iii, v, xxv.

Iroquois, Tab. ii; at New Orleans, i, 45; engages gun-boats, 52; at Vicksburg, 70-1; blockades *Sumter*, 145-6.

Iscodra, i, 289.

Island No. 10, i, 61, 66, 67.

Isle of Serpents, i, 293.

Ismail Bay, i, 300.

Itaipu, ii, 39, 42.

Itaipuru, i, 262.

Italy threatens Austria, i, 211; naval preparations, 212; state of fleet, 212-3; *personnel*, 213; gunners, 214, 215, 246; fleet arrives at Ancona, 215; want of guns, &c., 216; fleet ordered to sea, 218; attacks Lissa, 218-224; Austrians appear, 224-5; state of It. fleet, 225; no plans, *id.*; the battle, 231-247; heroism of *Re d'Italia's* crew, 237; of *Palestro's*, 241; mistake of It., 248-9; type of It. battleships, ii, 59.

Itasca, Tab. ii, iv; at New Orleans, i, 44; position, 45; passes forts, 53; retires, 54; at Mobile, 120, 128; torpedoed, 134.

Itata, ii, 33-4.

Ito training, ii, 53, 84; orders at Yalu, 87; why he drew off, 102-3, 115; after battle, 126; at Wei-hai-wei, *id.* at Port Arthur, 127; blockades Wei-hai-wei, 128-133; position in battle, 152; m. 56, 88, 95, 124.

Itsukushima, Tab. xix; loss xxi; Pl. xxii, ii, 58; des. ii, 58-9; at Yalu, 84; position in line, 88; loss, 105; hit at Wei-hai-wei, 133; m, 167.

Ivahy, i, 262.

J.

Jackson, Lieutenant, killed at Alexandria, i, 350.
Jackson, President, i, 38.
Jackson, Fort, *see* Fort Jackson.
Jackson, Tab. ii.
Jacob Bell, i, 148.
Jahde, i, 270, 273, 276, 277.
Jamaica, i, 155.
James River, i, 35, 105, 180.
Jamestown, i, 15, 19.
Janequeo attacks *Union*, i, 332-3.
Japan, revival of, ii, 51-3; fleet, 57-60; mercantile marine, 61; docks, *id.*; *personnel*, 53, 71 n., 94, 117; disregard of Chinese, 80, 82, 83; action off Asan, 67-71; sinking of *Kowshing*, 71-78; cruelty, 77-8; battle of Yalu, 85-106; heavy guns, 110-1; quick-firers, 111-2; signals, 119; captures Port Arthur, 127; blockades Wei-hai-wei, 127-133; torpedo attacks, 128—132; m, i, 169, 269, 278.

Japan, *see* Georgia.
Jardine and Matheson, ii, 72.
Jauréguiberry, i, 280.
Jauréguiberry, *see Amiral J.*
Java, ii, 174.
Javary, Tab. xvii; ii, 35; sunk, 38-9.
Jeff. Davis, i, 143.
Jejui, Paraguayan, i, 258; sunk, 260.

Jemmapes, des. ii, 266; m, 145, 271.
Jeune École. The followers of Admiral Aube in France. They hold that the torpedo-boat and cruiser have displaced the battle-ship; that speed is everything; and they have a great belief in bombardments, i, 321-2, 355, ii, 119; commerce destruction, i, 314, ii, 119; war of coasts, i, 321-2, 324, ii, 119; everything not changed, ii, 238.
Jih-tao or Ihtao, ii, 128.
John Elder, i, 306.
Johnston, on board *Tennessee*, i, 129, 131.
Johnstone, C., captain of *Camperdown*, ii, 201, 206.
Jones, on *Merrimac*, i, 6, 25.
Jouett, i, 128.
Juniata, Tab. v, i, 137, ii, 6.
Jupiter, English, *see Majestic*, ii, 237.
Jupiter, Brazilian, ii, 35.
Jylland, i, 226 n.

K.

Kai Chi, ii, 65.
Kai Koku, ii, 52-3.
Kaiser, Tab. viii, i, 227; position in Austrian line at Lissa, 231; movements, 234; rams *Porto-gallo*, 239; heavy loss, 240, 244, 246; m, 238.
Kaiser Maximilian, Tab. viii; des. i, 227; position in Austrian line at Lissa, 230; threatens *Palestro*, 240; part in battle, 242-3.
Kamiesch Bay, i, xxxiii.
Kansas, State of, i, 180.
Kansas, Tab. v, i, 137.
Kartali, i, 293.

Katahdin, gunboat, at New Orleans, i, 45; passes forts, 48; at Vicksburg, 70-1, Tab. ii.
Katahdin, ram, ii, 150, 229.
Kate, i, 193.
Katsuragi, ii, 84, 128, 133.
Kearsarge, des. i, 158-9; challenges *Alabama*, 159; the action, 160-163; sinking of *Alabama*, 163-4; m. 146, 157, ii. 137, 158.
Kelasour, i, 300.
Kelung, ii, 4.
Kennebec, Tab. ii, iv; at New Orleans, i, 45; does not pass forts, 53; at Vicksburg, 70-1; at Mobile, 120, 126, 128.
Kennon, i, 55.
Kentucky, State of, i, 65, 179.
Keokuk, Tab. iii; des. i, 90; attacks Sumter, 93; retires, *id.*; founders, 95.
Keystone State, Action with *Palmetto State*, i, 88-9.
Key West, i, 184.
Khedive, i, 336.
Khroumirs, ii, 1.
Kiel, M. xv, i, 274; position, i, 270-1; French dare not attack, 276; ships at, 278; difficulty of attacking, 282; ii, 275; m. 274, 277, 281.
Kifz-i-Rakhman, i, 287; in Danube, 289; attacked at Sulina, 295, 297.
Kilia, Mouth of Danube, i, 295, 296.
Kilidj Ali, i, 290.
Kimpai Narrows, ii, 12.
Kinburn, des. i, xxxiv; bombardment of, xxxv; captured, *id.*; English loss, xxxvi; m. i, 3; ii, 253.
Kineo, Tab. ii; at New Orleans, i, 45; collides with *Brooklyn*, 48, 49; at Port Hudson, 74-5.

Kingstown, ii, 189.
King Yuen, Tab. xviii; Pl. xx, ii, 14; des. ii, 64; with Chinese fleet at Yalu, 83; place in line, 88; moves out, 91; sunk by gunfire, 92; part in battle, 99; explosion, 98; hit by a big Canet shell (?), 111; on fire, 113; m. 101, 102, 120, 169.
Kobe Maru, ii, 61.
Kolberg, French prepare to bombard, i, 276, 277; exposed to attack, 284; ii, 276.
Kongo, des. ii, 57-8; m. 84.
König Wilhelm, Tab. x; des. i, 269; m. 272, 273, 278; flag of Germ. squad., ii, 192; rams *Grosser Kurfürst*, 193-4; damage to ram, 160, 195.
Korea, Chinese and Japanese land in, ii, 51; Chinese troops sent to, by sea, 72, 82.
Korea, Gulf of, ii, 72, 84.
Kotaka, des. ii, 61; at Wei-hai-wei, 127, 129.
Kowshing leaves Taku, ii, 72; stopped by *Naniwa*, 72; sunk, 74; violations of international law, 75-8; m. 51.
Krikun, i, 286.
Kronprinz, Tab. x; des. i, 269; m. 273, 278.
Kronstadt, i, 289.
Krupp guns at Yalu, ii, 56, 64, 65.
Kuang, &c., *see* Kwang.
Kuré or Huré, ii, 61.
Kuwan-shi or *Kwang Yi*, ii, 67, 71.
Kwang Kai or *Kuang Chia*, Tab. xviii; des. ii, 65; with Chinese fleet at Yalu, 83; position, 88; attacks *Akagi*, 91; runs away, 92; goes ashore, *id.*; destroyed, 93; did little fighting, 94; on fire, 113; m. 95, 102, 108.

INDEX OF NAMES. 337

Kwang Ping, Tab. xviii; des. ii, 65; with Chinese fleet at Yalu, 83; position, 88; engages Main Squadron, 91; engages *Saikio,* 96; loss, 105; m. 99.
Kwang Ting, ii, 126, 132.
Kwang Tsi, ii, 126.

L.

Labö, i, 270.
Lackawanna, Tab. iv; at Mobile, i, 120; *Tennessee* attempts to ram, 126; rams *Tennessee,* 129; collides with *Hartford,* 130.
Lacour, Col., ii, 275.
Lafayette, des. i, 73; passes Vicksburg, 78-9.
La Guira or Guayra, i, 199.
La Roncière le Noury, i, 272.
Laird, Messrs., shipbuilders, i, 152, 259, 307; ii, 183.
Laird Clowes, *see* Clowes.
Lai Yuen, Tab. xviii, xxv; des. ii, 64; with Chinese fleet, 83; position, 88; attacks *Akagi,* 91; set on fire, 95, 99; heat in engine-room, 165; loss, 105; disabled, 102; at Wei-hai-wei, 126; torpedoed, 132; m. 120.
Lalande, i, 268.
Lamb, Colonel, commands Fort Fisher, i, 135, 140.
Lambton, i, 349.
Lamoricière, i, 211.
Lancaster gun, ii, 245.
Lancaster, i, 68, 72, 77.
Lang, ii, 67.
Lanyon, ii, 199, 202; will not leave the admiral, 203.
La Seyne, i, 253; ii, 17, 58.
Latorre, i, 322-23.
Latouche-Tréville, des. ii, 267.
Lave, des. i, xxxii, at Kinburn, xxxiii-vi.
Lay torpedo, i, 322.
Lebedi, i, 286, 296.

Lee, Admiral, i, 108.
Lee, General, R. E., i, 135.
Lee, R.E., ex *Giraffe,* i, 188, 191.
Léger, ii, 268.
Lehigh, ii, 246.
Lemoine, q. i, 150.
Leonoff, Major-Gen., i, 294.
Léopard, ii, 1-2.
Lesina, i, 220, 222.
Leu-Kung-Tao or Tau, position, ii, 127, 128-9; captured, 133.
Levant, French ships in, 1870, i, 267.
Lévrier, ii, 268.
Lexington, des. i, 62; at Fort Henry, 63; at Shiloh, 65.
"Le Yacht," q. ii, 170.
Liberdade, Tab. xvii.
Lieutenant Poustchine, i, 296.
Li Hung Chang, order to Ting, ii, 81.
Lima, i, 314.
Lima Barros, i, 259, 263.
Lincoln, President, Ericsson writes to, i, 6; on "Black Sunday," 21; q. on importance of Mississippi, 37 n.; student of military matters, 83; on services of fleet, *id.* n.; proclaims blockade, 177, 181; moderation, 202.
Lindoya, i, 263.
Lion, ii, 75.
Lisbon, i, 166.
Lissa, M. xii, i, 220; Hoste's action off, i, 219; des. *id.*; Persano urged to attack, 217, 218; Albini against attack, 220; report of It. staff, *id.*; first bombardment, 220-2; second bombardment, 222-4; Austrian fleet appears, 225; telegrams to Tegetthof, 229; order of Austrians, 230; of Italians, 231-32; battle, 231-247; tactics of Tegetthof, 247-8; strategy of Italians, 248.

VOL. II.   Z

Little Rebel, i, 67, 69.
Liverpool, i, 152.
Loa, Peruvian, i, 255.
Loa, Chilian, i, 322, 324-25; destroyed by a mine, 333-34.
Long Island Sound, i, 168.
Lopez, Marshal, President of Paraguay, i, 257, 260, 264.
Lord Clyde, ii, 221.
Lord Warden, des. ii, 221 ; m. 185, 186.
Louisiana, State, i, 181.
Louisiana, Tab. ii ; des. i, 41 ; defects, 44; brought down to forts, 45; m. 51, 58.
Louisiana, powder ship at Fort Fisher, i, 138.
Louis, Phillippe, ii, 217.
Lousville, des. i, 62; at Fort Donelson, 64-5; passes Vicksburg, 78; at Grand Gulf, 79.
Lubeck, i, 277.
Lutfi-Djelil, i, 287; sunk, 289.
Lynx, Tab. xv; at Foochow, ii, 4; opens fire, 8; shells arsenal, 10.
Lyons, Admiral, i, xxxiii.

M.

Macedonian, ii, 174.
Mackau, Admiral, ii, 217.
Mackinaw, Tab. v, i, 137, 139.
Madeira Island, *Sea King* off, i, 166; strategical importance, 170-172.
Madonna Battery, i, 222-3.
Mafitt, commands *Florida*, i, 147-150.
Magallanes, des. i, 313; action with *Huascar*, 321; *Huascar* attempts to torpedo, 322; off Arica, 334; joins Congressionalists, ii, 16, Tab. xi, xvi, xxv.

Magenta, early battleship, Tab. x, i, 267, ii, 260.
Magenta, modern battleship, des. ii, 263; engines, 214; stern fire, 154; heavy guns, how mounted, 179; m. 271.
Magnanime, Tab. x, i, 267, 275.
Magnificent, see *Majestic*, ii, 237-8; induced draught, 254.
Mahan, q. ii, 103, 118.
Mahmoodieh, Tab. xxv, i, 287, 302.
Mahopac, Tab. v.
Maine, State of, i, 168.
Main Squadron, Japanese, ships composing, ii, 88; m. 90.
Maipo, ii, 20, 21.
Majestic, Tab. xxii; Frontispiece, vol. i; elevation, Pl. xxxix, ii, 232; des. ii, 237-239; wood on, 123; armour, 121, 241, 253; compared with foreign battleships, 242-244; with English cruisers, 141; m. i, 96, ii, 180, 212, 271.
Makaroff commands in Russian torpedo operations at Sulina, i, 293; at Sukkum, 298, at Batum, 301-2.
Malacca, Straits of, i, 171.
Mallory, Secretary Confederate States' Navy, q. i, 2-3; m. 6.
Manassas, des. i, 41; attacks Northern fleet, 47; rams *Brooklyn*, 50; disabled and burnt, 56; m. 49.
Manco Capac, Tab. xi, i, 312, 322.
Manego, see Porto M.
Manhattan, Tab. iv, des. i, 119; at Mobile, 120; guns disabled, 123; attacks *Tennessee* 130-1.
Marajo, Tab. xvii.
Maratanza, Tab. v, i, 137.
Marceau, Pl. xii, i, 270; ii, 263.
Marengo, at Sfax, ii, 2, 3, 261.

INDEX OF NAMES. 339

Maria Adelaide, Tab. vii, flagship of Albini, i, 245.

Maria Pia, Tab. vii; des. i, 213; at Ancona, 216; at Lissa, 219; off Comisa, 221; position when Austrians appeared, 225; place in line, 232; in the battle, 238; damage, 245.

Markham, second in command, Mediterranean fleet, 1893, on *Camperdown*, ii, 196; misgivings at Tryon's signal, 199, 201; court martial, 206.

Mars. See *Majestic*, ii, 237, n.

Marseilles, i, 275.

Marston, i, 24.

Martinique, *Sumter* at, i, 145; *Alabama* at, 153.

Maryland, State of, i, 179.

Mason, Confederate envoy, i, 201-3.

Masséna, ii, 264.

Matamoras, trade of neutrals with, i, 180, 185, 198.

Matthias Cousiño, with Chilian squad., i, 322; sent inshore, 323; blockades Callao, 333.

Mattabesett attacks *Albemarle*, i, 108-9.

Matuska, i, 286, n.

Maumee, Tab. v, i, 137.

Maurice, Col., q. i, 76

Maya, ii, 84.

Max., abbreviated form of name *Archduke Ferdinand Maximilian*, which *see*.

Maximilian, Archduke, i, 226.

McGiffin, commander on board Chinese ship, *Chen Yuen*, ii, 87; q. 79, 80, 85; puts out fire, 97; on *Tsi Yuen*, 109; on hits received by *Ting Yuen*, 111; on torpedoes fired, 114.

Mearim, Paraguayan, i, 260.

Mediterranean, United States' Ships in, 1861, i, 182; French fleet in, 1870, 267; ships retained there, 275; French communications there not threatened, 280; Russian squadron in, 1877, 280; French fleet in, 1881, ii, 2; hypothetical French and English fleets, ii, 116; loss of *Victoria* in, 196-207.

Medea, Tab. xxiii, ii, 257.

Medjemieh, i, 289; at Sulina, 295.

Medusa, i, 270; blockaded on Japanese coast, 278.

Mejillones or Mexillones, i, 323.

Melbourne, *Shenandoah* at, i, 167.

Mello, Admiral, revolt of Brazilian navy in his favour, ii, 35; lacks army, 37; suffers from shorthandedness, 49; collapse at Rio, 42; m. 46.

Melpomène, ii, 138.

Memphis, battle of, i, 68-9; Confederate works at, 61; railroad to, 64.

Mercedita, action with *Palmetto State*, i, 87-8; taken to Port Royal, 89.

Mercury, ii, 255.

Merlin, ii, 6.

Merrimac or *Virginia*, design, i, 3-6; discrepancies in accounts of, 4; the original *Merrimac*, 3-4, n.; designers, 4-5; des. 4-5; crew, 6; armament, 5; Commodore Smith on, 13; defects, 15; action with *Congress* and *Cumberland*, 16-19; results of action, 20-21; alarm at Washington, 20; not sea-going, 21; repaired, 25; action with *Monitor*, 25-31; bad projectiles, 27; results of action, 31-3; subsequent history, 35; scuttled, *id.*; m. 228, ii, 160, 168; type of ironclad widely reproduced in South, i, 41, 61, 71, 87, 97, 105, 116; in North, 99, 266; abroad, 255.

z 2

INDEX OF NAMES.

Mersey. Tab. xxiii, ii, 256.
Mesoodieh, i, 287.
Messagiere, Tab. vii, i, 220.
Meteor, in West Indies, i, 270; des. 279; action with *Bouvet*, ib.
Meteor, i, 286.
Mexico, i, 179-80.
Mexico, Gulf of, i, 183.
Miami, i, 108-9.
Milan Decree, i, 197.
Milne, Admiral, ii, 184-5.
Milwall, i, 212.
Milwaukee, i, 134.
Min River, descent of, by French, ii, 11-12.
Mina, i, 294.
Minnesota, in Hampton Roads, i, 14, 18; attacked by *Merrimac*, 19, 25, 29-30; m, 24; torpedo attack on, 104; at Fort Fisher, 137, Tab. v, xxv.
Minotaur, ii, 185, 221.
Mississippi River, strategical importance, i, 37, 81, 83; fighting upon, 37-85; open to Unionists, 81; gunboats, 62, 73. ii, 227; tactical lessons, i, 84-5; value of navy, 83; Grant upon, 82; difficulty of navigating, 39; *Sumter* in, 144-5; blockade of mouth, 181.
Mississippi Sound, i, 114, 180.
Mississippi, State of, i, 64, 181.
Mississippi, Confederate, i, 41.
Mississippi, United States'. Tab. ii; at New Orleans, i, 40; place, 45; passes forts, 47-8; tries to ram *Manassas*, 56; at Port Hudson, 74; runs aground, 76; burnt, *ib*. 77; m. 84.
Missouri, State of, i, 37, 179.
Mitchell commands Confederate naval force at New Orleans, i, 41-2, 45; does not send down fire rafts, 46; m. 58.

Mobile, M. vi, i, 122; Confederate defences, i, 114-5; mines, 115; flotilla, 116-8; Northern fleet at, 119-20; attack begins, 121; *Brooklyn* stops, 122; Farragut goes on, 124; fort passed, 127; action in the Bay, 128—131; *Tennessee* surrenders, 132; tactical importance, 133-4, 59, 70, ii, 30; battle, m. 59, 70, 282, ii, 151, 164; bay, i, 180, 185; town, 147, 187.
Mohican, Tab. v, i, 137.
Moll, Von, killed at Lissa, i, 243.
Moltenört, i, 270.
Monadnock, Tab. v.
Mona Passage, i, 170.
Monarch, English turret-ship, Tab. xii; elevation, Pl. xxxvii, ii, 220; des. ii, 224-5, i, 339; at Alexandria, 336; position, 342-3; share in the bombardment, 346-8; no damage, 351; m. ii, 184-5, 196; guns, 223.
Monarch, U.S., ram, i, 68-9.
Monitor, elevation, Pl. ii, i, 10; section, Pl. iii, i, 26; designed by Ericsson, i, 6-7; the turret, 8-9, 33; completion of, 9; des. 9-10; criticism of, 11-2; price, 13; name, 14; passage to Hampton Roads, 23-4; arrival, 25; action with *Merrimac*, 26-31; defects, 27-8; pilot house hit, 30; results of action, 31-4; founders at sea, 36; m. ii, 137, 220; deck, 227, 246, i, 86; earlier designs by Ericsson, i, 8, ii, 218; type Lincoln believes in, i, 91; later examples, 89-90, 119; defects, i, 136, 266; armoured funnels, ii, 164.
Monocacy, ii, 6.
Monongahela, Tab. iv; at Port Hudson, i, 74-76; at Mobile, 120; rams *Tennessee*, 126; does so again, 129.

Montauk, Tab. iii; at Charlestown in action, i, 92-4; damage, 94; attacks Fort McAllister, 89-90; destroys *Nashville* 90, 105; hits on, ii, 246.

Montcalm, i, 267.

Mont du Roule, i, 160.

Montéchant, French strategist, q. i, 209, ii, 156.

Montevideo, ii, 36.

Montgomery, Tab. v; i, 137.

Monticello, Tab. v; i, 137.

Montt, commander of Congressionalist squadron, ii, 16.

Montz, Von, commands *Grosser Kurfürst*, ii, 194; saved, 195.

Moore commands *Independencia*, i, 315, 319, 321.

Moraga commands *Condell*, ii, 21; leads torpedo attack on *Blanco*, 21-2; his account, 23; number of torpedoes fired, 27; fights *Aconcagua*, 29; at Iquique, 31.

Morgan, *see* Fort M.

Morgan, des. i, 118; run aground, 128.

Morlaix, i, 165.

Morris, gallantry of, i, 17.

Morris commands *Florida*, i, 149, 150, 151.

Morse, ii, 268.

Mosher, armed tug, i, 48.

Mouin-i-Zaffre, i, 287; at Sulina, 289, 295.

Mound Battery, at Ft. Fisher, i, 142, 190.

Mound City, des. i, 62; rammed, 68; captures Fort Charles, 69; passes Vicksburg, 78.

Mount Vernon, i, 137.

Mucangué Island, ii, 40.

Mukhadem Khair, i, 287; at Sulina, 289, 293, 295, 297.

Mustapha Pasha, i, 295.

Musashi, ii, 84.

N.

Nada, ii, 41.

Nagasaki, ii, 61.

Nahant, Tab. iii; action with Ft. Sumter, i, 93-4; damage, 95; action with *Atlanta*, 99-100; hits, ii, 246.

Naniwa, Tab. xix; loss, xxi; elevation, Pl. xxvi, ii, 74; des. 60; efficiency of, 53; in G. of Korea, 67; opens on *Tsi Yuen*, 68, 70; attacks *Kwang-shi*, 71; stops *Kowshing*, 72; sinks her, 74-5; fires on men in water, 78; at Yalu, 84; place in line, 88; reconnoitres, 93; loss, 105; hits on, 109, 110; at Wei-hai-wei, bombards, 128; hit, 133.

Nan Shuin, ii, 65.

Nan Ting, ii, 65.

Nantucket, Tab. iii, i, 95; ii, 246.

Napoleon III. and the *Monitor*, i, 8; ii, 218.

Nashville, destroyed by *Montauk*, i, 90, 105.

Nassau, *Florida* seized and released at, i, 147; coals at, 148; watched by U.S. cruisers, 186; centre of blockade-running trade, 187, 189, 191, 193; number of ships clearing from, 194, 195.

Naval Defence Act, 1889, ii, 234, 26

Navarino at Sukhum, i, 298-9.

Nedjem-i-Chevket, i, 287.

Nelson, Lord, his relations to his captains, i, 214, 228; on frigates and battleships, ii, 138; place of admiral, 151; m. i, 59, 133, 219, 282; ii, 79, 137, 216.

INDEX OF NAMES.

Nelson, Tab. xxiii, 255; m. 266; 273, 274.
Neptune, English turret-ship, ii, 224-5.
Neptune, French battleship, ii, 263; Pl. xliii, ii, 264.
Neuse River, i, 180.
Neustadt, i, 277; ii, 276.
Newfoundland Banks, i, 153, 170.
New Ironsides, see Ironsides.
New Madrid, i, 66-7.
New Orleans, instructions to capture, i, 38; forts, 40; flotilla, 41-2; boom, 42; boom breached, 42, 44; Northern attack on forts, 46-54; action above forts, 54-56; capture of New Orleans, 56; consequences, *id.*; garrison of forts, 57-8; compared with Mobile, 59, 127; loss of fleet, 60; blockaded, 183; blockade running, 187; place of Farragut, ii, 151, n.; sandbags on board fleet, i, 39; ii. 164; m. i, 281.
New Orleans, i, 41.
New York, threatened by *Atlanta*, i, 167-68; dockyard, 184; m. 171, 198, 212.
Niagara captures *Georgia*, i, 166; off Charleston, 183; declines to fight *Stonewall*, 311, n.
Nichols, on *Ting Yuen*, ii, 87; killed, 97.
Nictheroy, ex *El Cid*, des. ii, 40; pneumatic gun, 40-1; appears off Rio, 42; at Tijucas Bay, *id.*; m. 151.
Nielly, ii, 13.
Nikolaiev Fort, *see* Fort N.
Nikopolis, or Nikopol, i, 294.
Nile, battle of the; loss at, ii, 106.
Nile, English turret-ship, Tab. xxii; Pl. iv, i, 32; elevation Pl. xxxix, ii, 232; des. ii, 233; low freeboard, 240; descended from *Monitor*, i, 33; in Mediterranean fleet, 1893, ii, 196, 199.

Niloff, i, 294.
Nils Juel, Danish, i, 226, n.
Nippon Yusen Kaisha, ii, 61.
Nitrate Ports, ii, 18.
Nordenfelt gun, ii, 250; torpedo, 259.
Norfolk, Virginia, capture of navy yard, i, 4; guns from, 35; recaptured, 184-85.
Norman, H., q. on Ting, ii, 55-6, 79; Chinese *personnel*, 56; foretells Yalu, 57.
Normand, i, 129.
Normandie, ii, 260.
North Atlantic squadron, i, 185.
Northbrook, Lord, ii, 184.
North Coast squadron, ii, 62.
North Sea, French squad. in, 1879, i, 267; Willaumez off Jahde, 272; Fourichon in, 275; his difficulties, 275-6; blockade continued, 277-8; observation substituted, 278.
Northampton, ii, 255, 273, 274.
Northumberland, ii, 221, 185.
Novara, Tab. viii, i, 226, 227.
Novgorod, i, 286.
Numancia, Spanish, in Pacific, 1865, i, 252; bombards Valparaiso, 253-54; at Callao, 255-56; hits, 256.
Nunez, Spanish Admiral, i, 253, 256.
Nyack, i, 137.
Nymphe, i, 277-78.

O.

Océan, elevation Pl. xlii, ii, 262; des. ii, 261; in Baltic squadron, i, 272; withdrawn, 277; m. 272.
Ocean, ii, 221.
Octorara, Tab. iv, at Mobile, i, 120, 123.

INDEX OF NAMES. 343

Odessa, i, xxxiv, 211, 218, 293, 296, 300.
O'Higgins, Tab. xi, xvi; des. i, 313; searches for Huascar, 322; chases Union, 325; joins Congressionalists, ii, 16; shells Valparaiso, 20; at Pacocho, 31.
Ohio River, i, 61.
Ohio, State of, i, 61, 65.
Old Dominion Trading Co., i, 188.
Olinda, Paraguayan, i, 258; grounds, 260.
Olustee, see Tallahassee.
Oneida, Tab. ii, iv; at New Orleans, i, 45; passes forts, 47; engages gunboats, 55; at Vicksburg, 70; at Mobile, 120; raked by Tennessee, 126; Florida runs past, 148-9.
Onohama dockyard, ii, 61, 129.
Onondaga, purchased by France, i, 266-7.
Opiniâtre, i, 266.
Opyt, i, 296.
Oran, i, 275.
Oreto, see Florida.
Orion, purchased from Turkey, ii, 224; m. 272.
Osage, i, 134.
Osaka, ii, 61.
Osceola, Tab. v, i, 137, 139.
Osmanieh, i, 287, 300.
Ossabaw Sound, i, 90, 95.
Ossipee, Tab. iv; at Mobile, i, 120, 126.
Otchakov, i, xxxiv.
Otsego, i, 113.
Outka, i, 286, 296.
Owasco, i, 43.

P.

Pacheco, killed on Blanco, ii, 26.
Pacific, Russian fleet in, 1877, i, 286.
Page, on battle of Lissa, i, 248; on armour, ii, 217.
Pagoda Point, ii, 5.
Paixhans, General, inventor of shell-gun, ii, 217.
Palacios wounded on Huascar, i, 329.
Palestro, Tab. vii; des. i, 213; at Ancona, 216; off Lissa, 225; position in line, 232; engages Drache, 235; takes fire, id.; courage of her crew, 241; she blows up, id.; m, 249; ii, 185.
Pallas, English, ii, 221.
Pallas, Brazilian, ii, 37.
Palliser shells have a chilled point, ii, 252.
Palmer, gallantry of, i, 70.
Palmer, Surgeon, i, 132.
Palmerston, Lord, i, 202.
Palmetto State, des. i, 87; action off Charleston, 87-89.
Pamlico Sound, i, 186, 106.
Para, i, 263.
Paraguari, Paraguayan, i, 258; rammed, 260.
Paraguay, M. xiv, i, 280; Lopez, tyrant of, i, 257-8; seizes Brazilian ship, 258; enters La Plata, id.; battle of Riachuelo, 260-1; Itaipuru, 262; Humaita, 262-4; boarding attacks on Brazilians, 264; end of the war, id.
Paraguay River, i, 257.
Parana River, i, 257, 259, 261.
Parauahyba or Paruahyba, Tab. xvii, i, 260, ii, 42.
Pareja, Admiral, Spanish, on Pacific Coast, i, 252; commits suicide, 253.
Paris, i, 276; treaty of, "free ships free goods," 169, 199. Russian Black Sea fleet limited, 287.
Paris, i, xxxi.

Parrot guns. A species of rifled gun: burst, i, 139; dangerous, 141.
Pascal, ii, 267.
Paso el Patria, i, 261.
Passaic, Tab. iii; at Charleston, i, 92-94; damage, 94; hits on, ii, 246; m. 312.
Patapsco, Tab. iii; at Charleston, i, 92-94; damage, 94; hits on, ii, 246.
Patrick Henry or Yorktown, i, 15.
Pawtucket, Tab. v, i, 137.
Payne, drowned on David, i, 104.
Peacock, ii, 174.
Pearl, Tab. xxiii, ii, 257.
Pe-chi-li, Gulf of, ii, 66, 127.
Pedro Ivo, sent against Aquidaban, ii, 43; pressure falls, 45.
Pciho, i, 35.
Peixoto, Marshal, President of Brazil; revolt against, ii. 35; his resources, 37; buys a fleet, 40; its uselessness, 41; crushes insurrection at Rio, 42.
Pélereschine, i, 305.
Pelorus, ii, 257.
Pemberton, General, i, 74.
Penelope, Tab. xii; des. i, 339; twin screws, ii, 254; position at Alexandria, 342-3; shells Mex, 347-8; loss, 350; damage, 350-1.
Penhoat, i, 277.
Pensacola, i, 184-5.
Pensacola at New Orleans, i, 40, 45; passes forts, 47.
Pequet, Tab. v, i, 137.
Persano, Count Carlo Pellion di, Italian Admiral; his record, i, 211-2; faith in Affondatore, 212, 215-217; want of character, 214-5, 217; complaints of fleet, 215; declines battle at Ancona,
216-7; puts to sea and returns, 218; ordered to sea, 219; strategy, 220; attacks Lissa, 220-224; his negligence, 224; makes numerous signals, 231-2; dispositions for battle, 232; changes flagship, 233; does not ram Kaiser, 238-9; no plans, 246; rage against him, 249; his trial, 250; condemnation, 251; m. ii, 79, 85, 151.
Persine, i, 290.
Peru, quarrel with Spain, i, 252; Callao bombarded, 254-6; repulse of Spaniards, 256; and England, affair of Huascar, i, 306-312; and Chili war, i, 312; fleet, 312, Tab. xi; geographical position, 314; bad gunnery, 317, 319; cross raiding, 321-2; battle of Iquique, 315-321; of Angamos, 324-331; blockade of Callao, 333.
Peterhoff, case of the, i, 198-200.
Petropaulovsk, i, 286.
Petz, Commodore, commands wooden ships at Lissa, i, 230-1; engages Ribotti, 234; rams Portogallo, 238-9.
Phaeton. ii, 196.
Philadelphia, 184.
Philadelphia target, ii, 150.
Philippines, i, 169.
Phœbe, i, 150.
Phyong Yang or Ping Yang, ii, 83.
Piemonte, i, 224.
Pierola, i, 306.
Pifarefski at Sukhum, i, 298-300.
Pikysyry, i, 264.
Pilcomayo, Tab. xi; des. i, 312; at Arica, 322; captured by Chilians, 332; re-armed, id., m. 333.

INDEX OF NAMES. 345

Ping Yuen, Tab. xviii; des. ii, 64; at Yalu, 83; place in line, 88; holds aloof, 92-3; attacks *Saikio*, 96; loss, 105; at Wei-hai-wei, 126; m. 132.

Pinola, Tab. ii; at New Orleans, rams boom, i, 44; position in attack, 45; does not pass forts, 53-4; m. 56; loss, 60; Vicksburg, 70.

Pinzon, Spanish Admiral, i, 252.

Pique, ii, 2.

Pirabebe, Paraguayan, i, 258.

Piratiny ex *Destroyer*, des. ii, 40.

Pisagua, *Huascar* off, i, 307, 308; m. ii, 20.

Pittsburg, des. i, 62; at Ft. Donelson, 64; hit, 65; passes Island No. 10, 67; passes Vicksburg, 78; at Grand Gulf, 79.

Pittsburg landing, i, 65.

Plevna, i, 287.

Plymouth, i, 170, 172, 273.

Plymouth, U.S.A., i, 111.

Podgoritza, i, 289.

Point de Galle, i, 171.

Pola, i, 229, 251.

Poltio, gallantry at Lissa, i, 237.

Polyphemus, Pl. vii, i, 132; des. ii, 228; utility, 147, 150.

Pope, General, i, 66.

Popoffkas, circular floating batteries, i, 286.

Port Arthur, docks at, ii, 66; Chinese fleet retires to, 93, 104; withdraws from, 126; captured by Japanese, 127.

Porter, Constructor, i, 5, 107.

Porter, Admiral, on *Monitor*, i, 12; commands mortar flotilla at New Orleans, recalls *Itasca*, etc., i, 54; commands on Mississippi, 74; runs past Vicksburg, 77-8, 84; commands at Ft. Fisher,

137-142; on strategy of commerce destroyers, q. 170; on Port Hudson, 77.

Port Hudson, fleet passes, 74-6; *Mississippi* lost, 76; Porter on strategic importance of passage, 77.

Portland, ii, 134.

Portland, Maine, i, 149.

Porto Karober attacked by Italian fleet, i, 222-3.

Porto Manego attacked by Italian fleet, i, 220, 222, 224.

Porto San Giorgio attacked by Italian fleet, i, 222-3.

Port Royal captured by Dupont, i, 56, 184; m. 20, 89, 185, 281.

Port Royal, Tab. iv; at Mobile, i, 120, 128.

Portsmouth, i, 307; ii, 195.

Portsmouth, U.S.A., i, 184.

Pothuau, ii, 267.

Poti, i, 301, 302.

Powerful, Tab. xxiii; des. ii, 135, 216, 256; water-tube boilers, 254.

Powhattan, Tab. v, i, 137, 183.

Prado, General, i, 315.

Prat, speech to his crew, i, 316; skill and daring, 315-16; boards the *Huascar*, 317; killed, *id.*; a national hero, *id.*; reward of his courage, 320.

Presidente Errazuriz, ii, 17.

Presidente Pinto, ii, 17.

Preussen with German squadron, ii, 193, 195.

Prince Albert, ii, 224.

Prince Consort, ii, 221.

Prince George, see *Majestic*, ii, 237, n.

Principe di Carignano, Tab. vii;
des. i, 213; at Ancona, 216; at
Lissa, 219, 225; place in line,
232; opens battle, 234; attacks
Kaiser, 239; damage, 245.

Principe Umberto, Tab. vii, i,
222.

Prinz Adalbert, Tab. x; des. i, 269;
at Wilhelmshaven, 273; in Elbe,
278.

Prinz Eugen, Tab. viii, des., i, 227;
place in line at Lissa, 230; in the
battle, 243.

Prinz Friedrich Karl, Tab. x,
des. i, 269; retires to Wilhelmshaven, 273, 278.

Provence in Adriatic, i, 227; in
Mediterranean fleet, 1870, 267;
in North sea, 275.

Pullino, ii, 259.

Pulo Condor, i, 157.

Purvis on board *Chih Yuen*, ii, 87;
drowned, 98.

Q.

Quaker City, i, 137.

Queen, ii, 212.

Queen of the West, i, 68; rams
Lovell, 69; on Vazoo, 71; passes
Vicksburg, 73; captured by Con-
federates, 74; rams *Webb*, id.

Queenstown, ii, 189.

Quilio, ii, 275.

Quinteros Bay, ii, 21; Congressionalist forces land at, 32.

R.

Radetzky, Tab. viii; i, 226, 227.

Ragheb Pasha, i, 337.

Raleigh, English, ii, 254.

Raleigh, Confederate, i, 15, 19.

Ramillies, submarine attack on, i,
102.

Ramillies, see *Royal Sovereign*, ii,
234, n.

Rappahanock, ex *Victor*, i, 166,
174.

Ras-el-Tin, see Fort R.

Rattlesnake, ii, 257.

Rattazzi, Italian Minister, i, 211.

Razzetti, gallantry at Lissa, i, 237.

Read, i, 149.

Rebolledo, Admiral, i, 315, 322.

Re di Portogallo, Tab. vii; des. i,
212; at Ancona, 216; at Lissa,
219, 221; position when Austrians
appeared, 225; place in line,
232; rammed by *Kaiser*, 238-9;
damage, 245; praised by Persano,
249.

Re d'Italia, Tab. vii; des. i, 212; at
Ancona, 216; Persano's flagship,
219; at Lissa, 221; position
when Austrians appeared, 225;
place in line, 232; Persano leaves
her, 233, 234; hotly attacked,
235; rammed and sunk, 236-7;
gallantry of her crew, 237;
rudder damaged (?) 243; praised
by Persano, 249; m. 237, 242,
244, 246, 247, 250; ii, 160.

Redoubtable, elevation, Pl. xlii, ii,
262; ii, 262.

Red River, stores destroyed on, i,
73; stores collected, 74, 77;
Federal expedition up, 80.

Reed, Sir E. J., ii, 221, 225.

Reina Regente, Spanish, ii, 207.

Reindeer, ii, 174.

Reine Blanche at Sfax, ii, 2, 3.

Renard, i, 267.

Rendel gunboats, carrying one very
heavy gun forward, ii, 65, 83,
126.

Réné Adolphe, i, 278.

Rennie, Messrs., i, 259.

Renown, Tab. xxii; des. ii, 253;
compared with cruiser, 140-1,
142; m. 145, 271.

INDEX OF NAMES.   347

Renown, see *Victoria*, ii, 196.
Republica, Tab. xvii; joins Mello, ii, 36; passes Rio forts, 37; rams *Rio de Janeiro*, 38; captures *Itaipu*, 39.
Repulse, ii, 221.
Repulse, see *Royal Sovereign*, ii, 234.
Requin, ii, 265-6.
Research, ii, 221.
Resistance, ii, 220.
Resolucion, Spanish, in Pacific, i, 252; bombards Valparaiso, 253-4; Callao, 255-6.
Resolution, see *Royal Sovereign*, ii, 234.
Retribution, privateer, i, 143.
Revanche in North Sea, i, 275; m. 257; at Sfax, ii, 2, 3.
Revenge, ii, 234.
Rhind, i, 138.
Rhode Island, i, 137.
Riachuelo, ii, 36.
Riachuelo, Battle of the, i. 259-261.
Ricasoli, Italian Minister, i, 215.
Richelieu, ii, 261.
Richmond dependent upon Wilmington, i, 135, 142; inland, 179; on James River, 180; prices at, 196; m. 4, 5, 35, 106, 107, 185, 187, 205.
Richmond, Tab. ii, iv; at New Orleans, i, 45; passes forts, 52; at Vicksburg, 70-1; at Port Hudson, 74; at Mobile, sandbags, 119; place, 120; danger of collision, 125; passes forts, 126.
Rimac captured by *Huascar*, i, 322; hit, 333.
Rio de Janeiro, revolt of fleet at, ii, 35; desultory fighting, 36-41; collapse of revolt, 42; m. i, 171, 259; ii, 137.

Rio de *Janeiro*, ii, 38.
Rio Grande, Mexico, i, 185.
Rio Grande, i, 263-4.
Rio Grande do Sul, ii, 35.
Riojo, ii, 58.
"River Defence Fleet," i, 42, 58.
Riveros, Commodore, commands Chilian fleet, i, 322; action with *Huascar*, 323-331.
Roanoke River, i, 106, 110.
Roanoke Island, i, 184.
Roanoke at Hampton Roads, i, 14, 18.
Rochambeau purchased by France from United States, i, 266, 277; ii, 266.
Rodgers, G. W., killed, i, 101; ii, 168.
Rodgers, J., i, 101.
Rodimyi, i, 286.
Rodney of "Admiral" class, ii, 231.
Rodolph, i, 134.
Rodriguez killed, i, 327.
Rolf Krake, Danish turret-ship, i, 33.
Rose, i, 134.
Rossia, i, 286.
Rostislav, line-of-battleship, i, xxxi.
Rostislav, i, 304.
Rouen, i, 278.
Roumelia, i, 287.
Royal Alfred, ii, 221.
Royal Arthur, target practice, ii, 166.
Royal Oak, ii, 221.
Royal Oak, see *Royal Sovereign*, ii, 234.
Royal Sovereign, Coles' turret-ship, elevation, Pl. xxxvii, ii, 220; des. ii, 224; ancestor of new *Royal Sovereign*, 239; m. i, 33, ii, 183.

Royal Sovereign, English battleship, Tab. xxii; Pl. v, i, 56; elevation, Pl. xxxix, ii, 232; des. ii, 234-6; nickel steel decks, 253; crew, 213; engines, 214; unprotected surface in, i, 311; m. 96, ii, 123, 135, 144, 178, 239, 253.

Roydestvenski, i, 293.

Rügen, i, 277.

Rupert, des. ii, 226; deck, 227; m. 272.

Rurik, i, 311-4.

Russell, Lord John, afterwards Earl, i, 147, 165.

Russell, Scott, designs *Warrior*, ii, 219.

Russia, war with Turkey, i, 286; distribution of Russian fleet, *id.*; torpedo attacks in Black Sea, 290, 303; *Vesta* and *Assar-i-Chevket*, 303-4; defeat of Turks at Sinope, i, xxxi; bombardment of Kinburn, xxxiv-vi.

Rustchuk, i, 295.

S.

Sacramento, i, 311.

Saikio, Tab. xix; loss, xxi; torpedo affair, xxv; speed, &c., ii, 61; at Yalu, 87; not in line, *id.*, 88; hotly engaged, 91; torpedoes fired at, *id.*, 96; sent off for repairs, 93; part in battle, 95-6; loss, 105; hits on, 96, 109, 110; repaired, 126; escape of, 175; m. 112, 144.

St. André, Jean Bon, i, 218.

St. Augustines, seized by North, i, 184; distance from blockade-running centres, 187.

St. Bon, i, 223.

St. Helena, i, 171.

St. John, Knights of, lead-plated ship, ii, 217.

St. Lawrence at Hampton Roads, i, 14, 18, 20, 21.

St. Louis, see *De Kalb*.

St. Louis, ii, 264-5.

St. Marc, i, 278.

St. Philip, see Fort S.

St. Pierre, i, 145.

St. Thomas, i, 187, 195.

Sakamoto, killed at Yalu, ii, 95.

Salamander, Tab. viii; des. i, 227; at Lissa, 230; part in battle, 243.

Salto Oriental, Paraguayan, i, 258, 260.

Sampaio, see *Gustavo S.*

San Diego, ii, 33.

Sandri, i, 220.

San Francisco, ii, 33.

San Giovanni, Tab. vii.

San Jacinto stops *Trent*, i, 201; m. i, 153.

San Martino, Tab. vii; des. i, 213; at Ancona, 216; at Lissa, 219; position when Austrians appeared, 225; place in line, 232; in battle, 235, 237; damage, 245; praised by Persano, 249.

San Salvador, ii, 42; sent against *Aquidaban*, 42, 43.

Sanspareil, see *Victoria*, des. ii, 232; m. 174, 196, 271.

Santa Catherina, torpedo attack on *Aquidaban* at, ii, 43.

Santa Cruz, ii, 43.

Santa Rosa, i, 306.

Santiago, ii, 17.

Santiago di Cuba, Tab. v, i, 137.

Santos, ii, 35, 42.

Sdone, ii, 12, 13.

Sardinia, i, 211.

Sargente Aldea, Tab. xvi.

Sarthe, ii, 2.

INDEX OF NAMES.   349

Sassacus, Tab. v; engages *Albemarle*, i, 108; rams her, 109; disabled, 110; at Fort Fisher, 137, m. 168.

Saugus, Tab. v.

Savannah, *Atlanta* constructed at, i, 97, 98; fall of, 185; distance from blockade-running centres, 187; m. 190.

Savoie, i, 267.

Scharf, q. on U.S. Navy, i, 196; q. i, 44, 89, 100, 119, 124, 143, 144, 146, 185, 186, 188, 193, 201.

Schichau torpedo-boats, ii, 83, 129.

Schwartzenburg, Tab. viii, i, 226, 227.

Schwartzkopf torpedo, a species of Whitehead, ii, 45, 135.

Sciota, Tab. ii; at New Orleans, i, 45, 52; at Vicksburg, 70; destroyed by a mine, 134.

Scorpion turret-ship, i, 33, 168.

Sea Bride, i, 150.

Seaford, i, 208.

Sea King, see *Shenandoah*.

Sebastopol, naval attack on, i, xxxi, xxxvi, m. 288, 298.

Sedan, i, 276, 281.

Segurança, ii, 41.

Seifé, i, 289; on Danube, 290; sunk by torpedoes, 291-2; Tab. xxv.

Selma, i, 116.

Selma, i, 118; action with *Metacomet*, 127-8.

Semendria, i, 289.

Seminole, Tab. iv; at Mobile, i, 120.

Semmes commands *Sumter*, i, 144-6, 151; commands *Alabama*, 152-163; himself a prize-court, 153; eludes *San Jacinto*, 153; sinks *Hatteras*, 155; challenges *Kearsarge*, 159; his strategy, 170; Porter on, *id.*

Seneca, Tab. v, i, 137.

Serpent, loss of, ii, 208.

Serrano boards *Huascar*, i, 318.

Serras, ii, 275.

Sestrica, i, 286.

Seth Low, i, 23.

Seward, U.S. Secretary of State, i, 202.

Seymour, Sir Beauchamp, later Lord Alcester, commands at Alexandria, i, 336; ultimatum, 337; general order, 342-3; tactics, 355; m. 349.

Sfax, seized by Tunisian insurgents, ii, 1; bombarded, 2-3; captured, 4.

Sfax, ii, 267.

Shah, des. i, 307, ii, 254-5; pursues *Huascar*, i, 307-8; sights and attacks *Huascar*, 308, action, 308-9; cannot close, 308; not hit, 310; uses torpedo, 309; comparison of force, 310; m. 331, ii, 137, 239.

Shanghai, i, 169; squadron, ii, 62; docks, 65; m. 61.

Shannon, ii, 228, 255, 273, 274.

Sheipoo, torpedo attack on *Yu-yen*, ii, 13, 14, 137.

Shenandoah, commerce-destroyer, ex *Sea King*, cruise of, i, 166-7; English payment for, 174; purposeless destruction by, 176.

Shenandoah, United States, Tab. v, i, 137.

Sherman, General, i, 73.

Sherman, Confederate, i, 48.

Shiloh, battle of, navy at, i, 65.

Shogun, ii, 52.

Shopaul Island, ii, 75.

Shoppek, Cape, ii, 84.

Sicily, i, 211.

Silvado, ironclad, i, 263.

Silvado sent against *Aquidaban*, ii, 43; part in the action, 45; hit, 46.
Singapore, U.S. ships laid up at, i, 169.
Singapore, Straits of, i, 157.
Sinope, Battle of, i, xxxi.
Sinope, at Sukhum, i, 298, 299; at Batum, 301, 302.
S. J. Waring, i, 143.
Slatina, i, 289.
Slidell, Confederate Envoy, i, 201-3.
Smith, Commodore, i, 11, 12, 13.
Smith, Lieutenant, gallantry of, i, 18.
Smith, Melancton, i, 76, 108.
Smith-Dorrien, i, 337.
Smith's Island, i, 186.
Sokul, ii, 268.
Soley, Professor, q. i. 165, 206.
Solferino, Tab. x, i, 267.
Sonoma, i, 148, 200.
Soto, Col., Balmacedist, ii, 20.
Soto, Aspirant, Congressionalist, drowned on *Blanco*, ii, 27.
Sound, The, i, 277.
Southampton, i, 165.
South Atlantic Squadron, i, 185.
Southfield engages *Albemarle*, 108; rammed by her, *id.*; wreck of, 111.
South Sea, i, 269.
Spain, quarrel with Peru, i, 252; squadron despatched to Pacific, 252; loss of *Covadonga*, 253; bombardment of Valparaiso, 253-4; of Callao, 254-6; withdrawal of squadron, 257.
Springbok, case of the, i, 200-1.
Stag, i, 188.
Stanton, U.S. Secretary of War, i, 20.
Stchelinski, i, 302.

Stella d'Italia, i, 220.
Stenzel, q. i, 37.
Stevens, designer of floating battery, i, 3; ii, 218.
Stevens at Mobile, i, 123, 127.
Stettin, i, 277; ii, 62, 64.
Stimers, Chief Engineer, i, 26, 95.
Stockton, fitted with screw, ii, 211.
Stodder, i, 28.
Stonewall Jackson, Tab. ii; at New Orleans, 55; abandoned, *id.*
Stonewall Jackson, ram; des. i, 168; at Corunna, 311.
Stoney, T., i, 103.
Stowell, Lord, rule of "continuous voyages," i, 198-9.
Stralsund, i, 277, 278.
Suenson, Danish Commodore, i, 226.
Suffren, ii, 261.
Suffren's attack upon Hughes, i, 150.
Sulina, ships at, 1877, i, 289; Russian torpedo affair at, 293-4; attempt to capture, 295-6.
Sulina at Sulina, i, 296; sunk, 297.
Sullivan's Island, i, 87.
Sultan, Tab. xii, xxii; at Alexandria, i, 336; des. 338, ii, 222; orders to, i, 342-3; part in bombardment, 346; ammunition exhausted, 348; loss, 349; damage, 351.
Sumter, commerce-destroyer, ex *Habana*; des. i, 144; gets to sea, 145; at St. Pierre, 145; escapes, 146; sold at Gibraltar, *id.*
Sumter, *see* Fort S.
Sumter, ram, i, 67, 68.
Sunda, Straits of, i, 156-7, 171, 172

Superb, Tab. x; purchased from Turkey, ii, 224; at Alexandria, i, 338; des. 338-9; orders to, 342-3; part in bombardment, 346-7; loss, 349; damage, 351.
Surprise, ii, 6.
Surprise, privateer, i, 173.
Surveillante, i, 267; flagship of French Baltic fleet, 272; loses her rudder on North Sea Coast, 277.
Suzuki, ii, 131.
Symonds, Sir T., Admiral on *Captain*, ii, 185.

T.

Tabarka, seized by French, ii, 1.
Tacna, ii, 18.
Tacony, i, 149.
Tacony, Tab. v, i, 137.
Tacuari, Paraguayan, i, 258, 260.
Tage, ii, 267.
Takachiho, Tab. xix; loss, xxi; see sister ship, *Naniwa*; des. ii, 60; in Gulf of Korea, 67; engagement with *Chen Yuen* (?) 79; at Yalu, 84; place in line, 88; opens on *Lai Yuen*, 92; loss, 105.
Takao, ii, 84.
Taku, docks, ii, 65-6; transports leave, 67, 83; return to, 101; m. 73.
Takushan, ii, 89.
Talien Bay, ii, 83, 93.
Tallahassee, ex *Atlanta*, destroys commerce, i, 167-8; threatens New York, 168; m. 170.
Tamandaré, Admiral, character of, i, 258, 259; his battles, 261-2; recalled, 263.
Tamandaré, see *Almirante T*.
Tamplin, Mr., ii, 74, 75.
Tang commands *Chih Yuen*, ii, 98, 139.

Tangiers, ii, 207.
Taranto, i, 215, 229.
Tarapaca, ii, 18.
Tatnall commands *Merrimac*, i, 35; "blood thicker than water," 35. 349.
Taureau, Tab. x; des. ii, 265; m. i, 267.
Tavara, surgeon on *Huascar*, i, 328.
Tayi, i, 263.
Tchen Kiang, ii, 13; sunk, 15.
Tchesmé, line-of-battle ship, i, xxxi.
Tchesmé, torpedo-boat, at Sulina, i, 293-4; at Sukhum, 298; at Batum, 301-2; second attack, 302-3.
Teaser, i, 15, 19.
Tecumseh, Tab. iv; at Mobile, i, 119; des. *id.*; position, 120; opens fire, 122; crosses line of torpedoes, 124; sunk, *id.*; heroism of Craven, *id.*; loss on, 132.
Tegetthof, Baron Wilhelm von, his record, i, 225-6; off Ancona, 216; message to Lissa, 222; scruples as to Venetian sailors, 227; prepares his *personnel*, 228; ram and concentrated broadsides, *id.*; appears off Ancona, 229; again visits it, 229; doubts whether Lissa is the Italian objective, 230; instructions, *id.*; sails in battle order, 230-1; bad weather, 231; breaks Italian line, 234; rams *Re d'Italia*, 236; reaches Lissa, 241; why he did not renew action, 241; tactics 247-8; rewards, 251; death, 251; m. ii, 79, 102, 103, 107, 151, 155.
Telegraph Tower, i, 222.
Téméraire, Tab. xii; elevation, Pl. xxxvi, i, 220; des. ii, 223-4, i, 339; at Alexandria, 339; orders to, 342-3; part in bombardment, 347-9; no damage, 351; good shooting, 353; armour deck, ii, 227.

INDEX OF NAMES.

Tempête, ii, 265.
Teneriffe, i, 150.
Tengchow-feu, ii, 127, 129.
Tennessee River, forts on, i, 61, 63; importance of, 64, 83.
Tennessee, State, i, 64, 65, 178, 179.
Tennessee, des. i, 116-7; crosses flats, 117-8, 124; at Mobile Bay, 118, 119; encounters *Hartford*, 125; attempts to ram, 125-7; retires to Fort Morgan, 127; renews action, 128; rammed by *Monongahela* and *Lackawanna*, 129; attacked by monitors, 130-1; strikes, 132; damage and loss, 132; m. ii, 137.
Tenrio, ii, 84, 133.
Terceira, i, 152.
Terribile, Tab. vii; des. i, 212; at Ancona, 216; at Lissa, 219; off Comisa, 221, 222, 224; position when Austrians appeared, 225, 232, 238; comes up, 240; complaints, 250.
Terrible, English, Tab. xxiii; compared with battleship, ii, 141-3; des. 256; m. 214; water-tube boilers, 254.
Terrible, French, ii, 265-6.
Terror, i, xxxiii.
Terry, General, i, 140.
Texas, i, 37, 180, 181, 185, 190.
Thames Iron Works, ii, 58.
Thames, i, 212.
Thétis in Channel, 1870, i, 267; sent to Baltic, 272; m. 273.
Thomson, Messrs., ii, 58, 208.
Thunder, i, xxxiii.
Thunderbolt, i, xxxiii.
Thunderer, built, ii, 226; hydraulic worked guns, 247; accidents on, 208, 248; re-armed, 272.
Ticonderoga, Tab. v, i, 137.

Tijucas Bay, ii, 43, 46.
Timbo, i, 262.
Timby, an inventor, i, 8.
"Tinclads" described, i, 73, 182.
Ting, Ju Chang, Admiral, portrait, Pl. xxvii, ii, 84; his experience, ii, 55, 57; his strategical adviser, 66; knowledge of naval matters, 79; his wishes, 80; orders of Li Hung Chang to, 81; general orders to fleet, 79, 85; his strategy, 82-3; sees Japanese, 85; tactics, 84-5; at Wei-hai-wei, 126, 128, 133; commits suicide, 133; his mistake, 133-4, 86.
Ting Yuen, Tab. xviii; Pl. xxxii, ii, 122; elevation, Pl. xxiii, ii, 62; des. ii, 62-4; defects, 63; Ting's flagship, 80, 83; at Yalu, place in line, 88; Europeans on board, 87; opens fire, 89; *Hiyei* passes her, 91; terrible fire upon, 92; on fire, 93, 97, 113; military mast hit, 96; Nichols killed, 97; shots fired, 109; hits on, 111; 6-inch gun not disabled, 121; heavy guns disabled, 122; at Wei-hai-wei, 126; torpedoed, 131, 132, 135; m. 94, 105, 112.
Tiradentes, Tab. xvii, ii, 36, 42.
Tokio Arsenal, ii, 61; university, ii, 52.
Tonkin, ii, 4.
Tonnant, ii, 265.
Tonnante, i, xxxii, xxxiii.
Tonnerre, ii, 265.
Torpedoist at Sukhum, i, 298, 300.
Toulba Pasha, i, 357.
Toulon, line Toulon-Algiers, i, 275, 280; m. 219, 355.
Toultcha, i, 296.
Tourville at Sfax, ii, 1-2.
Trafalgar, loss at, ii, 106; duration, 115; place of admiral, 151.

Trafalgar, see sister ship, *Nile*, ii, 232.
Trajano, Tab. xvii, ii, 36.
Treaty of Paris, see Paris.
Tredegar Works, i, 107.
Tréhouart, ii, 266.
Trent, case of the, i, 201-3.
Trenton, i, 6.
Trident, des. ii, 261; at Sfax, 2; m. 272.
Trieste, i, 230.
Triomphante, Tab. xv; des. ii, 5; at Foochow, 8; sinks *Ching Wei*, 9; descends Min River, 11-12; m. 266.
Tri Sviatitelia, i, xxxi.
Trusty, i, xxxiii.
Tryon, Sir G., Commander-in-Chief on Mediterranean station, 1893, on the *Victoria*, ii, 196; his fatal order, 197-8; executed, 199; the *Victoria* rammed, 200; on the chart-house, 202; drowned, 204; blamed by court-martial, 205; system of following the leader, 152; maxim, 153.
Tsan Chieng or *Tsao Kiang* in G. of Korea, ii, 67; captured, 71, 72. Same as *Tehen Kiang*?
Tshao Yong, Tab. xviii; des. ii, 65; at Yalu, 83; slow in moving, 86; place in line, 88; set on fire, 90; disabled, 92; steering gear damaged, 99; men crowd in tops, 107; fire on, 113; m. 102.
Tsi Yuen, Tab. xviii; des. ii, 64-5; action off Asan, 67-71, 53; appearance of ship after, 70; Japanese version, 67, 75; at Yalu, 83; place in line, 88; European on board, 87; runs away, 92; cowardice (?) 100; collides with *Yang Wei*, 92, 107, 115; part in battle, 100-1; loss, 105; at Wei-hai-wei, 126; captured by Japanese, 153; hit on conning tower at Asan, 68, 168; m. 78, 79, 144.

Tsuboi commands flying squadron, i, 67; at Yalu, 84; leads, 88, 152; raises speed, 90; m. 101.
Tsukushi, ii, 84.
Tsung-li-Yamen, its strategy, ii, 56-7; orders to Ting, 80, 82, 103; corrupt, 54.
Tunis, French protectorate, ii, 1; insurgents seize Sfax, 1; bombardment, 2, 3; captured, 4.
Turenne, ii, 266.
Turkey, war with Russia, 1877, i, 286; state of Turkish fleet, 287; has command of sea, 287-8; task before Turks, 288; precautions against torpedoes, 293, 297; ineffective blockade of Russian coast, 288-9, 290; torpedo attacks, 290-302; *Vesta* and *Assar-i-Chevket*, 303-4; *Lutfi Djelil* sunk, 289, 290; little use of fleet, 287-8; defeat of Turks at Sinope, 1, xxxi.
Tuscaloosa, ex *Conrad*, i, 156.
Tuscarora, Tab. v; at Fort Fisher, i, 137; blockades *Sumter*, 146; *Nashville*, 165.
Tuscumbia, des. i, 73; passes Vicksburg, 78; at Grand Gulf, 79.
Tyler, i, 62; at Fort Henry, 63; at Fort Donelson, 64; at Shiloh, 65; chased by *Arkansas*, 71.
Tyler on *Ting Yuen*, ii, 87; describes torpedo attack, 135.

U.

Unadilla, Tab. v, i, 137.
Unebi, ii, 208.
United States, object of Federal or Northern party, i, 1; navy in 1861, 2, 181; measures taken to create a navy, 62, 182; ironclads ordered, 7, 9, 62; consequence of naval weakness, 2; the *Monitor*, 7-14; *Merrimac* defeats *Congress* and *Cumberland*, 15-

20; alarm in North, 20; *Monitor* faces *Merrimac*, 26-31; importance of Mississippi, &c., 37, 82, 83; New Orleans captured, 39-56; the Mississippi opened, 63-81; unsuccessful attack on Charleston, 92-102; the *Albemarle* torpedoed, 110-3; Mobile Bay forced, 116, 132; Fort Fisher captured, 136-142; Southern commerce-destruction, 143-168; effects, 168-9; strategy of North, 169-173; Geneva arbitration, 174; the blockade, 181; its objects, 177-8; occupation of Confederate ports, 184-5; blockade-running, 185-196; treatment of neutrals, 197-8; case of *Peterhoff*, 199-200; of *Trent*, 201-2; doctrine of contraband, 203; the blockade crushed the South, 196-7; war of 1812, i, 197; and Chile, ii, 33-4.

Union, Tab. xi; des. i, 312; with *Huascar*, cross-raiding, 314, 322; sighted by Chilian fleet, 323; escapes, 324; torpedo attack on, 332; Tab. xxv; bombarded, 333; hit, 334.

Urano, Tab. xvii, ii, 38.

Uribe, i, 318.

Uruguay, i, 257.

V.

Vacca commands a division of Italian fleet, i, 219; for attack on Lissa, 220; at Comisa, 221; enters San Giorgio Harbour, 223; quoted on Persano's conduct, 224; left unsupported, 234; wheels on Austrians, 234; his conduct, 246; complains of gunnery, 247; given command, 251.

Valeureuse, i, 267, 275.

Valmy, ii, 266.

Valparaiso, bombarded by Spaniards, i, 253-4; neutral property destroyed, 254; futility of bombardment, *id.*; no docks, 314; open port, 315; revolt of Congressionalists at, ii, 16; fighting at, 19-20; captured, 32; m. i, 322, 332; ii, 30.

Vanderbilt, Tab. v; at Hampton Roads, i, 35; at Fort Fisher, 137; sent in chase of *Alabama*, 156, 171; coals in British ports, 149.

Vanguard, des. ii, 191; with Channel Squadron, 189; collides with *Iron Duke*, 190; sinks, 190-1; court-martial on, 192.

Varese, Tab. vii; des. i, 213; at Ancona, 216; at Lissa, 219; off Comisa (?) 221, 222, 224; position when Austrians appeared, 225, 232; arrives, 234; in battle, 238; collides with *Ancona*, 240; Persano complains of, 250.

Vargus, ii, 23.

Varuna, Tab. ii; at New Orleans, i, 45; passes forts, 47; engages Confederate flotilla, 54-5; rammed and sunk, 55.

Vauban, ii, 266, 272.

Venezuela, i, 153.

Vengeur, line-of-battle ship, i, 17.

Vengeur, ii. 265.

Venice, cession to France, i, 227; m. 211, 213, 230.

Venus, Tab. xvii, ii, 42.

Verestchagine, i, 295.

Vergara, lost on *Blanco*, ii, 27.

Vesta, i, 286; action with *Assar-i-Chevket*, 303-4; m. ii, 137.

Vesuvius, ii, 151.

Vice-Admiral Popoff, i, 286.

Vicksburg, fortified, i, 61; Farragut passes, 69-71; a second time, 72; attempts to turn, 74; Porter and Grant pass, 77-8; fall of, 79, 81; importance of, 81; m. 77, 83, 282.

Victor, see *Rappahanock*.
Victor Emmanuel, Tab. vii.
Victoria, ex *Renown*, Tab. xxii; Pl. xxxiv, ii, 196; elevations, Pl. xxxix, ii, 232; xxxv, ii, 202; des. ii, 196, 232; with Mediterranean fleet, 1893, 196; dangerous manœuvre ordered, 197-8; executed, 199; rammed by *Camperdown*, 200; capsizes, 204; loss of life, 204-5; court-martial, 205-6; why she sank, 206; value of armour-belt, 207; m. 233, 239; triple-expansion engines, 254.
Victoria, Peruvian, i, 255.
Victorieuse, ii, 266.
Victorious, see sister ship, *Majestic*, ii, 237, n.
Vigo, i, 278.
Villa de Madrid, Spanish, in Pacific, i, 252; at Valparaiso, 253-4; at Callao, 255-6.
Villars, Tab. xv; at Foochow, ii, 4, 6, 9; descends Min, 11-12.
Villegagnon Island and Fort, ii, 37, 38.
Villeneuve, ii, 213.
Vincedora, Spanish, in Pacific, i, 252; at Valparaiso, 253-4; Callao, 255-6.
Vipère, Tab. xv, ii, 4, 10.
Virginia, State of, i, 180, 181.
Virginia, ironclad, see *Merrimac*.
Virginia Volunteer Navy, i, 188.
Vishnevetski, Lieut., at Sukhum, i, 298, 300.
Vladimir, i, 286, 292.
Vnoutchek, i, 286.
Volta, Tab. xv; at Foochowi, ii, 4; position, 6; torpedo attacks by her launches, 8; damage, 11; descends Min, 11-12.
Voltigeur at Sfax, ii, 2.
Voron, i, 286, 296.
Vulcan Company, ii, 62.

W.

Wabash, Tab. v, xxv, i, 137.
Wachusett attacks *Florida*, i, 150.
Walke, Rear-Admiral, runs past Island No. 10, i, 66, 67; at Fort Henry, 63; at Fort Donelson, 64.
Walker, Sir B., ii, 184.
Wampanoag, ii, 212, 254.
Wandenkolk, Admiral, ii, 35.
Wangeroog, i, 270, 278.
Warrior, English ironclad, Tab. xxii; elevation, Pl. xxxvii, ii, 220; des. ii, 219-220; compartments, 219, n.; armament, id.; speed, 212; superior to *Monitor*, i, 32; m. ii, 189, 221, 239.
Washington, Tab. vii.
Washington Commission, i, 204.
Wasp, ii, 208.
Waterwitch, ii, 258.
Watts, Isaac, ii, 219.
Waymouth, Capt., ii, 220.
Webb, Lieut., Confederate, i, 100.
Webb, i, 74.
Webb's Yard, New York, i, 212.
Weehawken, monitor, Tab. iii; action with *Atlanta*, i, 99-100; action with Sumter, 92; has bootjack fitted, id.; opens fire, id.; damage, 94; again attacks, 101; injury to, 102; founders off Charleston.
Wei-hai-wei, M. xxv, ii, 52; Chinese naval port, ii, 66; *Tsi Yuen* reaches, 69; fleet at, 82; Japanese off (?) 84; Chinese fleet at, after Yalu, 126; attacked by Japanese, 127-8; bombarded, 128; torpedo attacks, 129-133; collapse of Chinese, 133; m. 71, 80, 81, 137.
Wei Yuen, Tab. xxv; Pl. xx, ii, 14; ii, 132.

INDEX OF NAMES.

Welles, Secretary of Navy, U.S.A., i, 91, 202.

West, The, i, 37, 65, 83, 180.

West Africa, i, 269, 270.

West Gulf Squadron, i, 185.

West Indies, i, 156, 170, 171, 270, 279.

West Virginia, i, 172.

Whampoa, docks at, ii, 66.

Wharton, Lieut., i, 30.

Whitehead torpedo, cigar-shaped and divided into three compartments, containing explosive charge of 60-200lb. gun-cotton, compressed air, and propelling machinery, des. ii, 258-9; range, 259; defects, 161; employment in war, Tab. xxv; first used by *Shah*, i, 309.

Whitworth gun, an early rifle with hexagonal bore, ii, 251.

Wilhelmshaven, state of, 1870, i, 270, 273; ships at 278; practicability of an attack, 283; m. 281, 282.

Wilkes, Capt., i, 171, 201-3.

Wilkinson, Capt., blockade-runner, on risk of blockade-running, i, 191; runs, 195; m. 189.

Willaumez, *see* Bouët.

William, i, 199.

William, King of Prussia, i, 270.

Wilmington, works at, i, 2; importance of, 135, 142, 179-80; blockade-runners at, 183, 192, 194; blockading squadron, 185-6; distance from neutral ports, 187; run into, 190-1.

Windward Island, i, 148.

Windward Passage, i, 176.

Winnebago, monitor, Tab. iv; des. i, 119; at Mobile, 120; turret jammed, 123; saves *Oneida*, 126; engages *Tennessee*, 131.

Winona, Tab. ii; at New Orleans, i, 45; does not pass forts, 53-4; at Vicksburg, 70; off Mobile, 147.

Winslow, Capt., of *Kearsarge*, i, 157-165.

Winslow, Confederate privateer, i, 143.

Wissahickon, Tab. ii; at New Orleans, i, 45; passes forts, 48; at Vicksburg, 70.

Wivern, i, 33, 168.

Worden, commander of *Monitor*, i, 14, 24; attempts to ram, 28; wounded in pilot-house, 30; ii, 168.

Wörth, i, 275, 281.

Wyalusing, action with *Albemarle*, i, 108, 110.

Wyoming, i, 157, 172.

Y.

Yaeyama, ii, 75.

Yalu, Battle of the, or Haiyang, M. xxvi, xxvii, xxviii, ii, 88, 90, 92; fleets at, Tab. xviii, xix; compared, xx; loss of Japanese, xxi; value of deductions from, ii, 57, 116-7, 137; Chinese tactics, 85-6, 103, 118; battle, 87-101; compared with Lissa, 101-3, losses on both sides, 104-6; life-saving, 107; gunnery at, 108-110; structural damage to Japanese ships, 110; artillery preponderance of Japanese, *id.*; size at, 112; fires, 113; ram and torpedo, 114-5; signals, 119; armour at, 119-20, 121-2, 159; special circumstances restrained Japanese, 103, 123; speed at, 124; place of commanders at, 152-3; escape of *Saikio* at, 175; m. 56, 71, 126, 133.

Yalu River, ii, 81.

Yamato, Marshal, ii, 84.

INDEX OF NAMES. 357

Yang Pao, ii, 65.
Yang Wei, Tab. xviii; des. ii, 65; at Yalu, 83; slow in taking up position, 86; place in line, 88; crushed by Japanese fire, 90; rammed by *Tsi Yuen*, 92, 99, 100; fire on, 99; sunk, 102; riddled, 112; weak ship, 144; m. 113.
Yang Woo, Tab. xv, xxv; at Foochow, ii, 5, 6; torpedoed, 8.
Yantic, Tab. v, i, 137.
Yarrow torpedo boats, ii, 83.
Yazoo River, i, 71, 73, 80.
Ybera, i, 258.
Ye Sing, ii, 65.
Ygurei, i, 258, 260.
Ylo, i, 307; action between *Shah* and *Huascar* off, 308-9, 310.
Yokosuka, Japanese dockyard, ii, 58, 60, 61.
Ypora, Lake, i, 264.
Ypora, Paraguayan ship, i, 258.

Yorktown or *Patrick Henry*, i, 15, 19.
Yoshino, Tab. xix; loss, xxi; Pl. xxiv, ii, 68; des. ii, 59-60; action off Asan with *Tsi Yuen*, 67-71; bridge and conning tower hit? 69; badly handled, 53; at Yalu, 84; place in line, 88; sinks *Chih Yuen*, 92, 98; uses cordite, 112; at Wei-hai-wei, 130; hit, 133.
Yu Sing, Tab. xv, ii, 5.
Yu Yen, Tab. xxv, ii, 13, 14.

Z.

Z and Montéchant, q. i, 209; ii, 156.
Zalinski gun, ii, 20, 150-1.
Zatzarennyi, Lieut., at Sulina, i, 293-4; at Sukhum, 298-301; at Batum, 301-3.
Zealous, ii, 221.
Zouave, i, 15, 18.

INDEX III.

SUBJECT-MATTER AND TECHNICAL TERMS.

A.

Accidental collisions, in action, *Brooklyn* and *Kineo*, i, 48; *Nahant* and *Keokuk*, 93; *Lackawanna* and *Hartford*, 130; *Ancona* and *Varese*, 240; *Blanco* and *Cochrane*, 328; *Tsi Yuen* and *Yang Wei*, ii, 100, 115; discussed, ii, 158, 169.

In manœuvres, &c., during peace, *Iron Duke* and *Vanguard*, ii, 189-91; *König Wilhelm* and *Grosser Kurfürst*, 192-6; *Camperdown* and *Victoria*, 197-205.

Accidental hits, on *Cochrane*, i, 329; discussed, ii, 155-6, 181.

All-round fire, *Monitor* had, i, 27; want of, on *Huascar*, 326; on *Audacious* class, ii, 222; on French designs, 263, 264-5, 269; on *Dupuy-de-Lôme*, 267, 269.

Ammunition, expended at Lissa, Italians, Tab. vii; Austrians, Tab. viii; at Charleston, Tab. iii; at Fort Fisher, Tab. v; at Alexandria, Tab. xiv; heavy expenditure against forts, i, 221; in bombardments, 283-4; Italian, exhausted at close of Lissa, 240; fails Spaniards at Callao, 253; English, exhausted at Alexandria, 348-9; short supply at Alexandria, 342; runs short on *Kwang Yi*, ii, 71; Chinese and Japanese, runs short at Yalu, 103, 93; importance of full supply, 123; limits action, 175; danger of exposed, explosion on *Cochrane*, i, 335; on *Matsushima*, ii, 94, 180-1; discussed, 180-1; on *Palestro*, i, 241, ii, 180, n.; on *Tamandaré*, 181, n.; hoists to guns, ii, 165; 180-1, n.; difficulty of supplying, on *Akagi*, ii, 95.

Arc of fire, limited, compels end-on battle, on *Ting Yuen*, &c., ii, 63-4.

Armour, application to ships, early, ii, 217; adopted by Confederates, i, 3; by Napoleon III., i, xxxii; by Federals, 6; quality of early, ii, 253; improvement in quality of, 253; at Kinburn, defies perforation, i, xxxvi; on *Monitor* and *Merrimac*, 28; at Charleston, 94-5; laminated and solid armour, 94; perforated, on *Atlanta*, 99; on Eads gunboats, 62, 63, 70, 84; on *Tennessee*, 130-1, 132; at Lissa, defies attack, 243, 245, 246; perforated on Brazilian ironclads, 262; *Huascar's* defies *Shah*, 309; fails against *Cochrane*, 325-8; *Cochrane* and *Blanco* not perforated, 331; resistance at Alexandria, 350-1, 354-5; at Yalu, ii, 119-122; withstands long-range attack, 159, 170; summary, 246; protects life, i, 33, 93, 132, 246; ii, 104-6;

360 INDEX OF SUBJECT-MATTER, ETC.

assures flotation, ii, 179; great resistance at sharp angles, 155; methods of application, Plates xxxvii, ii, 220; xxxix, ii, 232; xlii, ii, 262; ii, 226-7, 240; keeps out high explosives, ii, 172; resists ram, 207. Thin, dangerous; Eads gunboats, i, 62, 63, 70, 84; *Huascar*, 331; on *Tsi Yuen*, ii, 68, 70; on Chinese fleet removed, 86-7; at Yalu, 170; cloth instead of, *id*.

"Armour-clad." A vessel carrying vertical armour, *i.e.*, armour on her sides and gun-positions, *e.g.*, the *Royal Sovereign*.

Army, co-operation of, with navy ineffective: Confederate, at New Orleans, i, 44, 58; on Mississippi, 81-2, n.; *Albemarle*, 113; Fort Fisher, 137-8; need of army for offensive naval warfare, i, 61, 185, 205-6, 281-2, 284; ii, 33, 50, 127-8.

Artillery, *see* Guns.
Progress of, ii, 245-252; Tab. xxiv.

Auxiliary battery or secondary armament, carried by all modern ships, and is midway in power between the heavy guns and the machine guns, ii, 229; development of, 230, 234, 235, 238; protection of, on older ships, *nil*, ii, 164; easily put out of action, 172; on "Admirals," 230; *Victoria*, 232; *Nile*, 233; *Royal Sovereign*, 235; *Majestic*, 238, 243; improvised protection, 164.

"Axial fire." Fire right ahead or right astern, parallel to ship's axis or direction: of French ships, ii, 154, 269; of English "écheloned" turret-ships, 228; of Chinese battleships, 63.

Bow-fire necessary to meet torpedo attacks, ii, 180; strong, of *Conqueror* and *Victoria*, 232; necessary if enemy fight stern battle, 118.

Stern-fire, powerful French, 154; allows stern battle, 118; deficient, of *Conqueror* and *Victoria*, 232.

B.

"Backing." The timber cushion upon which armour plates are generally mounted, of teak or oak. Improvements in, ii, 254.

"Barbette." A circular or pear-shaped armoured inclosure, inside which is a turntable carrying the gun or guns. The latter fire over the edge of the armour, not through port-holes, and are said to be mounted *en barbette*. Pl. xiii, i, 280, shows a gun so mounted, forward; ii, 242.

Descends from turret, i, 33; on *Téméraire*, i, 339, ii, 223; "Admirals," ii, 230; *Royal Sovereign*, 234; *Centurion*, 236; *Majestic*, 238; French type, 261.

Danger of unprotected bases, ii, 164, 230, 263, 269.

Bases, seizure of, by North on Southern coast, i, 184-5, 205, 210; the Athenian strategy, 184; Japanese, in G. of Korea, ii, 84.

Battle, forecast of, chap. xxiii, ii, 136-182.

Close action or *mêlée* at Lissa, i, 234-248, ii, 101; may be necessary to close, 117; end-on attack results in, 154; weaker side may close, 159; hits numerous in, 170; its features, 172; torpedo-boat in, 149, 172; artillery in, 180.

Long-range action, Yalu one, ii, 101, 103; will probably be followed by close action, 117, 159; hits few in, 120; fire discipline in, 159; damage done in, 162-3; guns in, 180.

See also Loss, Torpedoes.

Battle formations discussed, ii, 153-8; Farragut's, at New Orleans, i, 45; at Mobile, 120; Tegetthof's, 230-1, 247-8; Persano's, 232; Ting's, ii, 85-6, 88; Ito's, 87-8. *See* Map xxix, ii, 156, to the figures of which reference is made by number below.
Bow and quarter line. Fig. 1 shows two such lines; at Lissa, i, 230-1, 247; ii, 155-6.
Groups, Figs. 6, 9, at Yalu, ii, 85-6; Amezaga on, 118; discussed, 156.
Line-abreast, Fig. 2, weakness, ii, 153-4, 155; Ting's, 57, 118; leads to *mêlée*, 154, 160.
Line-ahead, Fig. 4, weakness, ii, 153; advantages, 155, 157; length of line, 145; at Lissa, i, 232; at Yalu, ii, 87-8.
Quincunx, Fig. 8, ii, 136.

Battleship. A ship designed to fight in the line of battle, carrying vertical armour and heavy armour-piercing guns. Differs from coast-defence ship in its superior sea-keeping qualities. The general features of such a ship are shown in *Frontispiece*, vol. i, the *Majestic*.
English, French, German, and American compared, ii, 242-5; small, 145, 270; antiquated, 145, 270; definition of standard or first class, 270; homogeneity of, 271; need of water-line armour, 177-8; ideal battleship, 177-180; proportion necessary for a blockade, i, 205; battleship *versus* cruiser, i, 311; ii, 138—144, 119—121. Mainstay of the Brazilian revolt, ii, 49; at Yalu, 120, 122. Development of battleship in England, ii, 209-242; in France, 260-269. Earlier French and English compared, ii, 269.

Belt, armour, why carried, ii, 178-9; protects against ram (?) 207; on cruisers, 143.

Blockade of Southern coast, 1861-5, why it was proclaimed, i, 177; when proclaimed, 181; terms of proclamation, *id*.; objects, to prevent exportation of cotton, 178, and importation of arms, 179; imperfect at first, 181; magnitude of task, 183; does not stop trade, *id*.; indifferent craft blockading, 184; bases necessary, *id*.; the work of the army in, 185; typical blockade, 185-6; great rise in prices due to, 196; effect of, 196-7; three periods of, 186-7; invalidated by escape of three ships, 183; ships too close hit one another, 192.
Questions arising out of. *See* Neutrals, England, International Law.
Northern, of Bermuda and Nassau, i, 186.
French, of German coast, 1870, i, 274; difficulty of coaling, 276; escape of *Arminius* and *Elisabeth*, 277; of *Augusta*, 278; becomes observation, 278.
Turkish, of Russian ports, 1877, i, 288; ineffective, 293.
Chilian, of Callao, 1880, i, 333.
Japanese, of Wei-hai-wei, 1895, ii, 127-133.
Blockade of French coast by England, discussed, i, 204-210; France and Confederacy compared, 205-6; English admirals on blockade, 204-5; types of ship necessary for, 208-9; isolated ships can run, 277, 280-1.

Blockade-runners, starting points, i, 187; typical run, 189-191; Northern precautions against, 186; three types, 187; sailing, 187-8; fast steamers, 188-9; these could do what they liked, 192, 203-4; how they might have been stopped, 206; magnitude of traffic, 188, 193-4; companies of, formed, 188; cargoes carried, 191, 193; enormous profits, 194; price of freight, 193; wages of captain and sailors, 195; risk of,

191-2, 194-5; insurance on, 195; audacity of, 192; number captured, 195; specialisation in trade, 203-4; illegal seizure of ships by North, 198-201; excuses for North, 203; number of claims against North, 204; m. i, 51, 136-7, 138, 146, 167, 168.

Boats, will be left on shore before battle, ii, 162; *Tsi Yuen's* on fire, ii, 69; left behind by Chinese, 87.

Boilers, water-tube or *tubulous*. In these the water is in the tubes, whereas in *tubular* boilers the gases of the fire pass through the tubes around which is the water. On *Brennus*, ii, 254; on *Powerful*, id.

Damage to boilers, &c., in action, *Itasca*, i, 53; *Essex*, 63; *Sumter*, 68; *Little Rebel*, 69; *Mound City*, 69; *Brooklyn*, 75; *Mercedita*, 88; *Sassacus*, 109; *Mackinaw*, 139; *Villa de Madrid*, 235; *Ivahy*, 262; *Huascar*, 328; torpedo boat No. 9, ii, 131.

Bombardment. *See* also High angle Fire, Forts.

Of open ports, Spaniards at Valparaiso, i, 253-4; no attacks upon, in American Civil War, 153; Bouët ordered to attack German, 276-7.

Danger from, i, 209, n.; *cf.* 334; strategically useless, 283-4; provokes reprisals, 209, n., 284.

Boom, at New Orleans, i, 42, 49-50, 59; at Charleston, 87, 93; round *Albemarle*, 112; at Kiel, 271; at Sulina, 293, 295, 296; at Batum, 297, n.; at Sukhum, 298; at Callao, 333; at Sta. Catherina, ii, 49; at Wei-hai-wei, 128.

Need of, i, 96; ii, 50.

Bow and quarter line. *See* Battle Formations.

Broadside, weak, of monitors, i, 136.
Return to powerful, ii, 154, 242.

Bunkers, advantage of full, in action, ii, 164.

C.

Cap-squares. The metal plates which fasten down the trunnions of a gun to its carriage. Most modern guns have no trunnions, i, 95.

Captain in action, injury to, ii, 168; protection of, 168-170. *See* Conning Tower.

Capture in war eliminated? ii, 176. [*Huascar* captured, i, 327-330.]

Casemate. An armoured inclosure on board ship in which guns are mounted, ii, 235, 236, 237, 238. Casemate-ships are vessels of the *Merrimac* type, des. i, 3-6.

Central pivot mountings. A form of gun-carriage which allows the gun to sweep a wide angle, generally from 60° to 270°. It is usually fitted with a shield, and sometimes has an ammunition hoist coming up through the pivot. Pl. xli, ii, 250, illustrates this form of mounting ; ii, 60.

Clearing for action, Farragut's fleet, i, 39, 119; at Alexandria, 342; *Tsi Yuen* not, at Asan, ii, 69; Chinese fleet, 80, 86-7; discussed, 162-3; time taken to clear, 162, n.

Coal supply, bad on *Sumter*, i, 144; runs low on Italian fleet, 240; poor of French ships, 273, n.; of *Royal Sovereign*, ii, 235; of various battleships, 244.

Smokeless, Russians use, i, 298; Japanese betrayed by smoke, ii, 86; blockade-runners use, i, 189.

Protection improvised on Chinese ships, ii, 79, 86.

Discussed, ii, 163-5.

INDEX OF SUBJECT-MATTER, ETC.

Coaling at sea, *Alabama*, i, 154; French in 1870, 275-6.
Coaling in neutral ports. *See* Neutrals.
Coaling stations, i, 173, 206.
Coasts, war of, demands special types of ships, i, 265-7; *Huascar* carries on, 321.
Colour of ships, at New Orleans, i, 30; at Lissa, 235; of Russian torpedo-boats, i, 298; of torpedo-boats at Sheipoo, ii, 13; at Yalu, ii, 89, 101; importance of distinctive, 181.
Commander-in-Chief, place of (1) in fleet; Farragut's views on, i, 46, 120, n., ii, 151-2, n.; Nelson's view, ii, 151-2; Tegetthof, i, 231-2; Persano, 233; Ito at Yalu, ii, 152; discussed, 151-2; (2) on shipboard, Farragut in rigging, i, 121; military masts (?) ii, 152.
Commerce, warfare against, Southern against Northern, i, 143-176; passengers of captured ships, how got rid of, 150, 154; surprise rendered it effective, 153, 157; strategy of Semmes, 170; defensive against him, 170-173; transfers of Northern shipping to English flag, 169, n.; strategic inutility, 175-6; causes which may have contributed to ruin Northern commerce, 169.
    Commerce-destroyers, Tab. vi; two types, i, 143-4; *Sumter*, 144-6; *Florida*, 146-151, *Florida's* tenders, 148-9; *Alabama*, 152-7; *Tuscaloosa*, 156; *Nashville*, 165; *Georgia*, 165-6; *Shenandoah*, 166-7; *Atlanta*, 167-8; should not fight, 158; Northern precautions wanting, 153, 157, 170-3; prizes taken by, Tab. vi.
    German *Augusta*, i, 278.
    Steamers hard to capture, i, 168, 175.
    Geneva rules on commerce-destroying, 174.

English will be assailed, i, 143, 169, 175, 176, 206, 209.
Compound armour, iron faced with steel, ii, 229, 253.
Compound engines, have two cylinders, one larger than the other. The steam is used at high pressure in the smaller cylinder, and when used passes at a lower pressure to the larger cylinder, where it is used again. Triple-expansion engines have three cylinders, quadruple four, ii, 254, 261.
Concussion, injuries from, i. 26, 28, 131, 327, 350; ii, 89, 96, 169.
Conning tower. An armoured structure, generally cylindrical in shape, from which the ship can be directed in action. Same as pilot house.
    Kinburn batteries, i, xxxii; *Merrimac's*, 5; *Monitor's*, 10; struck, 30; struck on Mississippi gunboats, 65, 78; in later monitors, 89; difficulty of vision, 94; Rodgers killed in, 101; struck on *Salamander*, 243; on *Assar-i-Chevket*? 305; on *Huascar*, 326, 330; on *Tsi Yuen*, ii, 68; on *Yoshino's* (?) 69; tops removed from Chinese, 86.
    Hits on, ii, 168; thick armour or duplication, 168-9; field of view, 153; on various ships, 244.
Continuous voyages, Lord Stowell on, i, 199, n.; U.S. courts, 199-200.
Contraband, U.S. hold Confederate Envoys, i, 201-2; Northern courts (U.S.) on, 203; rice held, ii, 15.
Crews, of modern ships small relatively, ii, 213.
"Cross-raiding." [Colomb, Naval Warfare, 3]. Warfare by retaliatory expeditions seeking to burn and destroy without an effort to control or command the sea. *See* Coasts, War of.

Peruvian in 1879, i, 314, 321, 322, 324; ii, 33.

Southern cruisers threaten Northern ports, i, 149, 150, 153, 167-8.

Cruiser. A vessel of less military strength than a battleship, in which offensive and defensive qualities are sacrificed to speed and coal supply. Generally without vertical armour, but invariably has a horizontal armour deck, which usually curves up amidships, descending below the water-line at the sides and ends.

Sketch of history, ii, 254-7; in naval battle, 138—145; compared with frigate, 139-140; cannot ram, i, 22; ii, 103.

Against battleship, i, 308-311; ii, 103, 119-121, 138-145.

Needs armour-piercing guns, i, 311.

Classes of, ii, 143; armoured cruisers, *id*.

Necessary to protect commerce, i, 175; for blockade, 205; place in blockade, 209.

D.

Deck armour, on early floating batteries, i, xxxii; on *Monitor*, i, 10, ii, 227; on Eads gunboats, ii, 227; adopted, *id.*; novel arrangement of, in *Renown*, 237; in various battleships, 244; cuts into rammer's bow, 205; discussed, ii, 177.

Injured on *Affondatore*, i, 244; does not save Chinese cruisers, ii, 120; danger to, 172, 179.

Declaration of war, attack without, ii, 68, 76.

Devolution of command, ii, 169, 181.

Differentiation of warships, ii, 215-216.

Of armament, ii, 180, 234-5.
Of armour, ii, 235.

Difficulty of distinguishing combatants, ii, 101; of estimating fighting force, ii, 216.

Dimension, failure of small ships at New Orleans, i, 54, n., 60; *Palestro* at Lissa, 135; at Yalu, ii, 112-3.

Increased by steam, ii, 212, 241-2; of battleship, 270; in Western fleets, 116.

Resistance of large ships, *Aquidaban*, ii, 49; *cf.* ii, 116, 120.

Disguise in general action, ii, 181.

Distance between ships in action, i, 344; ii, 145, n.

Distribution of armament, ii, 178; in French battleships, 179-180, 263, 264; disadvantages of, 269.

Division of command on the warship, ii, 169.

Division of fleet, ii, 144—147.

Docks, need of, i, 313.

Double on enemy. To turn the enemy's flank or concentrate a large force upon a small section of his command, ii, 118, 181.

Duration of battle, ii, 115, 175.

E.

Échéloned turret-ship. A vessel in which the turrets are placed not in the keel line, in the centre of the ship, but diagonally across it. *See* deck plan of *Chen Yuen*, Pl. xxiii, ii, 62; defect of, ii, 63-4; English, 154, 228-230.

Electricity, use of, ii, 247, 267, 268.

End-on battle, Ting orders, ii, 85; danger of, 153-4; French may fight, 154-5.

End-on fire. *See* Axial.

End-to-end collisions, ii, 160.

Ends, unarmoured, ships with, are technically called "light-enders."

INDEX OF SUBJECT-MATTER, ETC. 365

At Yalu, ii, 122; danger of, 174, 179, 207; adopted by England, 227; in "Admirals," 231; compared with French protection, 269-270.

Engines, breakdown (?) ii, 69, 88; numerous on board ship, 213-4.

England and the North, the Confederate cruisers, *Florida*, i, 146-7; *Alabama*, 152; *Georgia*, 165; sale of *Victor* to Confederates, 166; *Shenandoah*, 166-7; coals and refits at Melbourne, 167; ironclads built for Confederates, 33, 168; sympathy with South, 147, 173; acknowledges as belligerent, 181; visits of English cruisers to blockaded ports, 147; instructions concerning blockade, 183; and cotton supply, 178.
    Case of *Trent*, i, 200-3; *Peterhoff*, 198-200; *Springbok*, 200-1; interference of North with trade, 197, 200.
    Northern spies, i, 171.
    Interests antagonistic in peace and war, i, 197-8.
    Prize law, 199; commerce will be assailed, 143, 169, 176, 206, 209; national insurance, 175; our position better than that of North, 175; dependence on sea, 198, 205-6.
    And France, rice, ii, 15.
    Strategy in war with France, i, 204-210; fleet necessary to blockade French coast, 206-7.
    Want of reserve, i, 207.
    Naval strength of, i, 207; ii, 269—274.
    Personnel of fleet, ii, 173, 273-4.

Explosion on *Cochrane*, i, 334-5; on *Matsushima*, i, 335; ii, 94, 180; on *Palestro*, i, 241, 335; ii, 180; on *Tamandaré*, ii, 181.

F.

Fire discipline, ii, 159.

Fires in action, Sinope, i, xxxi; *Congress*, i, 19; *Minnesota*, 29; *Pinola*, 53; *Keystone State*, 88; *Kennebec*, 126; *Lackawanna*, 129; *Schwarzenburg*, 226, n.; *Palestro*, 235, 241-2; *Ancona*, 236; *Kaiser*, 239; *Affondatore*, 244; *Maria Pia*, 245; *San Martino*, *id.*; *Castelfidardo*, *id.*; at Lissa, 246; *Blanca*, 256; *Huascar*, 325, 329; on *Vesta*, i, 305; on Chinese ships at Foochow, ii, 8-10; at Yalu, ii, 113; *Tshao Yong*, 113; *Yang Wei*, 91; *Matsushima*, 94; *Hiyei*, 95; *Lai Yuen*, 95, 113; *Saikio*, 96; *Ting Yuen*, 96-7, 113; *Chen Yuen*, 97, 113; *King Yuen*, 99, 113; *Ching Yuen*, 92, 99, 113; *Kwang Kai*, 113; discussed, ii, 165-6.

"Fleet in being." [Colomb. Naval Warfare, 122-3.] A fleet not contained or masked by a superior or equal force. May be compared to an army acting on enemy's lines of communication or flanks. It suffices to bar territorial attacks? [*Quarterly*, clxxvii.] Tegetthof's fleet in being, i, 220, 222, 224, 248; German, 275, 277; Chinese, ii, 80-1; Japanese, 82-3, 102-3.

Floating batteries, Kinburn, i, xxxii-xxxvi, 3, 8, 207.

Food before battle, i, 231, 343; ii, 87.

Forced draught. Artificial current of air in boiler flues. To be distinguished from *induced draught*, ii, 254.

Forts, and ships, Kinburn, i, xxxiv-xxxvi; Duckworth, i, 38; New Orleans, 45-54; not silenced, 59; Vicksburg, 70-1; Port Hudson, 75-77; *Cincinnati* sunk by, 80; McAllister, 90; Charleston, 92-96, 101-103; Mobile, 124-127; forts surrender, 134; Fort Fisher, 135-142; Lissa, 220-224; Callao, 255-6; on Parana, 262-3; Alex-

andria, 336-357; Wei-hai-wei, ii, 127-133; at Rio, ii, 36-40; ships cannot silence, i, 57, 128; monitors' fire too slow against, 94, 136, 266; advantage of forts, 91, 96-7, 265-6, 282-3, 353-4.
  Ships can pass, i, 45-54, 70-1, 75-77, 124-127, 263-4, 282; ii, 11-12, 36-40, 49-50.

Freeboard, high, necessary for seagoing ship, i, 32-3; 325, ii, 178, 239.

Frigates at Lissa, i, 219, 221, 234; compared with line-of-battle ship, ii, 138-140.

Funnel, damage to, *Merrimac*, i, 30; *Arkansas*, 71; *Albemarle*, 109; *Tennessee*, 129, 131; *Chickasaw*, 123; *Kaiser*, 239; *Assar-i-Chevket*, 305; *Huascar*, 328, 329; *Tsi Yuen*, ii, 68-9.
  Effect of damage, ii, 163-4; protection of, *id.*, 178; i, 89.

Fuses, defective, i, 162, 247, 352; ii, 70, 172.

G.

Gunnery, changes in, i, 133; value of, 159, 165, 215, 247, ii, 123; bad, i, 213, 247; Chinese, ii, 96, 105, 108-9; English, at Alexandria, i, 345, 353, Western better than Eastern, ii, 117; quick-firer has improved, 166; influence of weather on, *id.*; errors in, 155.

Guns. A 6-inch gun is a gun of 6-inch calibre; a 68-ton gun is a gun of 68 tons weight; a 100-pounder is a gun firing a shot of 100lb. weight. Increased power of, i, 34, ii, 170; sketch of artillery progress, 245-252; supreme (?) 123; has beaten armour, 174; long range of, i, 97; moderate size best, ii, 124, 171, 179.

Monster guns useless against earthworks, i, 353-4; craze for, ii, 247-9; need of big guns, i, 311, ii, 111, 171; weak guns useless against armour, i, 309, ii, 111; destruction proportionate to size of shot, 139, 171; disabled at Asan, 68, 69; at Yalu, 94, 98, 100, 113-4.
  Early guns weak, ii, 240, 246.
  Northern (U.S.) guns, Tab. i; burst, i, 65, 139.
  Southern, i, 1, 40, 118, 179.
  At Lissa, compared, i, 246; at Yalu, ii, 108-9.
  English, at Alexandria, i, 339; Egyptian, 341; disabled, 351-2.
  French and Chinese, at Foochow, ii, 5-6.
  Heavy guns in long-range encounter, ii, 159, 171; at close quarters, 172.
  Carriages, improvements in, ii, 247.
  Hydraulic-worked, at Alexandria, i, 345-6, 353; on Chinese ships, ii, 63; on *Téméraire*, 223; on *Dreadnought*, 226; *Thunderer*, 247; general adoption, *id.*; in French fleet, 262.
  Light (Gatling, Nordenfelts, &c.), effective against *Huascar*, i, 326; little value against large ships, ii, 30; deadly at Foochow, 10; on *Aquidaban*, 48-9; little damage by, at Yalu, 125.
  Long, advantage of, ii, 248, 250.
  Pneumatic, on *Nictheroy*, ii, 40-1, 42, 49; accuracy of, 150-1.
  Quick-firing, des, ii, 109-110, 251; Pl. xl, n, 246; xli, ii, 250. Value against torpedo-boats, i, 303-4; ii, 48-9, 149. For armament of merchant ships in war, i, 175. Promote accurate shooting, ii, 166; in long-range engagement, 159; structural influence, 227, 232, 238; tactical influence, 156; development of, 250-1; 8-inch, 170-1; as secondary armament of warship, 178, 179, 180;

Thunderer class impervious to, 226. At Yalu, ii, 90, 108-110, 111-2.
Rapidity of fire, ii, 180; importance of, 110; of heavy guns, 249.
Smooth-bores, U.S., i, 26; low charge, *id.*; short range, 127; effective on water-line, 161, 165.

H.

Handiness, i, 308, 316.

High-angle fire, at New Orleans, i, 43-45; effect of, 57; Farragut's belief in, *id.*, 75, n.; use of, in modern bombardments, 96-7; at Sulina, 296-7; on *Vesta* against *Assar-i-Chevket*, 304-5; *Inflexible's* at Alexandria, 352; Egyptian, 354; value of, urged by Lieut. Goodrich, 355. Howitzer ships, i, 355. Howitzer, ii, 171.

High explosives. The name given to substances exploding with greater violence than gunpowder, as dynamite, melinite, and cordite. They are used as the bursting charge of shells, ii, 252.
Effect of shells charged with, ii, 163, 172; at Yalu, 113; structural influence of, 227, 241.

Hits in action, on *Monitor*, i, 32; *Merrimac*, *id.*; *Rolf Krake*, 33; gunboats at Fort Henry, 63; at Donelson, 65; at Grand Gulf, 79; on monitors at Charleston, 93, 95; *Atlanta*, 99-100; second Charleston attack, 101; *Tennessee*, 132; fleet at Fort Fisher, 139; *Florida*, 148; *Alabama*, 164; *Kearsarge*, 164; *Formidable*, 223; at Lissa, 245; Brazilian fleet at Humaita, 263; flat-boat action, 262; *Vesta*, 305; *Assar-i-Chevket*, *id.*; *Huascar* against *Shah*, 309; against Chilian fleet, 325-330; *Cochrane*, 331; Alexandra, 346;

fleet at Alexandria, 350-1; French fleet at Foochow, ii, 11; *Blanco*, 19-20; *Aquidaban*, 37—38; *Tsi Yuen*, 68, 70; *Matsushima*, 94; *Saikio*, 96; *Hiyei*, 95; *Akagi*, *id.*; Japanese fleet, 110; Chinese, 111-2.
By heavy guns, i, 240, 243, 255, 256, 318, 325-330; ii, 9, 74, 94, 96, 98, 109, 111, 172.
On torpedo-craft, ii, 26-7, 46, 96, 132.
Percentage of, at Charleston, i, 96; Lissa, 246, 247; *Shah's*, 309-310; Chilian against *Huascar*, 330; on forts at Alexandria, 352; gunboats against *Aconcagua*, ii, 30; at Yalu, 109-110; future battle, 166.

Homogeneity of structure in battle-ships, ii, 236, 271.

Hydraulics. *See* Guns, Hydraulic-worked.
Ships propelled by, ii, 258.

I.

Induced draught. *See* Forced Draught.

Insurance, on blockade-runners, i, 195; national, of commerce, 175.

International law. *See also* England. Commerce-destroyers; *Sumter* not allowed to coal, i, 146; *Florida* uses English flag, 147; coals at Nassau, 148; at Barbadoes, 149; rule in regard to coaling, *id.*, n; at Bermuda, *id.*; *Shenandoah* coals at Melbourne, 167; negligence of England in regard to commerce-destroyers. *See* England.
Geneva rules, i, 174; *Alabama* claims, 174-5.
Sumter at Cadiz, i, 146; *Alabama* at Noronha, 156; coals at Blanquilla, 153; French lay embargo on *Victor*, 166; attitude of France, 174, n.

INDEX OF SUBJECT-MATTER, ETC.

U.S. warships, breach of French neutrality, i, 145; infraction of English coaling rules, 149; seizure of *Florida*, 151; complaints anent *Deerhound*, 164; repairs in English ports, 167; precedents against, 173; in blockade treatment of neutrals, 197; case of *Peterhoff*, 198-200; continuous voyages, 199, n.; Slidell and Mason, *Trent* case, 201-3; doctrine of contraband, 203; English claims against North, 204; pre-emption, *id.*; illegalities in blockade, 183.

French not allowed to coal at Heligoland, i, 275; nor English pilots to act for, 273; but Danish pilots procurable, *id.*

Neutral property destroyed at Valparaiso, i, 254; French hold rice contraband, ii, 15; coal at Hong-Kong, *id.*; blockade of Formosa, *id.*

Neutrals stop blockade of Valparaiso, ii, 19; of Rio, 37.

Case of *Itata*, ii, 33-4; of *Kowshing*, 75-81.

Hostilities without declaration of war, ii, 76.

Ironclad against wooden ship, *Merrimac* in Hampton Roads, i, 14-20; action off Charleston, 87-9; *Albemarle* and gunboats, 106-110; *Kaiser*, 239.

Against unarmoured ship, *Assar-i-Chevket*, i, 304-5; *Shah* and *Huascar*, 308-11. See also Yalu.

L.

Landing party from fleet at Fort Fisher, i, 141; diminutive with modern ships, 283.

Lashing of ships in pairs, i, 74, 120, 263.

Laws of war, breach of, ii, 78-79.

Life-saving in battle, or after, *Kearsarge*, i, 163-4; Lissa, 237-8; duty of victor, ii, 78; at Yalu, 106-107; Tegetthof wishes for conference to settle, 107; life-belts, 163.

Light draught, necessary on Southern coast, i, 7, 35, 40, 90, 117, 125; French had no — ships, 266, 268, 273, 282; value of, to *Huascar*, 308, 310; to *Covadonga*, 316, 319; to *Esmeralda*, 316; English want of, at Alexandria, 339; value to *Condor*, 346; to gunboats, 347; at Sfax, ii, 2, 4.

Lights misplaced, not, i, 273.

Line-ahead. *See* Battle Formations.

Line-abreast. *See* Battle Formations.

Loss in naval action. Men, Table of six great battles, ii, 106; past and present, 106-7, 173-4, 181.

At New Orleans, i, 60; Donelson, 65; Vicksburg, 71; Port Hudson, 77; Grand Gulf, 79-80; Charleston, 96; Mobile, 132; Fort Fisher, 140; *Alabama* and *Hatteras*, 155; and *Kearsarge*, 163, 164; *Formidable*, 223; bombardment of Lissa, 224; off Heligoland, 226; *Kaiser's* at Lissa, 244; total at Lissa, *id.*, 245; Spanish, at Callao, 256; *Meteor* and *Bouvet*, 279; *Esmeralda's*, 319; *Huascar's* at Angamos, 331; English, at Alexandria, 349-50; French, at Foochow, ii, 11; Chinese, *id.*; *Blanco's*, 26; *Tsi Yuen's*, 69; at Yalu, 104-6.

Ships, Lissa, i, 244-5, ii, 102; at Yalu, ii, 102 and n.; in future naval battle, 174-6.

M.

Machinery, influence on dimension, ii, 210-212; on size of crew, 213; quantity of, on shipboard, 213-4.

Mantlets, i, 52; ii, 86.

Masts hit, *Tsi Yuen*, ii, 69; *Akagi*, 95; *Hiyei*, *id.*; *Ting Yuen*, 96; *Chen Yuen*, 98.
  Value of military, with tops, good field of view, ii, 153; faults of, 167.
  On various types of battleship, ii, 244.
  Farragut places pilot in top, i, 75; climbs rigging at Mobile, 121.

Matériel, influence on tactics, &c., i. 283; ii, 63-4, 86, 153.

Melinite. A preparation of picric acid, used as a bursting charge in shells. At Yalu, ii, 113; effects of explosion, 172, 252.

Metallurgy, progress in, ii, 253.

Metacentric height. The height, above the centre of gravity of a floating body with a list, of the point through which the resultant upward pressure of the fluid always passes. The higher the metacentre, the more stable the ship. In *Captain*, ii, 184.

Mines, none at New Orleans, i, 58; *Cairo* sunk by, 73; *De Kalb* sunk, 84; damage *Montauk*, 90; *Ironsides* over, at Charleston, 93; Northern losses from, 104, 113; at Mobile, 115; sink *Tecumseh*, 124; *Hartford* crosses, 125; loss of ships in Mobile Bay, 134; sink *Rio de Janeiro*, 262; sink *Sulina*, 297; attempt to destroy Chilian ironclads by, 334; *Loa* and *Covadonga* sunk by, *id.*; none at Alexandria, 342, 354; at Sta. Catherina, ii, 49; at Wei-hai-wei, 128.
  Influence of, i, 97, 122-4; at Fasana, 228; at Kiel, 271, 276; at Iquique, 316.
  Necessary in a channel for defence, ii, 50; ports may be closed with, i, 209, n.
  Weehawken's torpedo-catcher, i, 92.

Misunderstanding of orders, i, 28, 52.

Monitor. A class of vessels with low freeboard, carrying their armament in a turret, named after their prototype, Ericsson's *Monitor*, Pl. iii, i, 26. A modern ship of this type is the *Nile*, Pl. iv, i, 32.
  Small target, i, 26, 97, 308; invulnerable, i, 101; to quick-firer, ii, 226.

Mortars. *See* High-angle Fire.

Muzzle-loaders, *Shah's* fail, i. 311; *Cochrane's* perforate, 330; reason for difference, 331; in English fleet, ii, 246; why abandoned, 248.

N.

Naval strength of England and France, ii, 269-274.

Navy, English, sketch of, ii, 219-259.

Navy, French, sketch of, ii, 260-274.

Nickel steel. An alloy of steel and nickel of exceptional toughness, ii, 253.

Night actions, ii, 176-7.

O.

"Observation." The strategical plan of watching a hostile port with light and fast ships, the main squadron being at a distance. Opposed to blockade, in which the main squadron is close at hand.

Organisation, want of Italian, i, 248-9; want of French, 271; German, 283.

P.

Personnel, Southern, i, 2, 42, 99, 100, 129, 159; Northern, 62, 132-3; Italian, 213-4, 215; Austrian, 227; Chinese, ii, 55-7; importance of, i, 159; long service, ii, 274.

Philanthropy in war, i, 239-240; ii, 31.

Physical health of men, i, 209; ii, 239.

Plans, Italians have none, i, 224, 246; Chinese have none, ii, 86, 101.

Plunging fire. Fire delivered from an elevation.

Popular feeling, leads to attack on Lissa, i, 217; and French fleet in Baltic, 280-1.

Ports, open. *See also* Bombardment.
Threatened, i, 150, 153, 167-8.

Prizes, *Sumter's*, i, 146; *Florida's* tenders, 147; *Florida's*, 151; *Alabama's*, 157; *Nashville's*, 165; *Georgia's*, 166; *Shenandoah's*, 167; captured by North in blockade, 195.

Projectiles. Those most commonly used are armour-piercing projectiles, or shot containing a very small bursting charge, and common shell or shell containing a large bursting charge. In addition, shrapnel shell is used by most navies, and case shot by a few.
Bad, of *Monitor*, i, 26; of *Merrimac*, 27; Chinese supply of, ii, 125; early rifled, 246; improvement in, 252.

Protection, portions of ship which need, ii, 177-8.
English and French systems of, 270.
Improvised, cables, i, 39, 64, 67, 119, 159, 228. Sandbags, i, 39, 85, 119. Coal, i, 64; ii, 79-86, 163. Timber, i, 67, 78. Cloth, ii, 170.

R.

Railways, in South, i, 179; on French coast, 205; German, 281; torpedo-boats sent by, 289.

Ram. [Laird Clowes. Journal United Service Institution, 1894.]
Merrimac uses, on *Cumberland*, i, 16-7; attempts of *Monitor* to use, 28; of *Merrimac*, 29-30; of *Manassas*, 48, 49, 50, 56; of *Mississippi*, 56; *Cincinnati* sunk by, 68; *Mound City* rammed, *id.*; *Lovell* rammed, 69; *Beauregard* rammed, *id.*; *Arkansas*, 72; *Queen of West* rams, 73; *Indianola* rammed, 74; use of, on Mississippi, 84-5; *Albemarle* rams *Southfield*, 108; rammed by *Sassacus*, 109-110; *Tennessee* uses, at Fort Morgan, 125, 126, 127; rammed by *Monongahela*, 126; a second time, 129; by *Lackawanna*, *id.*; by *Hartford*, 130; latter collides with *Lackawanna*, *id.*; *Florida* rammed, 150; Austrian attempts to ram at Lissa, 234-235; *Re d'Italia* rammed, 236; *Ferdinand Max* rams, 238; *Affondatore* dare not ram *Kaiser*, 238-9, 244; *Kaiser* rams *Portogallo*, 239; *Kaiser Max* rams, 242-3; *Prinz Eugen* tries to ram, 243; *Amazonas* rams *Paraguari*, 260; Bouët prepared to use, 272; *Bouvet* rams, 279; *Huascar* attempts to, 309; rams *Esmeralda*, 317-8; *Independencia* runs ashore trying to use, 319; *Huascar* attempts to ram *Magallanes*, 321; *Bianco*, 328; *Cochrane* attempts to ram *Huascar*, 329; at Angamos, 331; *Republica* rams?, ii, 38; Ting thinks of using, 85-6; *Chih Yuen* attempts to, 92, 101; *Tsi Yuen* rams *Yang Wei*, 115; *Iron Duke* rams *Vanguard*, 190-2; *König Wilhelm* rams *Grosser Kurfürst*, 192-6; *Camperdown* rams *Victoria*, 199-205.

Value of, i, 22; Italian confidence in, 224; why Tegetthof used, 228; in Chilian-Peruvian War, 331-2; at Yalu, ii, 114; in general action, 159, n., 160, 172.
Shock of ramming, i, 236, ii, 190, 194, 200; injury to "rammer," ii, 160; *Ferdinand Max*, i, 245; *Iron Duke*, ii, 192; *König Wilhelm*, 195; *Camperdown*, 203, *Huascar*, i, 320.
Special vessels for, i, 22; *Manassas*, 41; on Mississippi, 67; *Polyphemus*, ii, 147, 228; *Katahdin*, 150.
Speed necessary for, i, 32, 125, 132; ii, 150.
Structural influence, ii, 241.
Resistance to, of belt, ii, 207.

Ransom of prizes. See Tab. vi; vessels bonded, i, 146, 149, 154, 157, 166, 167, 168.

Recessed ports, ii, 221.

Reserve of ships for a blockade, i, 205; of men, 207; English lack of, ii, 274; in general action, 145-7.

Rice contraband, ii, 15.

S.

Sails, speed with, ii, 213; value as auxiliary, 212.

"Scouts." Light and fast ships designed to watch or discover a hostile fleet. More lightly armed than the cruiser.

Screw, twin, ii, 254; triple, 265; quadruple, i, 119, n.

Sea power, its influence, i, 1, 13; victory of *Merrimac*, 20; virtual, of *Monitor*, 31; on Mississippi, 61; at Shiloh, 65; Porter on, 77. Grant on, 82; Lincoln on, *id.*; general in Civil War, 82-3. Fort Fisher brings fall of Richmond, 142; South exposed to, 37, 178-181; pressure on South, 185; blockade crushed South,

196-7; enables turning of Humaita, 264; action of, in Franco-German War, 280-5; in Russo-Turkish War, 287-8; Chili and Peru, 314, 332; in Chilian Civil War, ii, 18; in China-Japan War, 86, 102.

Search lights, use of, i, 337, 349, ii, 11; *Blanco* does not use, 25; *Aquidaban's*, 45; in night action, 177.

Shields on guns. *See* Armour, Thin.

Signals, smoke, i, 190; in blockading, 208; at Lissa, Persano's unseen, 233-4, 240; Tegetthof's not seen, 243; at Yalu, Chinese, ii, 85, 97; Japanese, 101, 119; in general action discussed, 152-3.
Mechanical, on shipboard, ii, 167-8, 201.

Size. *See* Dimension.

Speaking tubes, break down on *Monitor*, i, 28; on *Hartford*, 75; shattered on *Tsi Yuen*, ii, 68; mistakes due to, on *Sampaio*, 44-5; in battle, use of, 167.

Specialisation. *See* Differentiation.

Speed, influence in action, *Monitor*, i, 27, 32; *Albemarle* too slow to ram, 109; *Tennessee* too slow, 125, 132; *Alabama* and *Hatteras*, 155; and *Kearsarge*, 158; *Drache* cannot catch *Palestro*, 235; *Shah* avoids ram, 310; want of, in Chilian fleet, 314, 321, 322; speed in "war of coasts," 321; strategical value of, ii, 33; at Yalu, 87, 89, 112-3, 123-4; in stern battle, 154; may allow of surprise, 87; advance in, 212; normal, of sailing ship, 213; limitations to, 258; of various battleships, 244.

Springs to cables, use of, i, 346.

Stability, importance of, ii, 166; *Captain* unstable, 183, n., 184, 189; *Vanguard*, 191; *Hoche* reported unstable, 269.

Steam facilitates rapid movements, ii, 158; renders doubling difficult, *id.*; effect of, on naval construction, 210-4; increases size, 212.

Smoke, at New Orleans hampers Federals, i, 54; at Port Hudson, 75, 84; at Lissa, 234-5; at Alexandria, 344; at Yalu, ii, 100, 101; hides Japanese, 89; from fires on Chinese ships, 96; with weather-gauge, smoke blows on enemy, 158; used to escape, i, 192-3.

    Smokeless powder at Rio, ii, 39; in Japanese fleet, 58, 112; tactical influence, 252.

Steamers, two only captured by Southern cruisers, i, 150, 154, 157, 168; hard to destroy, 175; captured blockade-running, 195.

Steering gear exposed of *Tennessee*, i, 117; damaged, 131; exposed of *Re d'Italia*, 212; damaged, 235; of *Huascar* damaged, 327-8; of *Tsi Yuen* damaged, ii, 68, 69; of *Saikio*, 96.

Stern-battle, ii, 118, 154-5.

Stratagems, masts dressed with branches, i, 43; dummy monitor, 75; method of sending message, 77; coal-torpedo, 105; disguised powder-ship, 138; Semmes' at St. Pierre, 146; use of neutral flag, 147; dummy guns, 149; *Alabama's* against *Hatteras*, 155; *Tuscarora's* use of twenty-four hours' law, 165; use of smoke by blockade-runners, 192-3; false colours at Lissa, 229; false hail, 291; questionable Peruvian, 333-4; *Covadonga's* at Iquique, 316; *Imperial's*, ii, 32; cloth, use of, 170.

Strategy, of Southern warfare against commerce, i, 170-1; commerce - destruction useless, 175-6; suggested strategy for North, 170-3; strategical importance of blockade, 179, 196-7; Northern occupation of bases,

184-5; of blockade in war with France, 204-210; of Italians in 1866, 218-220, 224, 248; of Tegetthof, 229, 230; of Franco-German War, 271; blockade, 274; French force in Mediterranean, 275; Bouët's communications, 277; attacks on forts, &c., 281-285; of Russians against Turks, 288; of Peruvians, 321-2; of Chilian insurgents, ii, 32-3; of Mello, 37; of Chinese, 79-80, 82-3, 134; steam has not changed, 119.

Structural damage at Yalu, ii, 110-2; in naval battle, 162, 175. *See also* Hits.

Submarine navigation, i, 103-4; ii, 259, 268.

Sun in enemy's eyes, ii, 159.

Surrendered ship, position of, i, 88.

T.

Tactical diameter, ii, 197.

Tactics. *See also* Ram, Torpedo, Battle.

    Mississippi contests, i, 84; Mobile, 120-1; faulty, of Buchanan, 127; of blockade, 208-210; of Tegetthof, 228, 230-1, 247-8; of Persano, 232-3; of *Huascar*, 308, 310; of *Esmeralda*, 316; Chilians against *Huascar*, 326, 327-8; of bombardment, 342-3, 347-8; 354-6; French on the Min, ii, 12; orders of Ting, 85-6; of Ito, 88; his mistakes, 91, 118; features of Yalu, 114, 116-124; gun, ram, and torpedo, 123, 159-162; ulterior motives affect Japanese, 103; structure of Chinese ships affects Chinese, 64, 86; steam and, 158.

Telegraph, absence of, hampered North, i, 173; used at Lissa, 222, 229-30; cut by Italians, 230; used by Peruvians, 315; at Alexandria, 337-8; engine-room, ii, 167-8, 201.

Telephone, ii, 167.

"Tonnage." The old system of measuring, or builder's measurement, expressed the internal cubic capacity of the ship. Modern or displacement tonnage gives the weight of water displaced by the ship.

Top-hamper, ii, 162-3.

Tops, use of, i, 75, 244, 326, 345, ii, 10; for signals, 153. *See also* Masts.

Torpedo, instances of employment in Tab. xxv, with reference to page.
  Invented, i, 102; development of, ii, 258-9; range, 259; structural influence, 215-6, 241.
  Loaded on deck, danger of, ii, 27, 30, 69, 71, 94, 98, 122.
  Tactical value, prevents ramming, i, 22; in action, ii, 160-1; its defects, 161, n.; equalises cruiser and battleship, 142; useless against ship in motion (?) 49.
  Effect of explosion, *Blanco*, ii, 26; *Aquidaban*, 47; *Ting Yuen*, 135.
  Compared with pneumatic gun, ii, 150-1.
  In Russo-Turkish War, i, 303-304.
  At Yalu, ii, 91, 96, 114.
  At Wei-hai-wei, in low temperature, ii, 130, 131, 132.
  Controllable, ii, 41, 161, 259.
  Spar and Whitehead, ii, 258.

Torpedo-boats, Tab. xxv, early, Southern, i, 102-3; Northern, 110-1; Russian, 286; sent overland, 289; Chilian, 332; fight between boats off Callao, 333; French, at Foochow and Sheipoo, ii, 13; Brazilian, at Sta. Catherina, 43, 45, 46; Japanese, 61; Chinese, at Yalu, 83, 91, 93, 96, 125; Japanese, at Port Arthur, 127; at Wei-hai-wei, 129, 130, 131, 132; Chinese, make a dash, 130.

Value of, in blockades, i, 208, 284, 288-9, 333; stopped by obstructions, 293; difficulty of finding target, 301, ii, 44, 48; danger to friends, i, ii, 149, 302; effect of quick-firer on, i, 303-4, ii, 48-9; prevent Japanese pursuit at Yalu, 93, 115; in general action, 123; place, 148; should not lightly be exposed, 149; in *mêlée*, *id*., 172; ideal torpedo-boat officer, 149, n.; save life, 107.
  Precautions against, i, 297, n., 301, n.; ii, 31, 49, 128, 134.

Torpedo-boat carrier, i, 292-3, 304.

Torpedo-boat destroyer in blockade, i, 208; English, ii, 258, 273.

Torpedo flotillas, French and English, ii, 273, 274.

Torpedo gunboat, *Condell* and *Lynch*, des. ii, 16-7; sink *Blanco*, 22-9; attack *Aconcagua*, 29-30; *Sampaio* or *Aurora*, des. 36; sinks *Aquidaban*, 42-9; useless for open fighting, ii, 30; place in battle, 147; English, 257; French, 268, 273.

Transfers of shipping, United States to English flag, i, 169, n.

Triple expansion. *See* Compound Engines.

Turret, Ericsson's, i, 6-8; Coles', 8, 33; Timby's, 8; *Monitor's*, 10; defects, 27, 28; concussion on, 28; jams of monitor turrets at Charleston, 94; at Mobile, 123; *Rolf Krake's* works well, 33; accident in, at Callao, 256; *Huascar's* perforated, 327-329; jams for a short time, 325; early English turret, ii, 224.

Turret-ship. *See* Index II. *Monitor, Nile, Devastation*, &c.

Twenty-four hours' law requires the elapse of twenty-four hours before one hostile ship can follow another from a neutral port. Breach of, i, 145; blockade by use of, 165; mentioned, 158, 278-9.

U.

Uniformity. *See* Homogeneity.

V.

Ventilation, artificial, of floating batteries; i, xxxii; of *Monitor*, 11; damage to ventilators in battle, ii, 163-4, 178.

W.

Water ballast, ii, 165.

Water-line, hits on, *Cincinnati*, i, 81; *Keystone State*, 88; *Gaines*, 128; *Alabama*, 162-3; at Yalu, ii, 121; on *Ching Yuen*, 133; effect of, on modern ship, 174, n., *cf.* 207, n.

Weak ships in line. *See* Dimension, Yalu, New Orleans.

Weather, influences shooting, ii, 166; *cf.* i, 231.

www.ingramcontent.com/pod-product-compliance
Lightning Source LLC
Chambersburg PA
CBHW022111300426
44117CB00007B/668